Creators of Intelligence

Industry secrets from AI leaders that you can easily apply
to advance your data science career

Dr. Alex Antic

BIRMINGHAM—MUMBAI

Creators of Intelligence

Group Product Manager: Ali Abidi
Publishing Product Manager: Gebin George
Senior Editor: David Sugarman
Copy Editor: Safis Editing
Project Coordinator: Aparna Nair, Farheen Fathima
Proofreader: Safis Editing
Indexer: Rekha Nair
Production Designer: Aparna Bhagat

First published: May 2023

Production reference: 1280423

Published by Packt Publishing Ltd.
Livery Place
35 Livery Street
Birmingham
B3 2PB, UK.

ISBN 978-1-80461-648-2

www.packtpub.com

Foreword

One of the most positive aspects of social media is that each individual can search for, find, and connect with groups of like-minded people, even with groups that are small, specialized, or off the beaten path. Over the past 37 years, I have focused on developing and advancing the fields of data, analytics, and artificial intelligence to further the goals and objectives of large and small companies, universities, government agencies, US high schools, not-for-profit organizations, and individuals.

When I started my career journey, the vast majority of professionals, and people in general, regarded data as a nearly useless byproduct of transactions, tracking systems, or compliance requirements. From the very beginning of my career, it was clear to me that all the systems that we developed relied on data, and any insights that we might derive originated from data. I have always believed that data has incredible intrinsic value, and analytics is how we discover, unlock, refine, revise, and leverage that value. Back in the 1980s when I started my career, being a member of the data and analytics community was like being in a club of one.

In the early 90s, I worked at IBM. My focus was on helping consumer-packaged goods companies understand the value and use of data warehousing and business intelligence systems. As usual, I was constantly evangelizing the next wave of technologies and business benefits of collecting, integrating, and analyzing data. Years after I left IBM, I bumped into my manager from my time there. After we caught up on careers, family, and life in general, he turned to me and said, "When you were talking about and encouraging people to think more about data and advanced analytics, I thought that you were crazy." I smiled and replied, "Not crazy, just early to the party."

As I was reading the chapters of this book, I smiled as I was drawn in by the stories of the journeys of the analytics leaders that Dr. Antic interviews. I was reminded that we are still in the early days of data and analytics. In the grand scheme of things, the data and analytics community is a small group of people scattered around the globe.

Today, people say things to me like, "You can't have been in data and analytics for 37 years; the field has not been around that long." Well, yes, it has. The foundations of the data and analytics field were established in the 1930s and 1940s. The field attracted serious attention and began to grow in the 1950s and beyond.

Currently, we have significantly more people involved in the field of data and analytics, but we are just scratching the surface of how we can organize, clean, integrate, analyze, and exploit data and analytics. It may seem like we have traveled quite far, but in relation to what we can and have to do, we are just at the beginning.

The stories in this book – the stories from analytics professionals, practitioners, and leaders – are early signposts of what we have done and what we are doing today and a few talk about what we can do in the near and distant future. I feel a connection to each of these analytics professionals. I can see my journey in their words and responses. I am sure many readers of this timely and well-written book will feel it too.

This book helps our analytics community to connect with each other and to know that we are on a similar journey. Our global community is growing very quickly. We are finding each other on social media, we are connecting via books like this one, and we are learning and leveraging our individual and collective experiences to help our community to advance our craft, our best practices, and our understanding of how and where we can make a positive impact in the world.

I hope that you enjoy this book as much as I did. If you are reading this Foreword and wondering whether you will enjoy the book and gain value, I encourage you to buy the book and delve in; it is a fun and enjoyable read.

Also, I hope that you connect with as many data and analytics professionals as possible. We are a unique bunch! We need to support and assist each other in any way possible. I look forward to meeting online or in person very soon.

John K. Thompson, author of *Data for All, Building Analytics Teams*, and *Analytics: How to Win with Intelligence.*

Contributors

About the author

Dr. Alex Antic is an award-winning data science leader, consultant, and advisor, and a highly sought speaker and trainer, with over 20 years of experience. Alex is the CDO and co-founder of Healices Health, which is focused on advancing cancer care using data science, and co-founder of Two Twigs – a data science consulting, advisory, and training company. Alex has been described as "one of Australia's iconic data leaders" and "one of the most premium thought leaders in data analytics globally." He was recognized in 2021 as one of the top 5 analytics leaders by the Institute of Analytics Professionals of Australia (IAPA). Alex is an adjunct professor at RMIT University, UTS, and UNSW Canberra, and his qualifications include a Ph.D. in applied mathematics.

To my wonderful business and life partner, Dr. Tania Churchill, who, as a senior data leader herself, was always available to bounce ideas off, assisted with editing and contributions to the writing throughout, and very importantly, stopped me from procrastinating instead of writing.

To the team at Packt (Gebin George, David Sugarman, Farheen Fathima, Vinishka Kalra, Shifa Ansari, and Aparna Nair), who not only helped make this book a reality but also provided expert guidance and support.

I am incredibly grateful and indebted to all the data leaders who agreed to be interviewed for this book, and for taking the time to share their insights, advice, and personal stories.

Additional contributor

Dr. Tania Churchill is an experienced data scientist and data science leader. She is a co-founder of Two Twigs, a data science consulting, advisory, and training company. Tania's experience spans the public and private sectors, including a decade of experience within the Australian Federal Government. Tania is passionate about fusing data science research with domain expertise for social good - with a particular interest in Privacy Enhancing Technologies (PETs). She is an Adjunct Academic at the ANU, the recipient of an Australia Day Achievement Medallion for exceptional leadership, and her qualifications include a Masters and a Ph.D. in IT.

Table of Contents

Preface   xv

1

Introducing the Creators of Intelligence   1

2

Cortnie Abercrombie Wants the Truth   5

Getting into the business   5   Establishing a strong data culture   16
Discussing diversity and leadership   7   Designing data strategies   25
Implementing an ethical approach   Summary   35
to data   11

3

Edward Santow vs. Unethical AI   37

Developing responsible AI pathways   37   Responding to the challenges of
Applying ethics in practice   42   generative AI   49
Considering the broader impact of   Summary   50
AI on society   45

4

Kshira Saagar Tells a Story   53

The path to data science   53   Implementing a data-driven approach 55

Discussing leadership in data culture   62   Looking to the future of AI   68
Storytelling with data   65   Summary   70
Getting into the industry now   66

5

Consulting Insights with Charles Martin   71

Getting into AI   71   Measuring impact   85
Balancing research and consulting   73   Integrating data   86
Advising companies on their AI   Finding the limits of NLP   89
roadmap   75   Explainable AI and ethics   91
Understanding why data projects fail   78   Summary   93

6

Petar Veličković and His Deep Network   95

Entering the world of AI research   95   Using graphs for AGI   107
Discussing machine learning using   Bridging the gap between academia
graph networks   97   and industry   108
Applying graph neural networks   102   Getting into research   110
Pushing research boundaries with   Summary   112
machine learning   104

7

Kathleen Maley Analyzes the Industry   113

Pursuing a career in analytics   113   Overcoming roadblocks   123
Striving for diversity   116   Establishing an effective data culture 126
Becoming data-driven   118   Learning about analytics   127
Dealing with dueling datasets   123   Looking to the future   130
                                                   Summary   131

8

Kirk Borne Sees the Stars   **133**

Getting into the field	133	Why do AI projects fail?	142
Advising a new organization on becoming data-driven	136	Building an effective data culture	145
		Teaching data science	147
Structuring teams	139	Predicting the future of AI	149
Managing data scientists	141	Summary	150

9

Nikolaj Van Omme Can Solve Your Problems   **153**

Getting started	153	Measuring success	167
Assessing the progress of AI	154	Developing ethical AI in an organization	169
ML and OR	155		
Becoming data-driven	160	Starting out in data	172
Setting your project up to succeed	161	Looking to the future	173
Exploring leadership	165	Summary	175

10

Jason Tamara Widjaja and the AI People   **177**

Getting started in data science	177	The importance of data governance	187
Becoming data-driven	178	Discussing leadership	189
Managing data science projects	182	Advising new entrants to the field	191
Why AI projects fail	184	Generative AI and ChatGPT	193
Communicating a realistic expectation to clients and partners	185	Predicting the future	194
		Summary	195
Establishing a data culture	186		

11

Jon Whittle Turns Research into Action   197

Building a career	197
Translating research into real-world impact	199
Developing AI that is ethical, inclusive, and trustworthy	201
AI in Australia	206
Discussing leadership	207
Predicting the future of AI	208
Entering the industry today	209
Summary	210

12

Building the Dream Team with Althea Davis   211

Getting into data	211
Increasing diversity and inclusion	212
Working in consulting	214
Establishing a data service and culture	218
Managing projects	223
Why does AI fail?	225
Summary	226

13

Igor Halperin Watches the Markets   229

Coming to AI from another field	229
Applying ML to problems in finance	231
Making AI explainable and trustworthy	233
Planning for successful AI	234
Navigating hype	235
Discussing the role of education	236
Considering the future of AI	237
Summary	238

14

Christina Stathopoulos Exerts Her Influence   241

Becoming a data science leader	241
Observing changes in the field	243
Increasing diversity and inclusion in the field	245

Advising new organizations   246
Understanding why projects fail   248
Using data storytelling   251

Understanding the fundamental
skills of data science   252
Getting hired in data science   253
Progressing into leadership   254
Summary   255

15

Angshuman Ghosh Leads the Way   257

Getting into AI   257
Watching the field evolve   258
Becoming data-driven   260
Organizing a data team   262

Building a good data culture within
an organization   264
Understanding the value of data
storytelling   265
Hiring new team members   266
Summary   268

16

Maria Milosavljevic Assesses the Risks   269

Getting into analytics   269
Discussing diversity and inclusion   270
AI and analytics   271
Becoming data-driven   273
Ethical AI   277

Establishing a good data culture   279
Why do data science projects fail?   280
Discussing data leadership   281
Looking to the future   281
Summary   282

17

Stephane Doyen Follows the Science   283

Getting into data science   283
Becoming a leader   284
Becoming data-driven   288
Developing AI solutions for the
medical field   289

Putting the "science" in "data science" 293
Establishing a data culture at an
organization   295
Building the right team   296
Looking to the future of AI   298
Summary   301

18

Intelligent Leadership with Meri Rosich   303

Becoming a chief data officer	303	What makes a good data leader?	308
Improving diversity and inclusion	304	The importance of data storytelling	309
Discussing the high failure rates of AI projects	305	Making AI ethical and trustworthy	310
		Advice for aspiring data scientists	311
Becoming a data-driven organization	306	Looking forward	312
Establishing an effective data culture	308	Summary	313

19

Teaming Up with Dat Tran   315

Entering the industry	315	Discussing data storytelling	322
Discussing the high failure rates of AI projects	316	Hiring team members	323
		Advice for beginners	324
Setting up for success	318	Looking to the future	325
Establishing a good data culture	320	Summary	326
Being a data leader	321		

20

Collective Intelligence   327

Entering the field and becoming a successful data scientist	328	Scaling your data capability	335
Becoming a CDO and senior data leader	330	Structuring and managing data science teams	336
		Avoiding AI failure	337
Developing an effective data strategy	332	Measuring Success	339
Establishing a strong data culture	332	Storytelling with data	341
Becoming data-driven	333	Predicting the future of AI	341
Ethical and responsible AI	334	Striving for diversity and inclusion	342
Data literacy	335	The changemakers	343

Index   **345**

Other Books You May Enjoy   **352**

Preface

Amidst all the hype surrounding AI, how do you build a successful career in this rapidly evolving field, and develop the necessary skills and knowledge to lead impact and change for your organization?

This book answers these questions and many more through a series of in-depth one-on-one interviews between award-winning senior data science leader, Dr. Alex Antic, and 18 expert data leaders and CDOs.

Who this book is for

This book is primarily aimed at three broad audiences – leaders, practitioners, and educators within the broader data and analytics fields.

Senior data leaders, CDOs, and those aspiring to reach these positions will find value in the advice and insights shared on how to transition to senior levels, and the fundamental skills necessary for these roles.

Practicing data scientists, data and analytics professionals, those looking to reskill or upskill, and the younger generation who are contemplating a career in the field will be given invaluable advice and recommendations on the necessary skills and knowledge needed – and important insights on what employers look for when interviewing candidates.

Those developing and delivering university-level education, online courses, and executive training programs will benefit from the advice contained within about the relevant skills and expertise needed to meet the demands of industry, government, and start-ups.

The book makes no assumptions about the technical background in data, analytics, AI, or technology of the reader, but those who are technically proficient will find some of the more research-focused interviews of particular interest.

What this book covers

Chapter 1, *Introducing the Creators of Intelligence*, offers a brief introduction to the scope of this book.

Chapter 2, *Cortnie Abercrombie Wants the Truth*, provides an interview with an experienced C-suite leader with a focus on explaining the truth of AI.

Chapter 3, *Edward Santow vs. Unethical AI*, provides an interview with a former human rights commissioner and professor of responsible technology, who gives us a legal perspective.

Chapter 4, Kshira Saagar Tells a Story, provides an interview with an accomplished senior data science leader with a flair for storytelling.

Chapter 5, Consulting Insights with Charles Martin, provides an interview with an experienced consultant and active researcher developing novel AI tools, who doesn't pull his punches!

Chapter 6, Petar Veličković and His Deep Network, provides an interview with a DeepMind researcher pioneering the development and application of graph neural networks.

Chapter 7, Kathleen Maley Analyzes the Industry, provides an interview with a renowned senior analytics leader with a pragmatic focus.

Chapter 8, Kirk Borne Sees the Stars, provides an interview with Kirk Borne; from NASA astrophysicist to data science influencer – he's seen it all.

Chapter 9, Nikolaj Van Omme Can Solve Your Problems, provides an interview with a pioneer of combining machine learning with operations research – a hybrid approach to overcoming limitations with machine learning.

Chapter 10, Jason Tamara Widjaja and the AI People, provides an interview with a human-centered leader with expertise in the application of AI to biopharmaceuticals.

Chapter 11, Jon Whittle Turns Research into Action, provides an interview with an experienced research leader translating research into impact.

Chapter 12, Building the Dream Team with Althea Davis, provides an interview with an award-winning CDO creating business value through data.

Chapter 13, Igor Halperin Watches the Markets, provides an interview with a physicist turned quant, and expert in the application of machine learning to the field of finance.

Chapter 14, Christina Stathopoulos Exerts Her Influence, provides an interview with a data evangelist, advisor, educator, and experienced analytics leader.

Chapter 15, Angshuman Ghosh Leads the Way, provides an interview with a veteran data science leader with expertise developed across Fortune 100 companies.

Chapter 16, Maria Milosavljevic Assesses the Risks, provides an interview with an accomplished CDO focused on delivering public good with AI.

Chapter 17, Stephane Doyen Follows the Science, provides an interview on start-up success – applying AI to the medical field.

Chapter 18, Intelligent Leadership with Meri Rosich, provides an interview with an internationally recognized CDO experienced in leading large-scale data initiatives.

Chapter 19, Teaming Up with Dat Tran, provides an interview with an entrepreneurial-minded data and technology leader, leveraging Agile and lean practices for successful data science projects.

Chapter 20, Collective Intelligence, offers an exploration of the key insights, advice, and recommendations captured in the interviews.

Get in touch

Feedback from our readers is always welcome.

General feedback: If you have questions about any aspect of this book, email us at `customercare@packtpub.com` and mention the book title in the subject of your message.

Errata: Although we have taken every care to ensure the accuracy of our content, mistakes do happen. If you have found a mistake in this book, we would be grateful if you would report this to us. Please visit `www.packtpub.com/support/errata` and fill in the form.

Piracy: If you come across any illegal copies of our works in any form on the internet, we would be grateful if you would provide us with the location address or website name. Please contact us at `copyright@packt.com` with a link to the material.

If you are interested in becoming an author: If there is a topic that you have expertise in and you are interested in either writing or contributing to a book, please visit `authors.packtpub.com`.

Share Your Thoughts

Once you've read *Creators of Intelligence*, we'd love to hear your thoughts! Scan the QR code below to go straight to the Amazon review page for this book and share your feedback.

`https://packt.link/r/1-804-61648-6`

Your review is important to us and the tech community and will help us make sure we're delivering excellent quality content.

Download a free PDF copy of this book

Thanks for purchasing this book!

Do you like to read on the go but are unable to carry your print books everywhere? Is your eBook purchase not compatible with the device of your choice?

Don't worry, now with every Packt book you get a DRM-free PDF version of that book at no cost.

Read anywhere, any place, on any device. Search, copy, and paste code from your favorite technical books directly into your application.

The perks don't stop there, you can get exclusive access to discounts, newsletters, and great free content in your inbox daily

Follow these simple steps to get the benefits:

1. Scan the QR code or visit the link below

https://packt.link/free-ebook/9781804616482

2. Submit your proof of purchase
3. That's it! We'll send your free PDF and other benefits to your email directly

1

Introducing the Creators of Intelligence

When you tell someone that you're writing a book, the first (and most obvious) question you're asked is, "*What is the book about?*" Sometimes, however, you get asked an even more important question: "*Why are you writing it?*"

For me, the *why* was an evolution, rather than an inspiration that hit me like a lightning bolt. Some common themes began to emerge from the topics that I was asked to speak about at conferences and industry events. I started writing a blog to address these themes. The blog was also a fun way to share my learning, and to offer advice to the next generation of data scientists and those transitioning to the field. I subsequently launched my consulting business, and again, there were common themes in the challenges faced by my clients in their journey to unlock the value of their data, and in maturing their analytics capability.

This prompted me to consider writing a book to share my experience, and to offer key insights and advice on how to avoid common pitfalls and issues. I realized the value of "war stories" and advice from people walking the walk and overcoming obstacles to help drive impact and change for their clients or organizations. That was my inspiration – what if I could bring together the collective wisdom and knowledge of some of the leading data leaders from around the world and share it, in a way that makes you feel that you're part of the conversation?

In selecting a list of successful, experienced, and influential people to interview for this book, I reached out to the leaders I have been inspired by. I've purposefully tried to include a diverse group of people who are from different industries and sectors to gain insight into the ingredients for success – both where there are commonalities between industries and where there are differences. I also focused on individuals with a proven ability to build and lead a successful data and analytics capability from the ground up. Some of the leaders have backgrounds as hands-on technicians, while others have come to data and analytics leadership via other pathways.

Common themes were explored with each leader, eliciting pragmatic advice and insights that you can directly apply to your organization and career aspirations. The questions asked include the following:

- What skills and traits do you need to become a **Chief Data Officer (CDO)** or **senior data leader**?

- How do you become a successful data scientist?

- How can we increase diversity and inclusion in the data science/IT field?

- How do you avoid common pitfalls and challenges with your data projects?

- How do you develop pragmatic ethical and responsible **Artificial Intelligence (AI)** and data science solutions?

- As a leader, how do you attract, manage, and retain staff?

- How do you establish an effective data culture in your organization?

This book is primarily aimed at four audiences:

- CDOs and senior data leaders, or those aspiring to reach that position

- Practicing data scientists and data and analytics professionals

- Those developing and delivering university-level education, including undergraduate, postgraduate, and executive programs

- The general reader, including senior business leaders, who are interested in learning how data and analytics are used to drive decision-making in different domains and sectors, and wish to understand some of the challenges and opportunities in improving the data capability of an organization

In addition, there is a fifth audience that I have in mind who will also benefit from the conversations within this book – the younger generation who are contemplating a career in data, AI or broader technology sector, and working professionals who are considering a career change. Given the global talent shortages in the field, and the challenges with diversity, my hope is that these readers will be inspired and encouraged by these conversations to pursue a career that is intellectually challenging, exciting, (often) lucrative, and meaningful.

My conversations with the data leaders in this book address how to develop a successful career in the field. We also discussed diversity and inclusion issues within tech – acknowledging the lack of gender, racial, and socio-economic diversity in many areas of the sector. Several leaders that I spoke with recounted some of their own experiences with discrimination or unconscious bias, and I thank them for providing their insights and suggestions to tackle the lack of diversity.

The book makes no assumptions about your technical background in data and analytics. As such, I will define the three most common terms to set the scene:

- **Data science**: The typical and most common definition of data science frames it as an intersection of mathematics/statistics, computer programming, and business knowledge. This is a relatively accurate description of the key skills needed to become a data scientist; however, it doesn't clearly explain the practice of data science. Data science is the science of change, enabling an evidence-based data-informed culture. While technology is an integral and fundamental component, what drives data science success is encapsulated in the words "science," "change," and "culture," which all fundamentally depend on people, and not technology. From a technical perspective, data science often involves the creation of models – specifically, machine learning models – which is why I'm loosely using the terms *data science* and *AI* interchangeably in this book.

- **Machine learning**: Machine Learning (ML) is a branch of AI that focuses on developing computer programs that can learn from data, rather than explicitly needing to be programmed to perform a specific task. This can include learning how to summarize text (**natural language processing**), identifying individuals in a photo (**computer vision**), or identifying suspicious behavior (**fraud detection**). ML algorithms are trained on large representative samples of data from which they can "learn" key characteristics to perform a certain task.

- **AI**: Most commonly, when someone refers to AI, they're really speaking about ML. However, AI is a broader concept and refers to computer programs that can perform tasks that typically require human intelligence, such as reasoning and abstraction, which is beyond the scope of current AI systems.

These conversations were conducted between June 2022 and April 2023. They were recorded, transcribed, and then edited by myself and the team at Packt. The edited text was provided to each person I spoke to so they could revise it, ensuring that each person's views were conveyed clearly and fairly.

I'm honored and humbled to have had the privilege to discuss a range of topics that are of great interest to me with such an amazing group of data leaders. They've left me inspired and have sparked new ideas and insights for me. I hope you enjoy reading these conversations as much as I did recording them, and that you're able to apply what you learn to your own role and career.

2

Cortnie Abercrombie
Wants the Truth

With a wealth of experience as a C-suite AI and analytics leader, **Cortnie Abercrombie** is well placed to provide targeted advice to new and veteran data leaders alike, and to comment on the state of the AI field. She is the CEO and founder of AI Truth (`www.AITruth.org`), a founding editorial board member for Springer Nature's *AI and Ethics* journal, and the former **Chief Data Officer** (**CDO**) of IBM.

Cortnie is also the author of *What You Don't Know: AI's Unseen Influence on Your Life and How to Take Back Control*, which explains what companies are doing with AI that can impact your life, as well as the changes we should demand for the future. I was particularly keen to hear about her previous experience as a senior executive at IBM, which gave her a unique opportunity to see how Fortune 500 companies traversed successes and failures on their journey to becoming data-driven organizations.

Getting into the business

Alex Antic: You've completed an MBA and obviously have a strong strategic focus. You then pivoted and became an expert in the field of data science and AI. I'd like to better understand your career trajectory. Were there any pivotal points or people that influenced you along the way? What led to you becoming a global leader in the field of data science?

Cortnie Abercrombie: Thanks for the acknowledgment. There was actually a pivotal moment in my life. When I left college, I started in the marketing department of a rapidly growing internet start-up during the dot-com days. My business degree was concentrated on marketing. The founders of the start-up wanted to figure out how to sell services better than the next company – especially the big multinational companies – and how to take advantage of the unique growth and first-mover market dynamics.

I had this fantastic boss and he was like, "*You know what? We've all been relegated to marketing.*" That's not the fantastic part. He felt like marketing was a lesser-than type of function because he had been in sales, and 20-30 years ago, sales was the be-all and end-all. It was like the TV show *Mad Men* – they had all the power. There was an overarching and incorrect perception of marketing that it was just those two ladies over there who did some random thing that nobody knew anything about except for it having to do with advertising and tchotchkes: *"Go get me some more things with our logo on it,"* or whatever. But my boss said, *"No, we've got to rethink this whole entire function so that it is strategic and customer-driven, and I want to use databases and data to do that."*

At that time, that was actually pretty forward-thinking. I did not realize it then, but I was one of only a few marketing people in the market who knew how to sift through data to find insights. That was why he hired me – though my data skills were basic at first. This was almost 30 years ago and the people who knew how to do "data mining," as we called it back then, were few and far between. You would typically find them in actuarial fields or theoretical science fields, such as astrophysics, or even biological research areas where data collection and analysis were key.

In business, most people who knew anything about data had computer science backgrounds and worked as **Database Administrators** (**DBAs**) in the IT department. The IT department and the marketing department did not get along at all. The IT department did not consider marketing to be of strategic importance in that era, which meant that of all the project backlogs they had, we were the lowest priority. This meant our projects never got done. They saw marketing as the "swag" department, who gave out logoed stuff and threw parties. They did not take it seriously when I would ask for their help in getting more data to understand how we could grow our share of the wallet with customers or increase revenue by targeting specific customer segments.

Because of this extremely off-putting attitude, I was forced to take matters into my own hands, and I began amassing data directly for the marketing department. My boss, who was just as frustrated with IT as I was, said, *"OK, what do you need?"* I said, *"Give me a server,"* and he replied, *"OK. Done! What else?"* (Marketing had money, you know!) I said, *"Well, I'm probably going to need more data classes."* At that time, Oracle had a university – literally called Oracle University – where they offered week-long courses right down the street from the company. I took a bunch of classes there, and then I took a bunch of classes from SPSS. My thinking was that if the IT people could learn it, then I could learn it better. Then I befriended and bought pizzas for as many DBAs as I could on the IT side of the house to get their help in properly setting up the data pipelines to my server.

Then *voilà*! I was putting out strategic segment analyses based on product types and where we needed growth. I dug into which customers were the most profitable and how we could deepen our share of the wallet with each one – including the use of VIP programs. I focused on customer churn, including understanding the triggering events and root causes as well as what could bring them back. As that progressed, I realized I had executives flocking to me in the company, asking me more and more about the customers, how to sell to them, and how to create models – almost like the Amazon recommender system idea, but internal; if my client buys this, what else might they buy? What patterns are you seeing? My boss also added his own perspectives and we produced the analyses in what became an anticipated quarterly report called "the red book." Every major leader in the company wanted a readout and Q&A session of the analyses for their department, and the analytics from it ultimately helped the company to be acquired at top dollar by a major international internet group that you would recognize today.

My boss' persistence and insistence made me believe, at a very early point in my career, that you can learn anything – just get over there and learn it! That defined my career and the way that I look at data scientists now. The way I see data science as a profession is that it's all about asking the right questions and looking across the business at the business needs in the same way that I did when I was sitting there as a 20-something-year-old, trying to figure out what kind of analytics to put forward to an executive team who wanted to know everything about our customers. I needed to think about what we were trying to do as a company. I still take that approach to this day.

I think that many data scientists lose themselves – sometimes they put their blinders on or just find that it's an uncomfortable thing to get out and talk to other parts of the business. I think they forget that what they have to do is stick to the strategic plans of the business. It's been a struggle, with everybody I've worked with or that I bring on, to try and convince them to have that tenacious, persistent personality. I can teach a person any skill. Tech skills change with the times.

> The way that you think about solving problems and home in on the things that matter: that's going to determine the longevity and relevance you have in your career.

Discussing diversity and leadership

AA: The next question touches on how the industry has changed. We've come some way in regard to diversity and inclusion, but we still have a long way to go. I'm curious as to whether you've personally faced any challenges in that regard, or whether you've observed any things that maybe you weren't happy with, things that should have played out differently. Also, what should we be doing more broadly as leaders in the field to change things?

CA: This one's always been tough because there's always been that disparity when it comes to women in the industry. I think a lot of that is changing, as far as getting more women into STEM is concerned. A lot of women have a proclivity for math, or detailed thinking and connecting the dots. They are also amazing negotiators, which makes them some of the best data governance leaders. Anyone who has ever tried to convince a business leader to share "their" data – meaning they bought it with their budget and enhanced it with their expertise – will understand the nuanced negotiation skills it takes, akin to negotiating a prized piece of candy away from a child before dinner. Looking at connections between datasets too, having those thoughts of, "*Well, this could connect with that,*" and, "*Oh yeah, I've seen some data from this other team that could more clearly fit this model*" – that is a dynamic I've experienced more often when working with women.

Many of the guys I've worked with in data and analytics are very logic-driven – always thinking linearly and of validity before connecting the dots. But sometimes, something may fall out of the norm of the logic that they're used to. Take behavioral psychology, using an example many people could probably relate to. I say to my husband, "*Let's talk*." He'll be like, "*Oh my good gosh! What did I do? I don't want to have this talk because you're going to talk about squishy stuff like emotions such as happiness or disappointment that are not quantifiable and that I therefore cannot affect*." But today, in a lot of data-driven use cases, we're talking about data that we are trying to use to emotionally drive people to do things. We're trying to understand emotions, even in cars. We're trying to understand: *"Hey, is this person rage-driving based on data from the odometer, brake pad sensors, or steering wheel motions? Does this car need to be automated to slow down upon recognition of precursor behaviors? Is there some way we can intervene?"*

I see women as becoming more and more important in the data field. I'm not saying there aren't guys that can do all of these things too, but I just have seen – in my experience, anyway – that men tend to be very strong linearly and chronologically with thinking through problems. Meanwhile, women can comfortably think about patterns all over the place, all at the same time. I know men who seem to think as follows: *"I deal with this first, then I deal with this second, then I deal with this third, and then I check everything at the end."* Women are more free-associators: *"Oh, I see a pattern here, here, here, here, and here. Let's pull that all together. Let's see what that does. Let's run that. Let's try that 50 times to Sunday."*

When we marry these two styles together is when data science is at its strongest. It takes a lot of tolerance for each other's styles, and it takes a lot of giving each other grace and credit. Trying to look at many different patterns at once drives a particular type of person crazy. A lot of people in the data science field come from computer science backgrounds, where you think a little differently about things. In computer science, you are taught to think about the flow of the code: how does it work in terms of its logic? It's just a way of thinking. We have got to figure out how to marry all of this together so that women and men and any other groups can challenge that thinking process. What happens is code is set up that points only in one direction and there's no one divergent point from which you can stop and say, *"Wait! Did you think about these other five things that could be out here that you could have put into your algorithm?"*

Considerations for the types of data that should be included to get the best fit in a model are a good example of this dynamic – especially when you inevitably have to come up with a substitute or proxy for a piece of data that you don't have. For example, I have seen on many occasions data scientists wanting to turn to financial scores as an easy proxy to determine whether a person is "responsible" in general – not just with their money. They will then use this score to determine whether the person is a "responsible" driver. But someone's financial situation and how well they drive their car are most likely not causal. But if you are working with data scientists who all have good financial scores, they probably do not see how this data selection could be flawed for their model. In this case, they could benefit by having someone from a divergent socioeconomic background question this assumption and also inform the data science team of the personal impact on a person who is declared an irresponsible driver – for example, their insurance rates go up, which may cause them to have even less disposable income for gaining a higher education, continuing important maintenance on their car, or worse, triggering a choice between buying groceries or paying rent.

As leaders, we should do the following:

- Think about the types of models we want to build, their strategic importance, and their impact on society as a whole. We tend to think only of the direct impacts in front of our faces, such as *"Did we get a bonus for delivering the best revenue results or cost savings?"*

- Data science teams developing some of the world's most important algorithms can't all have the same mindset or they're going to miss important stuff that can affect people whom they have never even considered.

> **We need to strive to hire melting-pot teams who are as diverse as the people that the models could impact in society.**

- We need to make sure that our norms, processes, and tools allow for the time, space, and ability to push back. It's one thing to have a diverse team fully able to bring many perspectives to the table, but people need to be able to speak for themselves.

> **We need to give people the ability, in standup meetings and other time-sensitive situations, to push back when they need to, and to do so without fear of harm to their career.**

I have seen people be railroaded in the process so they don't get to voice their different views. I have also seen people be blacklisted from future high-profile data science initiatives if they dared to raise a red flag that could negatively impact the budget or time to market. We must be able to encourage speaking up; otherwise, we take on unnecessary risks that could cause the data science projects to fail, or worse, could bring down the company's reputation or regulatory compliance or cause a decline in the company's market value.

AA: I think that is a great way to summarize where some opportunities exist to try and really bring people together to work toward the common good.

The next question touches on your work at the leadership level. Say you're consulting or working with an organization that's just starting out in the field and they want to develop a capability, and they say to you, "*We want to get stuck into data science and AI. We don't know where to begin.***" How do you advise them to begin that journey?**

CA: Well, the first thing I always ask is, *"What is the strategy?"* It's the same question I asked myself in the role I was just telling you about: how can you be relevant to the company? How can you be the *most* relevant? How can you be *essential*?

Let's say there's an insurance company and their strategy is to reduce costs. One of the biggest areas where costs can be out of control is legal fees. Depending on the size, insurance companies can have hundreds if not thousands of law firms that they hire to take on all kinds of cases across state, national, and international borders. Let's say you want to use data to understand which law firms are overcharging and by how much. Each contract you have with every single law firm is probably massive. What you should think about is this: "*How can I use data to try and compare patterns of usage so that I get the most efficient contract from that legal firm?*" Also ask, "*Is that legal firm overcharging me based on the types of legal matters that we agreed they could take on? Are they overcharging according to the contract that I signed? Could I get the data to show how much money in total per law firm has been overcharged and then dispute the charges to save costs and thwart overcharging in the future?*" You can do that type of contrast-and-compare work using data and analytics, especially with machine learning capabilities.

I would start by understanding the strategic goals of the business by poring through statements made to investors by the executive board of the company. That's where you start: what is the most strategic thing that you need to accomplish, and how does data play into that? It's not the opposite way.

> You don't start with, "*I'm going to be data-driven.*" You start with, "*I'm going to achieve this strategic goal, and this is how I'm going to use data to do it.*"

AA: Once an organization has an idea of their business strategy and they can formulate a data strategy to support that, what are some of the next key steps? Would you recommend they first employ certain people, or should they first build a tech stack and then bring people on board? And how should they structure those teams? Do they need a CDO? Do they need various team leads? Do they centralize the data science capability or distribute it?

CA: That all depends on the company and the industry they're in, how small or big they are, and how much they're looking to streamline what they're doing. There are also legacy systems and technical debt to consider. But really, it goes back to the company's strategy and how much money it has to do these things.

Let's say that you're in a 1,000-person company, and you have the ability to hire a chief data and analytics officer who can then go and interview the CEO, find out what their strategy is, look across all of the available data, and get a good understanding of the inventory of data that exists in the company.

Then, you say, "*How can I use the existing data to take on some smaller projects that move us toward the goal, so that I can reach the finish line and have some wins?*" If you're just starting out with being data-driven, it's important to demonstrate small wins along the way. So, you take a long journey, and you say, "*I've got a 30-day deadline – here's what I'm going to deliver. I've got a 90-day deadline – here's what I'm going to deliver; all the way up to the 5-year plan! And this is how we're going to keep moving that trajectory up, depending on what the strategy is.*"

In that insurance example, you want to reduce costs. So, you might say, "*We're only going to take on law firms dealing with a certain type of legal matter.*" You divide things up. Have you ever heard the phrase *you eat an elephant one bite at a time*?

> Instead of trying to tackle a huge project all at once, you tackle it in bite sizes so that people can see results. As you progress, you start to grow as well; it's kind of like an internal start-up. You're an *intrapreneur*, so to speak, instead of an entrepreneur.

AA: Given that you provide advisory services to Fortune 500 companies, I'm curious to know: what are some of the common mistakes that you see organizations making when establishing or scaling their capability? What roadblocks do they often face, and what should they be doing to overcome them?

CA: The mistake I see the most is that there's a lot of politics involved in data and analytics, from the minute that you start looking into people's data. There's also a lot of politics in terms of **Chief Information Officers (CIOs)** versus data and analytics. We have CDOs and people who are trying to use data who consider themselves more data-astute than other groups. That means I see a lot of groups, before they even get the chance to run those bite-sized projects, getting mired down in the politics of who supposedly owns what data inside the company.

CDOs who come in from outside a company have to be really good politicians.

> CDOs have to be good at negotiating and influencing people who aren't their direct employees. They have to manage up and down and sideways. If they can't do that and they're just coming at things with a hardened attitude, they're doomed from the get-go.

You have to have supreme diplomacy skills when doing this job. A lot of times, I think the personalities involved are often what causes a company to have a failure in data and analytics, as opposed to not having the right data at hand. What's sad is that almost every company I know of has had all the data that they need right there at their fingertips within their own company, but unfortunately, infighting causes people to be delayed and not work together. That causes failure every time.

AA: I completely agree. I see the same thing: it's the people side that is often the real barrier, not the technology or the data.

Implementing an ethical approach to data

AA: I've also been seeing lately that a lot of organizations want to ensure they are responsible when it comes to AI: implementing ethics frameworks and having people accountable for the ethical and responsible use of AI and data. How do you typically guide them on this journey in regard to ethics and governance around AI? Do you have any particular frameworks or processes that you would recommend (across different organizations) in terms of what is really needed to become ethical in this space?

CA: This is more about culture. It's also about going back to the main high-stakes usages of AI.

People are going to hate to hear me say this, but not everybody needs to be concerned about ethics and AI. If they're just doing A/B testing on steroids, they don't necessarily need to be concerned.

When I talk about the ethical use of AI, the starting point is typically an impact assessment of the use case that you want to move forward based on.

Let's say it's some C-suite person who wants to make sure AI is being used ethically. Typically, the way you want to start with ethics is you want to look at the most impactful, high-stakes, high-competitive-advantage types of strategies that you're using AI for inside your company. You can determine that without going through some big, lengthy process. You can pretty much just tell. There's a red-yellow-green situation here. If it's going to be life and death – if a person steps into your self-driving car and can die, or someone outside of that car can die by being hit by that car – that's a high impact. That's a red-level threat. If someone's life, liberty, and safety can be affected, that's red. Health diagnostics, weaponry, or anything automated is going to fall into this category. Automated with humans involved is probably going to fall into the category at some point.

If you take a step down, the orange level would be along the lines of whether we are going to affect anyone's rights or happiness. Are we going to keep them from getting jobs, or anything that would fall into the UN framework of human rights? If we're going to limit their ability to have shelter, food, water, or education, or the pursuit of happiness, that's high stakes. We have to go through a much more rigorous process of evaluation: is the use case even an appropriate use of AI?

We can argue that some people in this field think more like computers than people. If you plotted a spectrum from computer to person, there are a lot of people who can fall into the thinking-like-a-computer part of the spectrum. You have to evaluate the use case considering that spectrum. Some people are OK with receiving, for example, a chatbot version of, "*Hey, you've got cancer.*" That, in my mind, is never going to be an acceptable use case for AI – never. But in other people's minds elsewhere on that spectrum, they'd say, "*It'd keep doctors from having to deliver that information, which would be hard on them.*" But what about the person on the receiving end of this news? Did you think about them?

I know it seems ridiculous to think a chatbot might deliver such horrific news to someone, but during COVID times, when doctors were overwhelmed and going down with COVID themselves, the idea was becoming more of a possibility. I'm so glad we didn't fully get there, but we were on the cusp. If you read my book, you'll see how close the UK **National Health Service (NHS)** was to implementing this.

AA: That's a great example. How effective do you think AI ethics boards are in helping organizations implement responsible AI?

CA: This one's a tough one. It's not because of the AI ethics boards. It's because of the way that the AI ethics boards are set up for failure or success by the group of people who is the most powerful inside the organization. We call this the **HIPPO**: the **highest-paid person's opinion** inside the organization. The HIPPO will be the one that is adhered to whenever there is a conflict between the AI ethics board opinion and the HIPPO. And this causes a natural contention that can make it hard for external AI ethics boards to function properly.

I just commented on this in *Wired* about the Axon CEO who openly announced he was thinking about developing taser-enabled drones for the inside of grade schools. It was a response to the Uvalde grade school shooting (`https://www.wired.com/story/taser-drone-axon-ai-ethics-board/`). The announcement caused his AI ethics board to resign because they had already discussed it with the CEO and disagreed on the development of taser-enabled drones for schools. They didn't want to be associated with that – rightfully so, in my opinion.

You need to consider how you set these boards up, how much power and visibility they will be given, and what you will use them for. In my mind, the best use of AI ethics boards is to bring in diverse sets of opinions, especially if you're a smaller start-up or a mid-sized firm that doesn't have access to people who are regularly researching in the field of ethics. Typically, you want to bring in an AI ethics board that's going to lend you their diversity in some way, shape, or form.

Diversity can exist in many different ways. It can be socioeconomic. I think at one point, Starbucks had considered doing away with cash, and what they found out real quick was that if you do away with cash, there's a whole group of people in America, about 40 million people, who don't have credit. They can't whip out a piece of plastic and pay for stuff because they get paid in cash. There's a whole cash world out there that's not necessarily criminal. The data scientists working on the project had never been without credit, and didn't realize this isn't the case for everyone.

You need those very diverse opinions to remind you that there are people out there that don't think like you. You need to figure out how to accommodate them or you may lose them, you may have public backlash, or you may face some kind of societal impact that you weren't expecting. I think that's where ethics boards are really strong, especially when you're moving into a new space that you haven't been in before.

Google's external AI ethics board was short-lived, but let's say Google had wanted that board because they were moving into weaponry, which they had not been doing. Let's say they wanted to think through fully what that decision entailed. They had a member of the board that was supposed to help them think through the ins and outs of that. I think those are really good ways to use a board.

Unfortunately, a lot of companies want to just virtue signal by having and announcing the board publicly and announcing when they take the board's advice, but not announcing when they don't take its advice. That's the rub right now: it's trying to figure out that balance, and that's hard.

AA: Speaking of ethics more broadly, there are so many different definitions around bias, fairness, and explainability. How do we reach common ground? Will we have a situation where different organizations are implementing different interpretations, and how does that affect the citizen or the consumer?

CA: That assumes that the decisions will be transparent at all, which is one of the things that I'm working on right now.

I've been promoting the **12 Tenets of Trust**, which is on the **AI Truth** website right now. I think you need to have some way for people to have a sense of agency and transparency in the process of what you're building. Otherwise, you will have bias that you can't account for, because we're all biased. We're only human, so we're biased. Unfortunately, we're also the ones who build all of the information, and the information comes from us, which is also biased. We also can't find data for every single aspect of our lives. As much as we produce data constantly, sometimes there really are areas of our lives where we can't provide data that will back up a decision about us that affects us.

> **12 Tenets of Trust**
>
> In Cortnie's fantastic new book, *What You Don't Know: AI's Unseen Influence on Your Life and How to Take Back Control*, she shares her insights and expertise to help everyone understand AI beyond all the hype. She also provides 12 tenets, or principles, to help creators develop trusted AI, which you can also find online: `https://www.aitruth.org/aitrustpledge`.
>
> You can read more about *What You Don't Know* on Cortnie's website: `https://www.cortnieabercrombie.com/`.

You really have to set up your high-impact machine learning capabilities with transparency in mind. Again, not A/B testing stuff, but the high-impact stuff that incorporates feedback from those who will be affected, if you can do that.

In the financial business, we know how to implement identity and verification. There should be no reason why we can't allow someone to verify their identity and then respond with, "*Hey! You are munging five different sets of data. I see that you had this one thing in my financial score. I see it here: you gave me access. I can actually see it online. I know where to go to get to it, and I can see the explanation of what was weighted, why I was given this score, and why it might have gone down or changed, and I want to dig into that because it's currently affecting my ability to get an education loan or a house loan.*" I should be able to click into the system, and then it should open up a way for me to answer questions such as, "*Where did that data come from? What was the sourcing of that? What was the lineage of that?*" Then, I can participate. Was that data correct or wrong? Did it have me living in some part of town that was maybe risky according to the model, even though I have a really nice house there?

That feedback loop and the ability to have some personal agency where we can weigh in is important. Even our music choices give us that ability. So why don't the most major decisions in our life? For music, I can say, "*I like that song; I don't like that song.*" On Netflix, I can say, "*I like that movie; I don't like that movie.*" Why can't we do that for the data that's affecting us the most? Why can't we have the explainability and transparency in place? Somebody on the backend should have an automated capability to compare, an ability to do this thumbs up/thumbs down per piece of information that's affecting our outcomes. They should have some sort of automation that can then affect another system to change data permanently.

Then there are notifications.

> **If something's high-impact, we owe it to people to give them notifications. It is the least we can do.**

If Netflix can do it, come on! If we're going to sabotage someone financially, I think that's the least we can do for them. Let them know and let them have the ability to weigh in, "*Yeah, that's my data. No, that's not my data. Oh, this person turned in a bad piece of information. It was a different Cortnie Abercrombie over here in this other part of town!*"

AA: That's one of the best arguments I've heard for explainable AI.

A lot of organizations fail in their endeavors in AI. I've seen statistics of around 85% failure rates. In your experience working with large organizations, do you think this is realistic, and if so, why is it happening like this? So much money is being spent on AI. It's relatively cheap and easy these days to develop machine learning models. Why are so many failing? Where are they getting it wrong?

CA: This is a hot point for me. This is a Cortnie Abercrombie statistic with absolutely no grounding, so I shouldn't even say it, but I think 90% of what goes wrong is in the data, and people just don't want to investigate the data because it takes time. It takes energy to come up with the right data that's fit for purpose. I think that we use a lot of scraped data. We beg, borrow, and steal whatever we have to because of the way that we have set up our processes. When we look at what is at the root of AI and machine learning models, there are three aspects. I'm sure every data scientist out there is going to say, "*You can't boil my whole job down to three things!*" I'm going to try anyway. It's data, algorithms, and training or implementation. I'm including the training of the algorithm in implementation. You could probably argue that it could go outside of the data side, or you could say it's part of the actual algorithm itself. But I think algorithm selection is its own beast, along with iterating constantly on that until you get the level of accuracy that you're looking for.

Additionally, we have another problem that nobody even acknowledges: what are the users going to do with this stuff? I was talking to someone who just bought a Tesla and had no instruction on how to use it. 20 to 30 years ago, even when we were just getting pivot tables in Excel, that was new stuff. People trained us on all those new things. Nowadays, we just hand stuff over and say, "*Here you go*," and we don't tell people anything about it. We don't tell them how it works or what it's been trained on. This friend of mine who bought the Tesla won't even use the self-parking feature on the car because she's like, "*What if a toddler runs out? Has this even been trained on toddlers? Can it even see a toddler with its camera?*" I think that's a legitimate question.

If you don't give people some level of training and understanding of something, they just won't use it, and that's how we get these failure rates. There's no trust. First of all, what data did you use to train it? That's probably the most basic question that everybody's going to have in their minds. The first question my friend had was, "*Has it been trained on toddlers?*" These are erratic little beings that can just dart right out behind a car. Someone else may be parking while we're parallel parking, and there may be a van approaching. Do I really trust this thing to take into account some erratic little being that's only 2 feet tall running out into the path? That question is legitimate in all cases of AI and data science. Where did you get the data, how did you train this thing, and do I trust that?

Think about scraping all those "labeled faces in the wild," which has been the most used open-data source for facial recognition – something like 35% of the pictures were of George Bush. Did you know that? That's ridiculous! Because it was a dataset created in the late 90s. It was mostly his face because we didn't have as much social media participation as we do now. Even the frequency of updates is so important. How many data scientists do you hear these days saying, "*Well, all of my 8,000 APIs are updated on this date*"? We don't know! We're just pulling this crap together, managing it 50 ways to

Sunday. There are 8,000 APIs coming in. I don't know when they all come in! I don't know where they came from! I don't know who put that together!

And yet you have the Cambridge Analytica situation where a company pulled Facebook user data, did a personality test on everybody, and used that data to target specific people with political campaigns.

You've got to know where this stuff comes from and you've got to investigate and interrogate it, especially if it's one of the major features driving your model. If it's actually something that's making a big difference, you owe it to yourself to know everything about those bits of data that are coming in. I'm not expecting people to know all of the 8,000 different APIs that they use, but they should know the major things that are affecting their models, and that's where I think things are going wrong: it's the data – and not understanding it. All of this leads to not having trust in the AI product or service. Lack of trust leads to non-use, and that leads to the failure rates we see today.

Establishing a strong data culture

AA: Data culture is a somewhat related issue. In your opinion, what does an effective data culture look like? How do you advise organizations on building a data culture?

CA: It goes back to the C-suite culture: the people in charge, how they view data, and how involved or how data-literate they are really affect the data culture. The reason that people have a poor data culture is usually that they have people who don't know anything about data at the top.

There can be problems at both the bottom and the top of the organization. I have a top 10 list on my website about this, in an article about when, as a data professional, you should just walk away from a situation because you're not really going to get anywhere (`https://www.aitruth.org/post/10-signs-you-might-want-to-walk-away-from-an-ai-initiative`). There can be this over-amorous feeling about data. The CEOs and C-suite-level people can sometimes think it's going to solve world hunger! They don't have a clue what it's actually supposed to be able to do within their bounds and their company, but they think that it does a lot more, and they think that somehow the data's going to do whatever analysis needs to be done itself. They don't think about the people who are actually performing the analysis, how long it takes people to get things done, and how much data needs to be cleaned up.

We used to laugh a little bit when I was at IBM about executives who would promise to get data solutions up and running within four weeks. We would say, "*Yeah, that's going to be just an initial investigation of your data.*" Anytime you're working with data, you have to understand the quality of the data that you have and so many other aspects, such as where you're going to get it.

At AT&T, they did projects for everybody on the planet and they had 1,000 people acting as "data librarians" – that's my term, not theirs. You could go to these expert data-sourcing resources and say, "*I'm on a computer vision harvesting project for John Deere tractors, and they want to know the varying stages of ripeness for a red cabbage. Do you happen to have pictures of red cabbages in varying stages of ripeness somewhere?*" There was a 1,000-person team that could say, "*Yes, there's a place over in Iowa that's been working on this. We will procure some datasets or an API for you.*"

Sometimes, the data is easy and readily available, depending on what you're trying to accomplish, but other times, it's a hard use case and you're going to be tapped to try to figure out where to get it. Where am I going to source this information, and is it even possible to do so? There's a whole investigation that has to happen. If your C-level leader doesn't understand what goes into it and doesn't trust the people that are working for them, it's not going to work. You're working with all these vendors that have been hodge-podged together, which a lot of C-suite people do because they just see the numbers: "*Oh, it's cheaper if I outsource this.*" But what you're dealing with sometimes is that they're just throwing bodies at the problem as opposed to actually having expertise – expertise can sometimes cost more.

The C-suite can have a great effect in terms of how much time they give to a project and how much leeway they give to people about finding data sources, investigating them, and pulling them together in the right ways. I've seen that when people are not given enough time or budget, they'll just go for the cut-throat version of things. They'll say, "*OK! I've got $1 per record to spend on something that should normally cost $100 per record,*" or, "*I need genetic information on this but I'm not allowed to have that, so I'm just going to make up some stuff.*"

You see all kinds of bad practices because the C-suite has unrealistic expectations. But then, you see bad behaviors going up too, from the bottom up. You see some data scientists that are just lazy. They don't want to do things the right way. They are collecting experience on their resume like baseball cards. They just want to go from the project that they just got offered to this project, to the next project, and then they're going to put that on their resume, and they're going to keep moving their salary up from $100,000, to $200,000, to $300,000, to $400,000.

There's bad data culture everywhere.

> The best thing you can do is be literate about the data issues, have some trusted people that you work with, and pay them well.

Look at the types of products people have taken on and how long they spend at a company. If they are one of those people that's just in it for 12 to 18 months and you only see one project per place on their resume, that's a pretty good sign that they're just going to rush through, not document anything, and then leave you holding the bag with no **Return on Investment** (**ROI**) at the end of it. That's my personal opinion.

AA: Yes, that all resonates with me. I love the way you also address the issue with data scientists themselves. People are often very guarded about speaking negatively, but let's be honest: there are many data scientists who are collecting experience, just trying to move up the ladder like any working professional. They're no different from anyone else. They can be quite crafty with what they put on their resumes.

CA: That's exactly right. My thought is, "*Go with the people who you trust, and if those people happen to be inside the company already, then just teach them the skills.*" Remember my boss from before, he said, "*I know you can do this. You're ambitious. We're just going to give you all the classes you need.*" I go for trust and the personality types that I think would do well, and then I just give them the skills.

That's how I approach it, as opposed to the opposite, where data science candidates have promising skills, and then they come in and then they're not really that fantastic, but you've paid a ton of money for them and possibly signed a long-term contract. You don't want to get to the point where you're at the end of the game and thinking, "*I've already sunk a million dollars, and now I have no idea what this person did.*"

There might even be no documentation because a lot of people see failing to document what they've done and how they've munged data together as a way to control the situation, provide job security for themselves, and increase their salary. Some people do that. I'm not saying everybody does, but some do.

Then, we also have the opposite situation. There are also abused data scientists out there too, who are truly trying to do the right things but the time frames and budgets that they've been given are just so unrealistic that they couldn't possibly deliver a quality end result. Every profession has good people and bad people and people in between just trying to survive.

AA: I'm sure you've seen many examples where things have gone awry in terms of poor data cultures.

CA: Let's face it: data culture is just so important. One other aspect of this that I'm learning through research right now is that within the pod structure of data engineers, junior data scientists, and senior data scientists, it is the junior data scientists who are the ones who are most likely to blow the whistle when they're not being paid attention to. When they're raising objections in a process, they need to be listened to. What we're not seeing in these big companies is the ability within the agile process to push back, to assume that there will be some red flags, and to slow down. We've become so accustomed to our agile processes delivering the **Minimal Viable Product** (**MVP**), even if it's just an API feed, that we've become used to delivering in six to eight weeks. Sometimes, that's just not possible.

Anytime someone pushes back or people are under pressure, it's important to know what they're going to do. Are they just going to drop everything and say, "*OK, fine, whatever. I'll just be unethical – it's fine. I'm just going to deliver this because my bonus depends on it,*" or do they actually care? Will they push back on this part of the process, going up against a senior data engineer and a senior data scientist? Can they hold their own there, and do you have support for them to hold their own? Will the lead data scientist interfacing with the chief marketing officer or digital officer give these people enough pushback ability? I think that's not happening at all.

I think what we have is a whole lot of abused data scientists who are trying to raise the flags, and others are saying, "*No, you're slowing down our process. We're not going to get that patent filed in time, or we're not going to get this thing out to market in time, so we're just going to push you down.*" If that's the case, you have a very toxic, very risky data culture, and you have to figure out how to address that so that raising red flags (and, more importantly, fixing critical issues that cause red flags) is the norm, not something that you're blacklisted for. You shouldn't be blacklisted as a data scientist for bringing up risks that need to be addressed before a product is released.

AA: You've touched on an issue that isn't normally voiced: the realities that data scientists face with unrealistic expectations and the culture they have to abide by to survive, which is often very toxic in nature.

What do you think practitioners (machine learning engineers and data scientists) should be doing to make sure they're contributing in a positive way to the ethical development of their models and products?

CA: The overarching way that it needs to be approached is from the top down and the bottom up. Practitioners definitely have to be that last stop. The buck stops here. The responsibility is in everybody's role. That's why I semi-hesitate when I see AI ethics being called out as a separate group within a company. It bothers me because I think that gives you an easy scapegoat when things go wrong.

> You just get to point over to a group that somehow failed you, when in reality, every single person that's involved in AI development should be feeling that they are responsible in some way, shape, or form for every part of what they're doing.

The thing that concerns me the most is this attitude in the practitioner groups of, "*Why does that even matter to me? That doesn't have anything to do with me.*" Those are the people that need to find the problems and bring them forward because they are the people who should be involved in actually investigating data and understanding how models are set up. Only they know what they chose to use, what they chose not to use, and the decisions that went into that.

Let's say they feel responsible, but they don't feel like they can raise their hand, raise the red flag, and say, "*Hey! I think we could do better on this data. I've seen a lot of flaws in it. I've seen that it doesn't have enough of this type of people. We just left things out in general, and I think that the proxy data was crap*" – that's a problem.

In the case of Amazon's hiring algorithm, we actually saw that the data scientists themselves were the ones who did the whistleblowing. They did it anonymously, and I'm so grateful that they did. Those executive sponsors (such as the chief human resources officer, for example) don't really know what they don't know, and so they would have continued forward with that hiring algorithm had the data science development team not raised the flag. The fact that we still have to call it whistleblowing means that we don't have a culture or a set of norms yet that is conducive to pushing back, and that's the problem.

AA: Yes, it's everyone's responsibility. I couldn't agree more.

With the growing influence of AI in our daily lives, trust is a big issue. How do we develop trust in AI? I think your book goes a long way by helping the layperson understand what is and isn't AI. Beyond that, more broadly as a society, what should we be doing? Also, how important is regulation in your opinion?

CA: We could strip those down, and each one has an answer.

In my book, I do talk about the *12 Tenets of Trust* because I think that all across the globe, we're in a time where we have the lowest levels of trust among people.

How can we hope to build and scale AI at this point when there's no explainability, no transparency, and no accessible information about what goes into models? I don't think anybody has earned the trust of anybody right now in the AI space. The typical answer is, "*Well, I have accuracy,*" but we all know that 100% accuracy on garbage is still garbage. I hate hearing, "*Well, my accuracy rate is so great.*" That is not a good, solid statement of whether or not we used sound judgment for the things that went into the model in the first place, the data that went in there, or even the way that we conducted the training of the model, and so forth.

People don't trust each other. There was a research study out here in the States from a group called Pew Research Center (`https://www.pewresearch.org/topic/politics-policy/trust-facts-democracy/`). Right now, we don't trust scientists, we don't trust the government, we don't trust the news, we don't trust social media, and we don't even trust each other as neighbors to do the right things. For all the many hundreds – possibly thousands – of ethical AI frameworks out there, I would say this. There's one simple thing that you have to remember, and that's the golden rule, which is to do unto others as you would have them do unto you. As long as you can remember that, you should do OK with trying to develop trust.

I think we've become addicted to moving fast and breaking things. That's why I recommend this 12-step program to break that addiction, which I'm calling the *12 Tenets of Trust*. Whatever you are doing in your data science, if you wouldn't steal $20 off of the ground if someone in front of you just dropped it, and you would instead go find them and say, "*Hey, here's your money back,*" then you should do that in your data science too. If your model's stealing money from people and you as a person would not do that in real life, and instead would chase someone down and give them their money back, you should adopt a similar mindset when you develop your models as well.

We tend to have this thought that we leave part of ourselves behind when we come into the office. We somehow think, "*OK, to move forward in data science, we just need a corporate mindset – we have to make money.*" That's our fiduciary responsibility to our stakeholders, but there's still a line. If you wouldn't do it otherwise, then don't do it just because you're working at a company because, at the end of the day, you still have to look your kids in the face. You have to look at yourself in the mirror and say, "*Hey, I did something good today,*" or, "*I did something terrible.*" So, the golden rule "do unto others" is the number-one way to make sure you've got trustworthy AI.

AA: Can you please elaborate on the 12 tenets?

CA: These are the 12 tenets that I think the public should expect all of us in AI development to adhere to in order to put their trust and faith in our products or services. The very first tenet we should meet is developing AI that is humane. People ask, "*What does being humane have to do with data science?*" But I think the very first thing we have to do when we come up with a high-impact use case to fund is ask, "*Is it a humane thing to do?*" The chatbot that delivers cancer diagnoses is probably not a good idea, for example. The same goes for that article in *Wired* that I was quoted in where a CEO wanted to use taser-powered drones in schools for children. That's not a humane use of AI. Do we need taser drones inside school buildings with small children? No, probably not. You've got to start with the use case, and say, *"Could this cause more harm than good? Is this an appropriate use of AI? Is this humane?"*

The second thing to ask is, *"Is this consensual? Am I taking some data from a person that is gathered for a whole different context from where it's being used?"* A prime example I use for that is a group called Clearview AI (`https://www.oaic.gov.au/updates/news-and-media/clearview-ai-breached-australians-privacy`). They scrape people's information and faces from social media and work with law enforcement to provide potential facial matches based on that data. When you find yourself begging, borrowing, and stealing information from people on social media sites, that is not the original intent of the information. If you want to violate people's trust, go right ahead and keep doing that. But if you want to build trust and make sure that you have a consensual approach where people know what their data is being used for and that it's being used in a way that is consistent with what they've agreed to, then you want to be transparent. We've already talked about that.

Also, transparency is not enough. It's not enough to say, *"Hey, I have the ability for you to know about these things."* If you don't inform people and make the data held accessible to them online so that they can see it for themselves, that's not what I consider to be transparent or accessible.

The other part of this is personal agency. I need to be able to go in and affect something to do with myself, or at least understand it. It should be explainable, which is the next tenet. Can I affect that that is the wrong address, for example? Can I say, *"That's not the right location. Please go find that information and rectify that for me"*? That rectification is actually part of the 12 tenets as well, which we don't often see. We see a lot of explainability right now, which is fine. It's hilarious to me, though, that everybody talks about explainability and nobody talks about accountability, traceability, or the ability to govern and rectify these situations, which are the other tenets.

We also need privacy and security. You can't just take the X-rays of people who have had COVID and stick them where any hacker can get to them, along with dates of birth and everything else that would reside with those medical records. That could happen if you come up with a fantastic X-ray-diagnostic type of machine learning capability. Healthcare is fraught with those kinds of fast-and-loose methods of just throwing people's information into a single place and then leaving it unlocked. It's like saying, *"I just threw all your stuff into a huge purse, and then I put a big advertisement on the outside of it, and I can't believe people went in there and stole your data. I don't understand why that happened."* People are not thinking about that. What people are thinking about as start-ups is, *"I've got to hurry up and get this product out there and claim first-mover status. I'm going fast and loose with people's information. I know there are some guidelines I'm supposed to be following, but within my own company I thought it would be OK if I just did this."* No. Just because you thought that would be OK within your own company, doesn't mean it's OK. Most of the breaches that we see today are from within, especially in the competitive environment among start-ups where infiltrations can be common and workers move from one competing start-up to another.

The final thing is making sure that your data's actually correct. You need to make sure you have fair, quality information going into your model, not just some random bits that you were able to procure from somebody who just played a game on Facebook, swears that they now have the psychographic information of people, goes out and labels people as persuadable or whatever else, and then sells the data to the highest bidder (true story: that's the Cambridge Analytica case). That's just wrong. I think that you have to know where your data's coming from. Does it have bias in it? Is it actually correct? If the data quality is really bad, you're going to have overfitting, which is also going to cause the model to put out weird results. It'll look like you're getting some level of accuracy, but when you go back and look at what you've got accuracy on, it's not going to come out right. You've got to pay attention to the data itself, and I've been amazed at how many data scientists don't actually go in and thoroughly investigate their data.

To recap, the first 10 of the 12 tenets are these: *humane, consensual, transparent, accessible, agency-imbuing, explainable, private and secure* (that's one), *fair and quality, accountable*, and *traceable*. Traceability is about whether you know *when* something went wrong and how it went wrong, and whether it can be traced back to a moment in time. People are using blockchain to do some amazing things with traceability, especially in credit reporting.

Tenet 11 is incorporating feedback. If something isn't working the way it should, there should be a way to input feedback. This is especially required for expert systems, such as AI trying to take the place of actuaries. Believe it or not, people don't find that a fun profession anymore, and we're finding that Gen Z and millennials don't really want to go into that field. Now we're training AI to try to do it, but if you don't have experts continuing to weigh in in some way, shape, or form, you'll have drift. Bias can also occur if you don't have ongoing feedback loops incorporated where humans are definitely in the loop.

The last one is *governed and rectifiable*. What's the point of having explainable AI if you're not intending in any way, shape, or form to fix anything that goes wrong? We're not even talking about that as an industry. We're so focused on bias and explainability that nobody's stopped to ask the question, "*Well, what do you do when you find bias? What do you do when you find out that your model has drifted all the way over here, or it's turned into a racist, like* **Tay**? *What do you do? Do you just shut it down?*"

> **Tay**
>
> Tay was an AI Twitter chatbot created by Microsoft in 2016. Intended to learn from its interactions with human users on Twitter, Tay ended up being taken offline after just a few hours as users taught the algorithm to make offensive statements (`https://www.bbc.co.uk/news/technology-35890188`).

Think about the Tesla example. We have self-driving cars that can't even be shut down from outside of the car. What do police do when they find a couple sleeping in their car that's just racing down the highway? None of us can do anything. Do we just have to hope and pray? No, that's not acceptable. We need to start considering how we're going to rectify situations when we build. It needs to be in the design from the get-go, and that's how people will trust us: they need technology that they find trustworthy. Can we shut things down? Can we govern them? Do we know when things actually went wrong? Is someone even accountable to fix this? That's my other pet peeve: you have model drift. Who's going to fix it? "*I don't know. That person already left the company.*" In what business are you allowed to say, "*I don't know,*" throw up your hands, and say, "*Well, anyway, on to the next customer.*"

AA: Have you seen that happen?

CA: Yes, absolutely. It's because of the way that AI gets implemented with all the different vendors and contractors. People are transitory in this space, and so what you see is that a project gets developed, and then it gets left behind with the people that are using it, but the people who are using it don't know how it works. So, there's nobody accountable.

AA: How important is regulation now in all of this at a government level and an organizational level? Are you an advocate for regulation?

CA: I usually am not an advocate for regulation. Not because of the intent of the regulation, but because of the government's ability to execute the regulation in a way that doesn't overreach to the point of creating barriers to entry for smaller companies who can't afford to lobby. That said, I do feel like some areas such as social media, autonomous weapons and vehicles, and healthcare urgently require regulations in order to prevent major disasters from happening. For example, we have major social media platforms that are influencing everything from teen suicide rates to how elections are won and the full-on destabilization of countries through the use of fake news and the amplification of hate through bots. This cannot be allowed to continue. For these firms to continue profiting from ads while society goes down in flames is akin to Nero watching as Rome burned.

The big question on my mind is, "*How can we get the regulations we need when we have congressional representatives whose average age is 70 or above, who don't even understand how these firms make their money or how the technology works?*". The Facebook (now Meta) and Google congressional hearings in the US, where a key congressman asked how Facebook makes its money, were a sad demonstration of just how little Congress knows about the tech industry. If they don't know how the business model works, how can they ever hope to understand what drives these companies to do what they do and under what circumstances they do them?

Facebook congressional hearings

In 2018, Facebook executives were required to speak at a congressional hearing in the US (`https://www.vox.com/policy-and-politics/2018/4/10/17222062/mark-zuckerberg-testimony-graham-facebook-regulations`).

While I have little faith in the actual lawmakers themselves understanding the problems, I do think there are enough of us out there advocating that we can educate them and the public at large. I wrote my book specifically for people who aren't in the tech industry to enable them to understand the issues and impacts better. I want to get my book into the hands of as many legislators as I can so that they can be brought up to speed on what AI is, how it works, why adverse impacts happen with AI, and more importantly, what we can do about them.

AA: So, you think education is important?

CA: Yes!

> I think the more that people – both inside and outside the tech industry – know, the better. That way, we can all exert pressure on both regulators and the businesses that use data science and AI. We need all people in society to be aware and understand the areas where these technologies affect our lives.

The areas people ask me about most often are related to jobs and being replaced by AI, such as helping a student pick a profession that won't be usurped by AI, or how ads track them around online. All these years, I've been collecting AI questions from lots of different types of people who I've encountered: ride-share and taxi drivers, church ladies, family members, postal workers, and parents in line with me at the grocery store. I've tried to understand the main questions and frustrations in the minds of people from a slew of different backgrounds. I've taken those main categories of questions and made them each into a chapter in my book. The topics include the following:

- AI in hiring – including AI interviews and social media flagging
- Job replacement and automation – which jobs and what skills
- Impacts on kids and teens – tech addiction, suicide, harm challenges, and trafficking
- Political polarization and radicalization – fake news, conspiracies, and bots
- Rights and liberties being usurped with AI use in criminal justice – predictive policing and facial matching algorithms
- Life-and-death AI decisions in healthcare

Globally, one of the areas I think people feel most frustrated with but can't quite pinpoint why is politics. We are all constantly outraged, pointing fingers back and forth between political parties. We cannot get beyond it. But what the average person doesn't understand is that there are actually bots out there keeping the angst going by magnifying outrageous salacious content designed to provoke visceral emotional reactions. These bots are intended to polarize and destabilize democratic countries. There are a lot more of these bots than there are moderate people. The theory is that there are a lot more moderate *people* out there but their content is boring, so it doesn't attract eyeballs. Because moderate content doesn't attract eyeballs, it doesn't get advertising dollars, and because of that, it doesn't get amplified.

If the standard person understands that this is how everything operates on social media, that in fact their next-door neighbor is just forwarding something that was put out there by a bot, or is fake information, I think it would change the dynamics of how much they invest their personal psyche into the hatred that goes into comments. If you know you're interacting with a bot, you're probably not going to be as hateful about it as you are if you think you're truly being somewhat usurped in your immediate society. I think that is what happens with bots. I think people see some of these comments online and think, "*Oh my gosh! This is a personal attack on me.*" These bots are so rampant, it's not even funny. They're there just to provoke hate in the most devious little ways.

AA: That's great context. It's something a lot of people just aren't really aware of. I think it's very important that data scientists are cognizant of how market forces and technology combine and are shaping society.

Designing data strategies

AA: I would like to discuss data strategies. I would love to know what you think are the key aspects of a data strategy. If you were asked by a company to develop a data strategy for them, what would you be thinking about? What aspects are front-of-mind for you, and how would you look at developing a roadmap for them to actually implement a strategy?

CA: I feel like I'm somewhat of a broken record on this, but I would always start with data, their business strategy, and then whoever the group that has the most data inside the company is. Maybe they have a big external project that they want to take on and they're going to need external data, but most companies start with some internal data that they can ratchet up to produce things.

It's hard to talk about a data strategy if you can't talk about the actual use case, what the strategy of the company is, or the industry of the company. You need to know what the market levers are and what data is available, but first and foremost, what the actual business strategy is and who has it, and how much money they have to go after what they're trying to accomplish. You also need to know about all the people inside the company that need to be at the table. I think the key aspect is having a business strategy and data strategy to support that, rather than something that's separate and distinct, which I've seen happen, and which is definitely a mistake.

AA: When organizations are going from proof of concept to suddenly having to productionize that model and capability, I often see them hit roadblocks. They either don't have the right processes in place or don't have the right people. I'm thinking of MLOps and things like that to scale that capability. Have you seen some of these issues, and how do some of these organizations overcome this? What do you advocate for when helping them go from dev to prod at a large scale?

CA: The first part of this is understanding the executive sponsor and what they're really trying to do, because somebody somewhere in that company is paying for the development of this AI or ML product or service. It's usually a one- or two-person type of situation.

> They may come with a whole team that thinks that they're the ones who are dictating the project, but in reality, it's one or two major executives who will be the ones that set the tone for the entire development project.

There are many reasons why data scientists can't get past the proof-of-concept stage. One is that they don't have thorough communication between them and the actual executive sponsor. When there is a breakdown in communication from the get-go as to what the executive sponsor is trying to accomplish – say you produce a proof of concept of some sort (maybe an MVP), and the sponsor says, "*Hmm,*" or "*I'm not sure*" – that means communication between you and them has already broken down.

The second thing that I see happen is the executive sponsor may not understand what the heck the data scientists have done. Data scientists need to clearly answer questions such as, "*What did you do? What data is in there?*" Explaining things in technical jargon is not helpful to a business sponsor. That is not communicating well.

> **Data scientists have to find a way to actually communicate, which means that the person on the other side of the table understands what they say.**

Throwing in all the different models, all the different mathematical processes, and all the things that were thought through but didn't go into the model will just make the business sponsor not trust the data science team. Instead, they will feel like the team is trying to snow them. The best thing the data science team can do is spell things out. If there's something that the executive sponsor says that the data science team doesn't understand, then the data scientists have to get under the hood of that. They need to keep asking questions of the sponsor until they think they have a good hold on the request or concerns, and then ask the sponsor, "*OK, could you write that down?*"

> **When people have to actually write things down, I find that things get a lot clearer for both the person writing and the person receiving the writing.**

Those are some of the reasons. The biggest one, though, in my mind, is still trust: not having trust about what you've done and what went into the process, or whether you begged, borrowed, and stole to get there. A lot of teams feel like the overarching thing is that they've got to meet that MVP stage, which has to happen in anywhere from four weeks to eight weeks. That's the typical thing, but that doesn't usually give you enough time to truly do what you need to do with the data. I think that's where things fail because you do all kinds of horrible things in this rush to get to the MVP, and then you think, "*OK, I'm going to scrap all that and start over because now I'm going to have the funding that I really need to move forward.*"

However, what happens is that once the sponsor or funder sees the MVP, they say, "*Oh, no. We're just going forward from here.*" You can't say it out loud, but you want to say, "*Wait a minute, no. We built all this stuff based on chewing gum that we just stuck together. This isn't going to scale.*" They continue, "*Yeah, this is great! I'm going to schedule a meeting. You're going to show this to the CEO next week.*" And before you know it, it just keeps getting bigger as more people get involved. You can't go back, and then you're stuck with what you did in the first place, and now you don't get the proper investment to do it the "right" way or the way you had hoped.

You really have to set the expectations and have that first sit-down with the executive when you're planning these things out. You need to say, "*Look, I'm just going to be honest with you. The reason none of these things have scaled for you in the past is that you didn't give the team enough time to build these things the right way and to truly invest in it.*" People just don't want to be honest about that. They're afraid that the first answer will be, "*No,*" but I actually think it's different if you're honest and forthright and you show a legitimate plan. A lot of data scientists don't want to do that; they don't want to show a legitimate plan because it'll show their steps. It'll document things, and that causes accountability. But if you're honest from the get-go, you might actually be able to scale your thing, make it outside of the proof-of-concept stage, and reduce the failure rate.

AA: One thing you mentioned really resonated with me when you talked about trust for data science teams. Something I always advocate for in my teams and mentees is that your goal is not to be seen only as a technical expert but as a trusted partner to the business. I think that is a really important differentiation for people to make in their minds: that they need to be trusted by their peers and senior executives, and not just be seen as someone who speaks technical jargon. I completely agree with you there. They have to speak and understand the language of business and translate their findings into actual outcomes: profit increase, efficiency gains, and so on. That is pivotal in building their careers, especially as they progress up the ladder, but we'll get to some of that in a moment.

You talked about time pressures, which I think is a really big one. Sometimes, they're very unrealistic, for various reasons. What are some of the best ways to help teams prioritize their demands, as a leader of a data science team? If you're inundated with requests from senior executives, how do you prioritize what to take on? Do you follow any particular processes or methodologies around trying to figure out how to triage all of this, continue to add value, but still manage expectations?

CA: It's always going to depend on the specifics; every business case is different. It depends on what you're trying to accomplish, the relationships that you have, and how much trust you have between yourself and your team. That's hard because a lot of us deal with lots of contractors and vendors, too. You're not always going to have access to 100 data scientists just sitting there all the time. Sometimes, you have to say, "*OK, let's pull from outside resources,*" and those people are hard to trust because you don't really know what their levels of understanding of your operating environment are. When you use internal data and you're working with vendors, there are a lot more checks that come with that.

That being said, the relationship that the data science team needs to have with sponsors and champions is everything, and it's everything when it comes to prioritization too. Let's say you're working with the chief marketing officer and they have been going to the news and publicly touting this strategic initiative that they're about to do. Let's say the feature that they keep mentioning – that you haven't even built yet – is at the top of their list. You talk to their people and you say, "*Oh, yeah! They're so excited about this, this, and this.*" You might have something that's not on that list that you're working on, because we all get scope creep, and nobody's worse for scope creep than inquisitive data scientists. We're always saying, "*Wow! Why did that finding come up?*" You have to stop and say, "*OK, wait a minute. Is this actually contributing to all of those things that this sponsor keeps mentioning?*"

There are things that people tell you are important to your face, and then there are things that people will reinforce are important by the things they say to the public, investors, or other executives. There are also things that people don't want to admit they actually find important because they're things that are – for lack of a better term – cosmetic.

They're not deep and meaningful and purposeful to the business, but they're things that might give extra oomph to their campaign. You can get little wins like that with your sponsors. That's how you build trust: by prioritizing the things that matter the most to sponsors and the business.

> The last thing you want to happen is that you put all your eggs in one executive sponsor basket and then that person walks out the door. Who's left as your big sponsor within the company that knows what you can do?

So, a lot of it is politics. Who are the sponsors that you need to attach to, and what are their priorities? What is it within a project that will get the response, "*Ah! Let's keep going so that we can build out the other features*"? You have to keep them going. If the excitement fades to nothing because you're producing things that are a cool feature but not something that was ever asked for, then stop.

AA: Yes. That's great advice on the practicalities of tying your mission to that of the key stakeholders and decision-makers and their priorities. I think that is vital and may not be what a lot of people want to hear, but that's the reality of it, isn't it? If you're not going to get funding, you're not going to survive.

CA: Yes.

> You've got to act like you're a start-up.

AA: Often, I see issues around how success is measured. What metrics should we be using? How do we know what impact we're creating with our models? Do you tend to think about any specific metrics? I know it's a general question because it depends on the organization, but are there any particular types of metrics you tend to use to measure the impact of the solutions they're developing?

CA: What was the strategic goal the project was designed to impact? What business metrics are measuring progress against the goal?

Let's go back to the example of the insurance company from earlier in the chapter. The executive sponsor, let's say it's the **Chief Legal Officer (CLO)**, is concerned that the costs of all the different legal firms they're doing business with are eating them alive. It is becoming a detriment to the company. You build a model – and an automation capability – with the sole purpose of flagging and refuting legal fee overcharges. To understand whether your model is successful in this example, the question you should ask is, "*How much money did my model save the company in refuted legal fees?*" You should be able to go back to the original goal of the business sponsor (in other words, the CLO) and measure against that. You shouldn't have a separate metric for your model. You have business metrics. Period. Did it have an impact or did it not, and by how much? If it wasn't that much of a reduction (for instance, half a percent, equating to a few dollars, when you're spending 100 million dollars on data sources and all kinds of other things), it's not worth it. You need to stop the project.

AA: Something we touched on earlier that I'd like to delve a little bit deeper into is data literacy. It's something that I see as a big blocker in many organizations, especially at senior executive levels: their lack of data literacy can make or break a project. Is this something you're seeing? How much do you believe in enterprise-wide data literacy and development? How do you normally try and increase data literacy? Through training programs, such as off-the-shelf training programs or bespoke training programs?

CA: Data literacy is paramount.

I think data literacy is one of the most important things a company can work on to be competitive. Companies that aren't data literate – if they're not using data to drive every business discussion or decide how they're going to strategically fund themselves – are probably not going to be in business for a long time. They probably have a very limited future versus companies that are data literate. I would say the same is true of senior-level executives who are data literate.

They will adapt their strategies more quickly to marketplace uncertainty – which is something we have seen a lot of since the pandemic and we will continue to see a lot of in the future.

Data literacy comes in degrees. Some senior-level executives may not understand the inner workings of a data science team, but they most likely have an idea of the data they would want to see to understand financial metrics associated with the performance of the company. If you want to help your team by helping senior-level executives become more data literate, then forget about training programs.

> Instead, while in the midst of tackling a big data project for the senior executive, take extra care to communicate about the processes, tools, and key members of the team. You are the trainer of the senior-level executive that sponsors your major AI initiative. Educate them.

Assign one person from your team whose sole focus is to invest their time in educating that sponsor about how the data science team works, what data they use and how they get it, and the process they undertake. That is the only and best way to train senior-level executives. Because that is the only way they will be vested enough in an outcome related to data, to get them to care. Hopefully, that doesn't come off as harsh. But senior-level executives have a lot to do in a day. Being data savvy only for data savvy's sake isn't going to be on the agenda. But being data savvy so they can understand the risks (meaning, risks of a project's failure to launch or no ROI) associated with an AI project they are investing in…well, that's worth it from their perspective.

I don't encounter a lot of non-data-literate companies anymore – meaning the entire company of employees has no idea how to use data. If you're in a high-impact industry and you don't know data, I don't think you're going to be in business very long, and we have honestly probably seen that weed-out happen quite extensively during the pandemic. Especially to be in a world where everything is digital, you need customer data, market data, connected-device data, and data of all kinds. This is an essential part of doing business in the world these days.

AA: Yes, I think the definition of data literacy has expanded somewhat. It's also about having a conceptual understanding of analytics, machine learning, and AI, at least at a very high level. I think that's become important because some organizations, as I'm sure you see, get swayed by a lot of vendors who are still on the Kool-Aid.

CA: Right.

> You're talking about degrees of data literacy. There are the basics, and then there's how not to get snowed by a vendor.

I think the procurement part of that is definitely a problem. It's so much of a problem that I helped the World Economic Forum put out a piece for senior **Human Resources(HR)** leaders to help them identify the things they should be asking vendors who come in and try to sell them an AI hiring capability of some sort. I think those are good ways to tackle that level of data literacy. If people are having problems understanding what they should be asking vendors, I think that's a whole different thing. It's a different level of data literacy, but it's definitely a problem.

AA: Yes. A great point.

I'd love to hear your thoughts on what makes a great data leader: someone at the senior level – say, the CDO level or the data team leadership level. What are some of the aspects that you see in good, strong data leaders?

CA: I had to do this back in 2013. I was asked by IBM to put together the very first CDO community to promote getting a centralized data analytics function up and running inside of the organizations we did business with – we didn't even know what to call that role at the time. Was it a CDO, was it a chief analytics officer, or was it a chief data and analytics officer? There was no real precedence yet as there were only three CDOs that existed in the marketplace at the time.

Think about the Fortune 500 companies that you know out there. For any given company, IBM was selling data and analytics capabilities to the finance, marketing, operations, and IT departments of these companies. The IBM sales reps saw some overlap in what departments were purchasing. They finally just said, "*We need one person. We're doing some of the same types of sales over and over, but we have little bits here and little bits there in different departments. We need to create a role that can oversee larger, strategic data and analytics implementations.*"

I thought this was massive hubris on IBM's part at the time. I said, "*What? You don't just go create a role in the marketplace!*" They replied, "*Oh, yeah! We've done it before. The CIO, back in the mainframe days: we had to create an information officer to run all of these systems and mainframes within the financial institutions and NASA.*" I said, "*OK. I guess we can do that, then.*" They said, "*OK. I want you to treat the CDO – the chief data and analytics officer, whatever you're going to call it – like your product.*"

At that time, I thought about all the different data leaders that I knew from working at Verizon, at Citi, and across all the different functions within a company, from marketing to finance. I really had to look at what makes a great data leader, because I knew I needed to find those who were doing the best possible things and hold them up as examples to the marketplace. Then, I'd have to get the CEOs of all the biggest companies, pull them together into round tables, and show them the great things that these people were doing that would actually make them want to consolidate data and analytics inside their company and not just have 50 other people trying to access data. Otherwise, we'd reinvent the wheel for data access for marketing, finance, and every other department inside their companies. That's nuts! They all want the same types of stuff. There should be one person in a company doing this stuff.

The answer to your question is that the best data leaders are the ones that are asking the highest-level questions in a company and are attached to the incentives and strategy of the company as a whole. They think strategically at the CEO level about how they use data to reach the company's goals. That's what they do. Plus, they're politicians.

> The worst data leaders are really good at their actual data job, but terrible at managing the minefields of relationships with CIOs or CEOs.

That or they'll drop the data package at the doorstep – real or virtual – of senior executives and run in the other direction, as opposed to having full conversations and welcoming and educating people, almost as business counselors.

You know school guidance counselors? Well, a cross between a business guidance counselor and a politician is what I would say most good data leaders are. They just happen to also know how to wield data to meet strategic ends. The best CDOs manage to carefully migrate past political minefields, form relationships with all the different C-suite roles, and then come away with sponsors and champions to undertake strategic data initiatives – because usually, they're not the ones with all of the money to do all the projects. They make friends and influence people so their data analytics teams – federated with other departments or centralized under them – can shine.

AA: What's your transition into being one of the global data leaders been like? Have you found that challenging? I'd just like to get a better feel for what it's been like for you.

CA: I was incredibly frustrated because I had a vision. Since CDOs were my product, I had a vision for what CDOs needed to be and when they needed to be that. I started with two data officers in the beginning. They would have a career progression that started with data governance. They'd pull all the data together inside companies (this was back in 2013) and try to make all of that different data work. That was the reason why we were even making the role in the first place. It was why we were trying to make this role in the market and trying to get the CEOs to demand the role. We were putting out job descriptions around what CDOs should be doing, so some hiring firms could go out, hire the people, and start them in an organizational structure at the company.

With that in mind, the maturity progression was supposed to be from data governance to business optimization and then to market innovations. For instance, business optimization would be looking across the business and doing more operationally sound analytics to help businesses cross-sell to customers and things of that nature. A lot of companies do that now. Back in 2013, that wasn't necessarily happening at the ubiquitous level that we see now. Finally, you'd get to the level of market innovation where the company's data could become its own source of revenue through the release of data products and related services. You get to a point where your product is internal information that can then be wielded as some sort of a monetization plan for your internal data and practices. That market innovation part of the career progression was and still is an area that many data analytics executives just couldn't reach, or by the time they finally got there, it had been usurped by a different C-level role, such as the chief digital officer, chief innovation officer, or chief data scientist.

On a second level, I feel like there's been a split of duties. There's a group that stays focused on data. Sometimes they stay in the IT department, and sometimes they're separate. Then, we've got chief analytics officers doing more on the business optimization front for the business. Market innovation got taken over somewhere along the path by data science pods. That's what I call the little teams that work underneath each sponsor as needed on different initiatives. They can often work for digital officers. Sometimes, they're straight-up working for IT, but a lot of times, they're vendors being contracted from outside the company to come in and work for specific sponsors, such as a chief marketing officer for a specific project such as an AI social media command center.

I thought that this would be a good progression, but now I see my peers are stuck with data governance or internal business analytics. I've moved into AI and I want them to come with me. I'm trying to bridge that gap currently to show them why they're relevant to AI, because a lot of data and analytics professionals are still not willing to move into that market innovation space.

AA: That's a great point. What advice would you give to someone who's transitioning from a technical role to a senior data leadership role? Say that their career aspiration is to go from being a hands-on data scientist to eventually becoming a CDO. What advice can be given on that trajectory with regard to skills, development, or education?

CA: Networking is vital.

> Go out and start networking and politicking with as many of the senior leaders within your company as possible. Start forming relationships and understanding what drives them and what's most important to them.

Also, understand where the landmines are as far as data goes inside the company. What are the types of things that senior executives will shut down? What will cause people to start pulling into themselves when you push too far? It means being a politician, having data charisma, knowing where you can help and where you can't help, and knowing the CEO and their strategies. Every single one of those C-suite members is going to interpret that main corporate strategy into a strategy for their individual departments, and then you've got to figure out how you're going to be relevant to those departments and recommend projects based on their goals.

AA: As for formal studies, you have an MBA in strategy and business development. Do you think extra, formal studies are important for career progression to leadership roles, or can a lot of it be done through intuition, hands-on practical learning, and maturity in terms of skills development?

CA: That's a tough one because I think everybody learns differently. What comes naturally to me may not come naturally to others. I've definitely learned that. I'm more of an extrovert. I think CDO roles are made 10 times harder if you're an introvert and you do just enjoy working with the data. I will also say that maybe you don't always have to be a CDO to enjoy and be fulfilled by your work. For some people, it'll stress them out too much to try and move into that type of role. If it's stressful like that, maybe the role's not for you. One of the hardest things to do is to figure out what makes you happy, because you may get to that chief data analytics officer spot after learning all these things and doing the activities and just think to yourself, "*I am so freaking miserable. Just get me back on the team. Plug me back into the matrix.*"

To answer the question more directly, if you are intent on becoming an executive in the data, analytics, and innovation field, you will definitely need to understand how business leaders think. I do not think you need to go to business school if you have a computer science or other degree…but you will definitely want to go and talk directly with business leaders in your company to understand their goals and any political struggles they are having within the company. You will have to become highly adept at building relationships, so the development of interpersonal skills will be key. The book *How to Win Friends and Influence People* by Dale Carnegie will probably be something you will want to read if you feel you have no natural intuition or proclivity when it comes to building relationships inside organizations. I would also rely on employees who have a high emotional IQ and a lengthy tenure inside the company to help me navigate the political landmines. These don't have to be your direct employees, but set up a regular lunchtime or something of this nature with them to understand more about what they are seeing.

AA: Final question: when you're hiring (although you're probably too senior to be involved with data scientists and data engineers at the hiring stage), what skills and attributes do you normally look for in people? What do you expect them to excel at in terms of the technical stuff, such as coding skills, curiosity, and attention to detail? Is there a list you go through in your mind of what you look for in people?

CA: I think this is going to be consistent with what I've been saying, but I can't teach you the personality aspects that I'm looking for. I can just teach you a skill – Python skills, for instance.

> The main thing is having people who are flexible and adaptable in their mindset, because some data engineers and data scientists can be very inflexible.

What I mean is that once they are told what they will be working on, then if something changes, they will become frustrated if they cannot work on the project as outlined at the beginning. Data science is highly experimental, and internal and external clients of data science projects change their minds constantly about various aspects of what you might be building. You need people who are comfortable with being flexible and adaptable and also willing to roll up their sleeves and pitch in with every other kind of job, such as data sourcing and pipelines, investigating data, cleaning up the data, and putting the data into various formats. The more data and analytics skills they have, the better for being flexible.

> **Data scientists need a positive and experimentation mindset.**

In data science projects, you're going to be flinging things at the wall. Much of what you do will fail as you experiment. You're going to be going back and forth, trying different things until they produce outcomes the team likes. You have to be comfortable with this sort of frenzied, iterative style of working. Even when you're frustrated, you have to be able to articulate what you're frustrated about. A data scientist can't just stew because they're not going to get very far and they're probably going to burn out on the project.

> **Data scientists have to have a team player mindset, but also can't be a pushover in the process either.**

They are going to be in what I call the "AI pod" environment. This is where everyone's work from the time the data is sourced and extracted to the point where the model is developed and released through an app or an API (data stream) can be thought of in a circular process structure. The work you do in the model testing part of the process may depend on the work of the person whose process came before you, such as the data engineering process. To go fast as a team member, you will have to work iteratively with people both directly in front of you and behind you in the process. They will need your help and you will need their help as you iterate to bring the project to successful outcomes. You'll be saying, "*That didn't work. Let's try again on this training data. Can you reset this such and such for me?*" Then, someone may come to you about your work and say, "*The model you built isn't working with the application environment; could you try such and such?*" I also wouldn't want to work with data scientists that don't stand up for themselves. If they truly believe that they are doing something the correct way, I don't want them to back down when challenged just to save hassle or time or because they do not want to be confrontational to a peer or team leader. This is especially true if there are ethical concerns that could put the validity of a project at risk. You have to be able to stand up for yourself.

So, it's all personality stuff. Very rarely would I turn someone away on skills. If you have the personality and the skills, that's best. I can toss you right in and you can get going. But I would expect that people can learn the math, Python or R, and data science skills, all of which can be taught. At IBM, we did a nine-month program with Galvanize.

> **But what you want and can't teach is a personality of being inquisitive, constantly learning, being able to work in teams, being interested, and connecting the dots constantly.**

I can't teach data scientists to want to investigate data, people, or the places they come from. I can't teach that.

Summary

I greatly enjoyed hearing about Cortnie's career, and her trajectory to becoming a leader in the field.

In developing successful data science capabilities, I agree with her advice on needing to align data and analytics to the broader business strategy.

Cortnie made it clear that successful data leaders need to be politically savvy in a business context, and that CDOs need to be "really good politicians" – and have strong negotiation and influencing skills . The people and politics elements can make or break a data science initiative in my experience. Cortnie and I agree that senior leaders also need to have **data charisma** – the ability to influence and negotiate with data – beyond strong data literacy skills, which we also discussed.

We discussed the challenges organizations face developing responsible and ethical AI and the reasons that AI ethics boards can fail. Cortnie pointed out that it's difficult to find people with the appropriate skills and diversity of opinion and background to sit on these boards. That difficulty can set a board up for failure.

Cortnie outlined her 12 guiding principles (The 12 Tenets) for how organizations can develop ethical, fair, and trusted AI solutions. One of the key themes that emerged was that people need agency in the data-driven decisions that affect them, with the ability to review and provide feedback on the data used in the decision. It's the responsibility of everyone – not just the tech developers – to help build responsible and socially aware AI solutions. I encourage you to think about how you might apply the 12 Tenets to your own work.

She also stressed the importance of listening to staff who raise objections about processes or projects, with staff at all levels being empowered to speak up and raise concerns, such as ethical issues.

One of the most important topics Cortnie raised was how to work out what is important to key stakeholders so you can ensure you develop solutions that meet their requirements. This can be non-trivial and challenging. Her advice was to try and work out what they actually want and need from the following:

- What they tell you
- What they tell others
- What they won't admit – such as their career motivations and political machinations

She also suggested identifying who you need to get on board as key sponsors, and figuring out how to attach and align your work to their priorities.

Cortnie also offered some valuable advice for designing business metrics that measure progress against strategic goals. In my experience, this is context specific, as designing good business metrics in the public sector, for instance, can be more challenging than in the private sector. One reason is that sometimes the metric you're optimizing is a second-order effect – for which you may not have suitable data – such as providing actionable and timely intelligence to another government agency.

For data scientists, Cortnie suggested not making the mistake of thinking that career progression only means moving up into leadership roles – and thus becoming less "hands-on" with the technology.

She looks for the people she hires to have attributes such as a flexible and adaptable mindset, enjoying experimentation and exploration, and being a team player.

Edward Santow vs. Unethical AI

As Australia's former Human Rights Commissioner, and a lawyer, **Edward Santow** is uniquely placed to provide advice on developing responsible and socially aware AI, and giving suggestions for tackling some of the challenges in creating practical ethical frameworks and policies.

Ed is an industry professor of responsible technology and the co-director of the Human Technology Institute at the **University of Technology Sydney** (**UTS**), where he plays a leading role in helping shape policy and governance that applies human values to new technology.

I was keen to hear Ed's take on how we can continue to innovate with technology while ensuring fair and equitable outcomes for all citizens – including maintaining basic human rights.

Developing responsible AI pathways

Alex Antic: I'd like to begin by learning a bit about you and your career trajectory. What made you pivot into human rights, and then ultimately into AI and ethics, and then become an academic in responsible tech?

Edward Santow: I specialized in human rights very early in my career as a lawyer. I'd always had this sense that I really wanted to help people if I can put it as vaguely as that, and that pathway toward human rights felt like a very natural one for me. In terms of the connection to technology, there were two crucial moments that took place two jobs ago.

I used to be the chief executive of an organization called the **Public Interest Advocacy Centre (PIAC)**. PIAC exists to provide free legal advice and assistance to people who can't afford lawyers. There was an enormous amount of work we did with Indigenous communities and others who had suffered very significant injustice. There were two specific cases that really made me stop and think about the direction of technology. I'll mention one which was quite unsettling.

We received a series of complaints from clients who all happened to be young – and by young, I mean between the ages of about 12 and 18. They all had this very similar story, which was that the police here in New South Wales were essentially harassing them. They were being very, very specific. They were saying that several times a week, say, between the hours of midnight and 6 a.m., they would have the police knocking on the door of their home and waking up everybody in their home to check on them, just to see where they were. Or, the police would come to their school, **Technical and Further Education** (**TAFE**), or their workplace, and they hated it, understandably. I would too.

So, there were a few things we started to notice: the first thing was that their story was consistent. The second thing we noticed was that they were not serious offenders, on the whole: they'd maybe been on a train without a ticket, that sort of thing. It didn't make sense to us that they had been identified as necessitating this close scrutiny. The third thing we noticed was that they all started to say the same thing: that the police, if they were asked, would say: "*Well, the reason we're checking up on you all this time is that you're on this list,*" and the list is called the **Suspect Target Management Plan**, or **STMP**. And the final thing that we noticed, which was unmistakable, was that all of those young people – children and young people – had dark skin. A lot of them were Indigenous but all of them had dark skin, and that was again, as I say, unmistakable.

We did a lot of work trying to get to the bottom of what on Earth was going on, and to cut a long story short, what became clear – because it was revealed in the New South Wales parliament – was that the police were using an algorithm to identify young people who were at greater risk of getting into a life of crime, and that was the basis on which individuals were selected to go on that list. 56%, for example, of those young people were Indigenous, even though less than 3% of the New South Wales population is Indigenous, and so clearly, something seemed to be going wrong.

The police officers themselves were not individually empowered to do something about it. They were just following whatever the computer spat out and told them to do, and it was causing real injustice. That really set me on this path of inquiry, and I was then able to pick up that strand when I became Human Rights Commissioner, to see how technology can cause real injustice, harm, and human rights violations, and that we need to do something about it. And if we do, then we can be excited about the positive things that new technology – such as automated decision-making AI – can do, and those positive things are genuinely exciting as well.

AA: That's fascinating, and I can completely understand the impetus behind your wanting to pursue a better understanding of the implications of technological abuse and misuse. Some of it, of course, is just through ignorance, as opposed to anything malevolent or nefarious. It touches on something we'll talk about later in relation to the application of AI in criminal justice in particular.

Speaking of your time as Human Rights Commissioner, I'd like to better understand from your perspective the outcomes of the fantastic *Human Rights and Technology Final Report* that you commissioned, which I found incredible to read through. First and foremost, the emotional side: how did you feel being part of the process leading up to the publication? Also, when you were uncovering some of those findings, before you even thought about recommendations, how did you feel, considering the position you were in and what you were discovering?

ES: I felt concerned. I felt that there were risks and threats that were not abstract, that were very real, and that we were not giving them the attention that they deserved, but I also felt strongly motivated to do something about it. What I mean by that is that we are at this moment at a crossroads where we can choose a path that will address those risks and threats and give us the things that we want and need from new tech, but we're not going to be at this crossroads forever. Failing to zero in on these issues is in itself a decision that we will, I think, ultimately come to regret.

Human Rights and Technology Final Report

The Human Rights and Technology Final Report was produced for the Australian government to consider the impact of technology on human rights. You can read the final report here: `https://humanrights.gov.au/our-work/rights-and-freedoms/publications/human-rights-and-technology-final-report-2021`.

The key findings include Australian citizens wanting safe, reliable, and fair technology, and that stronger laws and policies are needed to protect our basic human rights. The report provides 38 timely and pragmatic recommendations, including the Australian government commissioning an audit of all current or proposed uses of AI-informed decision-making by or on behalf of government agencies, and also legislating the requirement that any affected individual is notified where AI is materially used in making an administrative decision.

AA: **I think it's a pivotal point in history in some ways, where we have an opportunity to take control of how we use technology, which is ultimately there to serve us. People get very scared of robot overlords taking over, but it's much more insidious than that. There are many ways it can potentially harm – and is harming – us as a society that we're not aware of. So, reports like that are absolutely paramount to us, as a society, having a better understanding and taking some control back.**

What were some of the key findings, then, and some of your subsequent recommendations to try to address some of these concerns?

ES: Essentially, the report followed a problem-to-solution arc. In terms of the findings, the first was that there are enormous opportunities – economic opportunities, as well as opportunities to advance human rights – when it comes to the development and use of new technologies such as AI.

The second was that there are also risks and threats, and we've focused much more on the first than the second, and those risks and threats to our human rights go way beyond the narrow focus we've had to date on privacy. In fact, our sense, having done this major public consultation, was that privacy was often not even in the top 10 of people's concerns. What we heard over and over again from the public was that people were essentially saying to us, "*I've just realized that my personal information can be used against me.*" The sorts of human rights violations that people were concerned about were things such as discrimination and failure to accord fairness in the justice system, like the example I gave before, which are fundamental and go much more deeply into what it means to have your human rights upheld than a minor privacy violation such as having stuff marketed at you when you don't want it. So, that was really important in terms of the findings.

In terms of the recommendations, what we were saying was four-fold – that is, there are four key principles (*Fair*, *Accurate*, *Fit for Purpose*, and *Accountable*) that we thought really needed to infuse into everything when it comes to how we develop, use, and regulate new tech. The first thing was that the applications have to be fair.

> It's not OK to say, "Well, it works on people that look like me – white, middle-aged men – but it causes harm to others." That's not acceptable.

The second is that it has to be accurate. That sounds like it's so obvious that it need not be said, but it clearly does, because a lot of the time, we're talking about very early experimentation with data science and other new types of tech, and often, the results are highly inaccurate. For instance, using new technology, we might gain a more efficient way of achieving a particular activity, but if the method is less accurate, that's not acceptable: accuracy is crucial.

The third is fitness for purpose. That is really important because we often have AI tools that work really well in a specific context, but when you try and transplant them into a new context, that's when they start to cause problems.

Finally, the fourth is accountability. It's not OK to have an AI system that operates as a black box. You need to be able to get to the bottom of that. If you don't, then you can't ensure that the law has been followed in terms of how those tools have been developed and used.

AA: Business leaders are becoming much more cognizant of the fact that they need to take this seriously. There's the approach of regulation, which of course is becoming stronger and more overarching overseas. Australia, I believe, will follow suit. Business leaders often don't know where to begin. They reach out to people like me and say, "*We want to establish this capability. We've only invested in the tech. What do we do about the ethics?*" **For instance, I've had cases where they think just being seen doing something is enough to make the shareholders and customers happy, but most of them are becoming a lot more cluey in thinking,** "*We need a roadmap. We need something that's very pragmatic and practical.*"

What would be your recommendations for how they actually get started in this? What should they be doing to actually establish capabilities, both responsible and ethical in nature?

ES: I think there are a couple of problems that I want to unpack very quickly in answering that question. The first is that there's been a huge proliferation of ethical statements and ethical frameworks that have been developed for AI. A lot of them have been developed by companies – some by governments and other kinds of bodies, such as the **Organisation for Economic Co-operation and Development (OECD)** – and the vast majority of them are completely unobjectionable. They say things like, "*Do no harm,*" and "*Treat people fairly,*" and so on. That's fine, and I agree with all of those things, but the problem is that they're expressed at such a high level of generality that they provide no practical assistance. They articulate a set of principles that any purpose-led organization would already subscribe to, but there's a missing middle, which is how do you actually achieve this? And so, in practice, the emerging research on this – unsurprisingly – shows that those ethical frameworks almost always yield no discernible practical results.

So, I guess step one is to apply those general principles. You have to have a set of tools, processes, and so on that will help the organization apply those principles in a practical context.

The second thing is that there's often a gulf in companies, as well as in government, between the people on the technical side of developing a piece of an AI application and those who are procuring it and ultimately using it and overseeing it. The people in that second group – the non-technical people on the strategic side of that equation – often have very little understanding at all about AI. Often, AI is almost a synonym for magic, and so those processes often go terribly wrong from the outset, because if you don't know what you're buying, then you will find it very difficult to instruct an organization or an internal team on how they should develop that application in a way that will be fair, accurate, accountable, and fit for purpose. Then, when you come to implementing and overseeing the application, you'll have very little by way of expertise to make sure it does what it's supposed to do and nothing more.

The solution there, I think, is not to try and turn everybody on the strategic side of that equation into a PhD-level data scientist. Rather, they need to skill up to the level where they are at least cognizant of what AI is, its risks and threats, and its opportunities so that they can competently go through procuring, implementing, and overseeing it.

AA: I agree. Something positive that I'm seeing is many organizations are much more willing to commit to a broader enterprise-wide uplift in data literacy training, especially among the senior ranks, which helps allay some of these concerns.

I think it's also vital to create cross-collaboration between different groups and not just have the technical group as a silo, which is very problematic. I've seen that go wrong many times throughout my career.

When it comes to the public sector versus the private sector, do you think there are any differences in how they should go about developing these capabilities and solutions in a responsible way? Is there anything you've seen that works better for one than the other?

ES: There are some differences, but they're fairly subtle. Governments – not always, but often – are subject to more onerous requirements when it comes to human rights, and those have to be taken very seriously. Perhaps an example would be that a government generally is subject to an accountability requirement. It has to show its reasons for decisions on the whole, whereas corporations generally don't – or at least not to the same level. So, that's something that should be taken into account.

Broadly speaking, I actually think in a jurisdiction such as Australia, the corporate world and the governmental world have got a bit closer, in the sense that if you're a big corporation, you'll still be likely to outsource the development of a piece of AI, and then have some technical people internally who'll make sure that it's implemented effectively, and so on. If you're a big corporation, you're not likely to do it yourself, and it's similar for governments. Governments tend to outsource all of this.

So, those issues of procurement, integration, and oversight are quite similar. Obviously, there are some functions that only government fulfills – for example, when it comes to the court system – where we need to have regulators and others who are more skilled and knowledgeable than they are now about AI so that they can ensure that things don't go off the rails.

Applying ethics in practice

AA: The next question applies directly to my background in the technical space, and it's where I see a lot of peers and colleagues struggling in trying to understand the role they play in implementing responsible technology. Given that they don't tend to come from an ethical or legal background, it seems like they often struggle to understand what we can do apart from just trying to think about bias in data from a statistical point of view. Is there anything that they should be or could be doing to make sure that the solutions they're developing are developed in a conscious way, from data ingestion and model development to deployment and engaging with internal stakeholders?

ES: It's a great question. There's a cliche that is pretty applicable in this context: the technical people and the strategic people are all speaking English, but not the same language. That's really common – worryingly so. So, what can you do about it?

On the technical side, I think there are a couple of things. The first is to really go deeper into understanding the outcomes that the customer wants more precisely. I think that means trying to understand their context by doing more listening. Exploring the context in more detail means that ultimately, what you come up with is a product that is more likely to get out of beta and be very effective and durable, such that it won't need to constantly be tinkered with. So, that's the first thing.

The second is to try and see a lot of these ethical problems as essentially socio-technical. I think a lot of the mistakes that we've made in this area tend to be where we look at a socio-technical challenge only as a technical challenge or vice versa, or only as a social policy challenge.

There was a great experiment when I was at the Commission with **Gradient Institute**, **Data61**, and the **Consumer Policy Research Centre**. I'm not going into the details of it, but it was about the discrimination and unfairness that can result from the use of algorithmic decision-making systems. We kept hitting this brick wall in some of the conversations with the technical people where they would say, "*Well, we're sick of talking about this. Can you just give us the number?*" What did they mean by that? They said, "*If you're telling us that 50% of the customers have to be male and 50% of the customers have to be female, we can fix that problem almost immediately. If you tell me that 60% of the customers have to be female and 40% have to be male, I can do that too. I don't want to talk about it anymore.*"

That kind of reduction of what is actually quite a complex analysis (certainly as a legal question) to something that might have an arithmetic solution is dangerous and just plain wrong. There are a lot of legal questions that simply can't be reduced to that, and unless people on the technical side are willing to confront that, then they're living in a fantasy land and they're going to continue to get things wrong.

Now, I actually think most of the fault lies on the other side, with people like me, not with people on the technical side, but that's a real attitudinal problem that needs to change. Whenever a problem can be solved arithmetically, that's great, but it's unfortunately just not true of all problems.

AA: Often in those cases with the technical experts, if they're trying to mathematically define concepts such as fairness, bias, and explainability, then it can get quite difficult. That's especially true when you're trying to balance the accuracy of a model with, say, fairness when these are sometimes mutually exclusive.

The question is, how do we get common definitions? Do we need common definitions between organizations, industries and sectors, and countries? What are some of the definitions you tend to use that resonate with you, given your background?

ES: I think a lot of the common definitions are already there. Part of the problem is that we often start from the wrong spot. If we start from our ethical framework, which says things such as "*Do no harm*," you're going to think, "*What does that mean?*" You have a vague sense of what it means – you treat people fairly – but those aren't words that the law would ever use. Those are words that we use in vague marketing statements.

So, we need to start by looking somewhere else in order to find that common language, as you call it. For me, it's the law! The law's really clear. If you're designing an algorithmic system for a bank, the law generally doesn't say that you have to use a particular methodology to do so. The law says something quite different. What the law says is this: "*Your outcomes must not be unlawfully discriminatory, and this is what unlawful discrimination means.*" I'm just saying orthodox stuff: it's not a question of finding a new common language. It's about speaking the common language we already have available to us that we, for whatever reason, have often ignored. I think that that's a significant challenge but it's a very different one from inventing something new from scratch.

AA: So, you think the conversation ends up moving more toward some theoretical discussion around ethics rather than legalities, which are clear and well documented and we all have to abide by?

ES: Yes, that's right. An example I often give is this. Let's imagine a human bank manager has this view of the world that women make bad home-loan customers, and so this manager decides that he (I'm going to give him a gender) won't give home loans to women. We wouldn't be scratching our heads about that. We'd be saying, "*That guy is clearly acting unlawfully. That is blatant discrimination.*"

So, let's change the story and say that there's no human bank manager in this story. Instead, it's just an algorithm that's been poorly written. Maybe it has some bad training data and it's having the same outcome. It's making it incredibly difficult for women to get home loans as compared to men. It's baffling to me that we then don't say, "*That's clearly a legal problem, just as much of a legal problem as the first example.*" Instead, we say to ourselves, "*Huh – isn't this an interesting ethical dilemma?*" It's not an ethical dilemma at all! It's a breach of the law and it needs to be punished.

What I'm trying to say is that framing is not semantic. It's not just mere words. It really matters which category you put things in because if you put something in the category of "*this is a legal requirement,*" then that brings to bear a seriousness of purpose, which is a very important starting point. But secondly, it brings to bear hundreds of years of case law and the working through of exactly what things mean so that you can then benefit from that in understanding precisely what you have to do. Without that, I think that's where you end up flailing around.

AA: The next question touches on something you've just discussed: this notion of explainability, accountability, and transparency around models, both to determine whether the outputs are lawful and to give people a better understanding and appreciation of what's happening. I think there are often a lot of issues around how you define explainability – explainability to whom? It's a very technical explanation. Is it to the person who's debugging it? Is it to a manager, an auditor, or the end customer? Once again, do we need to have different definitions? What definition do we revert to, and for what reason do we want it to be explainable?

ES: I would say that the law already worked this through literally hundreds of years ago. This is not a new problem. The solution is hundreds of years old.

Let's forget that we're talking about a machine. Let's apply this to a human. When a human makes a decision, we can't saw open that person's brain and look at how the neurons are firing in order to get to their true motivation that way. Instead, we ask them to provide reasons for their decision, and then we interrogate those reasons and other contextual evidence to get as close as possible to the bottom of why they made the decision. We're not trying to get to a perfect explanation – we know that that's impossible and we don't need it. What we really need to know is something a little bit grosser, which essentially is this: has the person acted lawfully or unlawfully, and are there merits to the decision (which is the question of accuracy)?

We can apply that to the AI context. It's a much less difficult task to achieve. All we need to do is ensure that our system yields an explanation that is at least no worse than a human explanation.

The final point here is this: we know that humans and machines can each provide plausible explanations that are not the true explanations of why they made a decision. In other words, we know that humans and machines can each provide a smokescreen to hide their true motivation for a decision. Now, that is really interesting. We tried to spell this out in our report. When it comes to AI in decision-making, you need both things –you need an explanation that a layperson can understand – either as a court or a regulator, or just as an ordinary citizen, but which is not necessarily going to be a true reflection of what motivated the decision. You then also need a technical explanation, which only a small number of experts will be able to use in order to determine whether that layperson's explanation is an accurate reflection of the decision-making rationale.

Again, none of what I have just said is new. It's simply applying long-existing legal principles to this relatively new context of algorithmic or AI-informed decision-making.

AA: I love how you frame it through the history of the law. That is a great way to summarize the key concepts and get right to the heart of the matter without getting caught up in the ethics, which can be less direct.

Considering the broader impact of AI on society

AA: Bringing together a couple of things we've already discussed and that you've already raised – the human rights element, and also the outputs of some of these models – what are some of your biggest concerns with how AI and technology affect human rights in specific cohorts and generally? Also, what more do we need to do as individuals, as a society, and as organizations? How should we ensure we protect human rights with the ever-increasing growth of AI and its proliferation throughout society?

ES: I'll deal with those questions in reverse order.

The second question I can deal with quite quickly because I talked about it a moment ago. That is, you start by applying those four principles: fairness, accuracy, fitness for purpose, and accountability, and thinking through what that means – so while going through what's sometimes referred to as the AI life cycle. That's a really good way of baking in human rights protections.

Then, turning to the first part of your question, if we were having this conversation in Europe, it would be very different from the conversation we have about privacy in Australia. Privacy in Europe means something quite different. It's a much bigger concept than it is here in Australia, and that is true – to some extent – of North America as well.

> In Australia, privacy is narrow in the sense that it's the extent to which you are able to live a life that is secluded from the public domain, whereas in Europe, it accounts for that concern that I referred to before about your personal information being used against you.

That's significant because where your personal information is used against you, anything can happen. For instance, as we talked about earlier, you can be subject to unlawful discrimination. If the AI that goes wrong is a policing tool, then you could end up being unlawfully detained. You could have all kinds of justice rights violated.

What I'm trying to say is that the human rights that will be engaged – and possibly threatened – will usually depend on the context in which the AI is being used. That brings to bear on all human rights because we know that AI is being developed and used in pretty much every conceivable context.

AA: We spoke briefly about privacy before. When it comes to privacy and AI, can the two coexist? What are some of the main challenges that are posed regarding privacy when we talk about AI adoption? We've seen some of them already in society. Are there others that we need to be more aware of?

ES: There are three things for me that immediately leap out. There's privacy at a very individual level: your personal information is being used against you, which we talked about before. There's privacy at a more societal level: the cumulative effect of a whole bunch of AI, such as facial recognition AI, which I'm particularly interested in at the moment, as it can lead us toward changing the nature of society into one that essentially is governed by a mass-surveillance state, which is worrisome.

The third thing is that the way in which we use information, particularly personal information, is changing. Privacy law has traditionally protected that specific species of information known as personal information. When my personal information gets pushed together with yours and that of others, it ceases to be personal information because it's not referable back to me, and yet the insights can be used to affect me, sometimes negatively, and privacy law doesn't pick that up.

AA: We often hear about this notion of a large majority of AI projects failing – something like 85%, depending on which report you read or who you speak to. People often talk about the results in terms of profit and loss or brand damage for an organization. That's all well and good, but from a human rights perspective, what are some of the implications and consequences when failures occur and they're not being properly addressed or mitigated?

ES: From the perspective of the company or government agency that is using AI in a way that causes failure, they've got regulatory issues, reputational issues, commercial issues, and so on that they have to consider, but from an individual perspective, it's violations of human rights, and it really runs the full gamut. We've seen that already, in the example that I started with, about the STMP here in New South Wales. We've seen it in the justice system in the United States, with things such as the **Correctional Offender Management Profiling for Alternative Sanctions (COMPAS)** program – which is an algorithm that predicts the likelihood of recidivism for defendants. We've seen it with algorithms going wrong in banks and in recruitment, and so on. I guess "*it's everything*" is the short but perhaps slightly annoying answer.

AA: You just touched on an important element: ill-fated use of AI, such as recidivism and other cases. Overall, though, when cases like that become much more publicized, how do we ultimately build trust in AI? Ensuring human rights is a really positive step, a crucial step. More broadly than that, is that enough for people to trust these solutions in terms of proper automated decision-making rather than augmented decision-making, or are there other things we need to do so that people trust AI?

ES: I think trust is something that is earned. Everything you've said is important, but maybe you're right – the current methods are not sufficient. Often, the starting point is humility. You should not assume that your system will be right in every case, but rather assume that it will be frequently wrong. That way, you will have a series of checks in place so that you can kind of identify those errors and correct them in a way that will minimize harm – and even inconvenience and cost – to individuals who are affected.

AA: That gets me thinking about this example: say there are a million deaths caused by car accidents globally every year. Then, every car is replaced by a self-driving car. The toll drops down to 100 deaths per year. Is that enough for people to trust self-driving cars? How we think about trust is much more nuanced – there's a lot of complexity behind how you develop trust. Do we expect too much from AI systems, and should we?

ES: I think that that question is a really important one, and it takes us into a realm that is beyond my expertise.

There's a lot of psychology and anthropology involved in giving a really well-considered answer to that question, but I can make quick observations. For instance, for some activities (and driving might be one of them), society will expect more than just the technology to be as good as or even a little bit better than a human; they might expect a lot more than that, and we can't ignore that. That might actually be rational. There may be good reasons for that. It may be completely irrational, but either way, we need to confront that, work through what that means, and consider whether this is the community pushing the technology enthusiasts to prove themselves. It would have been highly rational in the early days of airplane flight for anyone to say, "*There's no way I'm going to get in one of those! That's crazy!*" But there was also a hangover that continued into the jet age, where people felt – incorrectly but understandably, perhaps – that getting into a plane was far more dangerous than getting into a car. That shifted over time, but it shifted over time largely because it was a highly regulated industry and remains so. Essentially, the manufacturers and operators of aeroplanes were forced to show proof of how the way in which they operated their planes was safe, and indeed safer than many other forms of transport, but that takes a bit of time. Plus, even if the public is wrong at a certain moment in time in thinking that one thing is more dangerous than another, I don't think we should ever ignore that. We should try and get to the bottom of what that might suggest.

AA: I think that raises a good point. It'll be interesting to see over time how society's appraisal of and appetite for risk changes as AI pushes the boundaries of some of these technologies and their adoption in certain sectors. Will our risk behavior change?

ES: I think it will. In fact, it almost always has as new technologies have become more widely adopted over time. You have the Maven class who are very comfortable with using new technology, but that in turn allows the rest of us to observe how those early adopters experienced the benefits and risks of the new technology before we choose whether, and if so, how, to use the technology ourselves.

AA: Regarding regulation, do we need government regulation? Can we expect organizations, such as big tech companies, to be trusted to regulate themselves?

ES: Yes, we need government regulation. There's no other field of activity that meaningfully affects people and is left completely unregulated in the long term. I hasten to say that's not true of tech either. There are a whole bunch of laws that apply; they may not be rigorously applied all the time, but they apply, and of course, they should be applied. We are having this debate at the moment regarding cryptocurrency, for example. That is, we are questioning whether existing law will suffice to regulate this relatively new phenomenon or whether new regulation is needed. In the meantime, there is an apparent regulatory vacuum, but if history is any guide, that vacuum is unlikely to persist.

AA: Is regulation being taken seriously, from what you've seen, especially during your time as Human Rights Commissioner?

ES: Good question, and a hard one to answer. I don't think I have a considered view about whether individual penalties are enough. I think the problem is more that they don't accumulate to be enough. Even through contract law and tort law, as well as direct regulatory action, it is still uncommon for someone who's done the wrong thing using AI to get caught. That's a moral and legal hazard because if people think they can get away with something, that may not necessarily mean that they deliberately go out and do the wrong thing, but it could also mean that they just don't pay it enough attention. The latter is a more common problem and a subtler one, but it can actually end up with exactly the same endpoint.

AA: You've previously spoken about this notion of dual affordance, which I find very interesting in relation to AI. Can you please explain what it means and provide some examples?

ES: The idea of dual affordance is not really one that applies specifically to new technology. A knife is a dual-affordance technology in that it can be used to do innocuous things such as cutting bread. It can also be used to stab people and cause harm. There's nothing specific to technology that makes it dual-affordance, but it's much more common when it comes to new technology, where it can be incredibly powerful. Whether that power is used for benign or malign ends is something that is not controlled by the technology itself. In one sense, that gives individuals enormous power to push technology in a particular way or just let it loose, and to me, that's dangerous.

AA: What should we be doing to make sure that graduates, as they become professionals and practitioners in the field of AI, are empowered and educated enough to make sure that from the time they hit the ground running they're aware of these concepts and know how to bake some of these important issues and conceptual designs into how they develop models and solutions?

ES: I think you've already pointed to one, which is making sure that people on the technical side are ethically literate. More important than that, they should be legally literate – it's not very helpful to see these things as primarily ethical questions. They're more legal.

There's also something that I think really applies to what we discussed earlier, which is the idea that a lot of these issues, when you try to take AI out of the laboratory and into the real world, are socio-technical in the sense that you apply an AI application to a specific situation. You use some human intelligence, and you use some machine intelligence, and that integration is really important – having people with a technical bent who see the socio-technical challenge so that they can value the input, expertise, and experience of others who do not have technical expert input, and so that they can better shape the way that they build and develop AI as responsive to those needs. A short way of describing all of this is a multidisciplinary approach. I'm hesitant about using that term because it's a bit of a buzzword and it can lose all meaning, but it does have genuine meaning here and is really important in this context.

Responding to the challenges of generative AI

AA: Given the popularity and advances in generative AI tools, such as ChatGPT, I'd like to get your thoughts on how generative AI has impacted ethics frameworks. What complications has it added?

ES: In one sense, it hasn't, as the frameworks are broad enough and apply to AI generally, and their application depends on adapting to the specific context in which they're being applied.

One of the great advantages of this is that generative AI is included within its scope. It may be a newer form of AI, as compared with analytical AI, but existing AI ethics frameworks already cover a range of privacy and human rights issue, so they *are* applicable. The previous work to create those frameworks has made it easier and faster to adapt to the specific aspects of generative AI from an ethical perspective.

One of the main complexities is the relatively low community understanding of how generative AI actually works and, particularly, the science behind it. Very few people can distinguish between analytical and generative AI. Most people in senior roles haven't made the distinction yet, or identified the true impact.

> **Generative AI**
> Generative AI is a type of machine learning which creates new data or content. Generative AI models are trained on vast amounts of content captured from across the internet. ChatGPT is the most prominent recent example of generative AI.

The issue is, if you don't understand the underlying technology well enough, then it's difficult to make the frameworks work in practice.

Analytical and generative AI share similar core science. However, generative AI can pose greater risks than simple classification AI. But the nature and scale of those risks generally hasn't been worked through in most organizations. Simply setting black and white rules – such as you can or can't use generative AI – isn't usually the best answer. You need to understand how to safely use it.

AA: How will organizations need to adapt their ethical frameworks in response to generative AI?

ES: First and foremost, they need to understand that skills and knowledge are vital. They need to upskill their staff and develop a better understanding of the technology and its implications – and this applies at all levels of the organization.

Second, they need to set a nuanced policy framework, outline how to use such technology safely, and develop appropriate risk mitigation procedures that can flag when it's not safe to rely on the outputs of generative AI applications. Most AI ethics frameworks don't go into this level of detail.

Finally, consideration needs to be given to how generative AI can be used lawfully. For example, entering confidential client data – or proprietary company data – into ChatGPT is likely to be unlawful, yet we also know this is happening.

> ChatGPT and confidential information
>
> Read about an incident involving ChatGPT on *Mashable*: `https://mashable.com/article/samsung-chatgpt-leak-details`.

AA: What advice can you offer CDOs and senior leaders in relation to navigating some of these challenges?

ES: There are simply no short cuts. People can't assume that even though others in their industry are using generative AI, that their organization can use it without considering the legal and ethical ramifications.

They also need to be able to experiment safely with such technology. For example, a new chat bot based on generative AI shouldn't be simply unleashed on customers. They need to first test and validate it in a controlled environment to understand all the risks – including the ethical and legal ramifications.

> **Technology such as generative AI is too dangerous to let loose without controls.**

Leaders need to ensure that an appropriately safe test environment is established to mitigate any risk of harm to staff or customers.

Summary

It was an honor to speak with Australia's former Human Rights Commissioner. Understanding the legislative frameworks within which we and our organizations operate is critical for all data leaders and practitioners. Ed provided practical advice on how organizations can establish responsible AI solutions.

Our discussion of the key findings and recommendations of the important *Human Rights and Technology Final Report* that he commissioned was of great interest. The report highlighted that human rights violations are of greatest concern to individuals in Australia, with privacy violations also being a cause for concern, but with potentially less impact. I believe that it's important for all business leaders to read the report, and try and implement the recommendations in their own organizations. Ed shares pragmatic suggestions on how to implement the recommendations and associated principles, in order to create ethical AI. This is incredibly valuable given that, as Ed explains, emerging research shows that most ethical frameworks include high-level (motherhood) statements, which in practice yield no discernible results.

Leaders should also take note of Ed's advice on the importance of upskilling to become cognizant of what AI is and isn't (including developing an understanding of generative AI), and the risks, threats, and opportunities it involves. This knowledge is essential to being able to competently manage procurement, implementation, and oversight activities.

I enjoyed hearing Ed's views on explainable AI – he pointed out that some of the issues being cast as ethical issues are not new, and are already addressed by existing legislative instruments such as the Privacy Act. This highlights the interplay between the philosophical and legal aspects of developing and managing fair and responsible AI, and the importance of being legally literate to help address socio-technical issues. This highlights the importance of taking a multi-disciplinary approach.

There are some pertinent points raised that should help anyone who's grappling with some of the nuances and challenges of ethical and responsible AI, such as ensuring that AI systems yield explanations that are "at least no worse than a human explanation." Specifically, two levels of explainability are required:

- A layperson's explanation
- A technical explanation

Furthermore, it's important to have a series of human checks in place so that errors in AI systems can be identified and addressed to minimize the potential harm to individuals.

Ed also offers some sage advice to organizations and leaders in navigating the ethical complexities and nuances of technology such as ChatGPT. He urges all leaders to ensure that they and their staff are educated to better understand the technology – not only how it works, and its possibilities, but also its limitations and potential negative impacts.

He also suggests establishing a test environment to help identify and mitigate risks prior to using such technology, which still isn't understood well enough. He believes the technology is too dangerous to be let loose without appropriate controls being put in place – a clear and timely warning.

4

Kshira Saagar Tells a Story

Kshira Saagar is head of international data science at DoorDash and the former **chief data officer (CDO)** of Latitude. He's been recognized as one of the top analytics leaders in Australia and has a wealth of experience developed across several industries and sectors. Kshira is generous with sharing his knowledge and expertise, and I was eager to discuss his recommendations on developing successful data and analytics teams from the ground up and how to manage the expectations of senior executives.

I enjoyed learning about his leadership style and his thoughts on what it takes to develop a successful career in the field.

The path to data science

Alex Antic: How did you become a leader in the field of data science, and were there any pivotal moments, mentors, or people that inspired you?

Kshira Saagar: As one falls in love, I fell into the data science space. I did not actively pursue data science when I started. There was no such *thing* as data science back then. It used to be business analytics, data analytics, and stuff like that. If there is one inspiration that I would definitely put a finger on, it would be my high-school math teacher. It was the way he tried to teach us math when we were in our formative stages of trying to figure out what integral calculus is and how it even works! He taught it so beautifully and tried to explain it using physical concepts.

We had this one whole session on statistical distributions and hypothesis testing: he taught us how log-normal works and how to identify the distribution. The growth of hair is a log-normal distribution because it can only start from positive, and it's a normal distribution. When you learn things like that, or how a Poisson distribution works, or how a Poisson process works, you get so excited.

After finishing high school, as I was going into university, I thought, *"I'll always work for a company that uses math in some form."* That was the mindset. Now, I didn't know exactly how that would work. I was lucky to be in the right place at the right time, as I came across a company whose tagline was *"Do the Math."* It doesn't get more obvious than that. I didn't know *what* to do with math, but I was going to figure it out. I majored in statistics, computer science, and math, so I thought, *"I'll figure this out."* It turned out that this company worked with data (they used statistics to interpret and analyze data), and that's how I got into it. I was more of a statistics-based data analyst, and over time, I've picked up the skills to be a data scientist and an engineer.

AA: I can completely understand the importance of having a good teacher who inspires you, especially in areas that can be a bit more abstract, such as mathematics. I had the same thing early on, and that really inspired me to pursue mathematics as a career. That resonates with me.

You're an expert in the data science field. You've been doing this for quite a while now, and you've worked across many different industries and sectors. What are some of the common challenges you've seen for organizations that cut across sectors? Also, what are some of the challenges when it comes to implementing data science and becoming data-driven that are specific to certain industries, such as retail and telecoms?

KS: There are several things that come to mind. One thing that definitely stops companies from actually benefiting from data is what I call the **data action loop**. We don't have a data action loop in a lot of places, so let me elaborate. We have a lot of businesses building data tools that have assets and dashboards, but nothing is done off the back of it. It's all lying there, but they can't make a decision using it because either they don't know whether it's right, they don't know how to use it, or they don't know what it is trying to tell them. They have these wonderful tools, segments, and models, but they just don't have the ability to take action using them. It could be something as simple as, "*I can see this number, but I don't know what action to take,*" meaning that that person uses their bias to do something rather than letting the data truly influence their decision.

So, that's one. The second thing is a lack of awareness of what exactly different data components do. For some people, it's one big box. Even people in the biggest companies I have worked for come to me and say, "*You're the data person, right? Can you just build this algorithm to find out why our business is not growing?*" It would be the same thing as going to a computer engineer and saying, "*Can you build a computer to write my book?*" People don't take the time to understand what each part of data science does and what parts of data science exist. They treat data as one big black box.

> The biggest challenge is not asking the right questions.

Sometimes we have the best tools and know what exactly to do, but we don't know what questions to ask. Asking the right questions is a skill. You have all the data, and you have all the dashboards, but what questions need to be asked?

I remember once, when working for a fashion retailer in Australia. The head of menswear came to us and said, "*I have these questions. If you can find the answers in six months, we'll never need any more data work.*" We were asked 10 questions on what kinds of people purchased menswear, what their traits were, and how they influenced future purchases. A teammate and I quickly typed away, wrote SQL queries, figured out the answers to everything in 25 minutes, put them in a document, and said, "*Here are all the answers. Can you ask more questions?*"

The head of menswear responded that those were all of the questions and they had nothing more to ask. My point to them was, *"No, you have to think of more questions and think of your business even more deeply to ask the right, tough questions."* So, they went away, conducted an offsite meeting, and thought of more gnarly questions. The outcome was a deeper understanding of their own business and some intellectually stimulating problems for the data team to solve.

I think those are all typical challenges for almost all businesses: not being able to use data to make decisions.

AA: Have there been any sectors you've worked in where the challenges they faced were specific to a particular industry?

KS: I think one big challenge with finance specifically, or any regulated industry, is the fact that you can't go into the level of detail that you want. More often than not, you're road-blocked by the regulatory nature of the work being used as an excuse not to innovate. That's definitely a problem, but there are so many other problems, inside and outside of finance, that can be solved but go unsolved because we don't know what questions to ask. The retail world is the Wild West: we capture everything, we track everything, and nobody asks any questions. Innovation is still missing in many industries.

Implementing a data-driven approach

AA: In trying to overcome some of these challenges, if you were to give some advice to an organization that's beginning its journey to become data-driven, what would you tell them as regards how to begin, how to prioritize the different stages (people, tech, culture, and so on), and how to implement their approach? In addition to that, how should they structure the team?

KS: When I address start-ups, or I work for bigger boards, I only advise one thing:

> *"Place the data team closest to the decision-maker."*

So, wherever the person who makes the biggest decisions in the company is, that's where the data team sits. It's as simple as that. If the **chief financial officer (CFO)** is making decisions, place the data team in their team. If it's the **chief executive officer (CEO)** making the decisions, place it in their team. If it's a tech company and the **chief technology officer (CTO)** is making the decisions, put it in their team. What we end up doing is trying to force-fit an approach that works for one company but not another. A CFO might say, *"I think all the analysts should sit in my team,"* but maybe that doesn't work because the CFO doesn't have the decision-making power: say they don't help with merchandising if it's a retail company. It makes more sense for the data team to sit within merchandising, with the creative director's team (which is actually making the decisions), or the operations team (which is actually running the business).

A second bit of advice I typically give teams is about team makeup. If the team in question is smaller, I do an exercise where I ask them who they want. Everyone thinks they want a "data scientist" because that's what their investors tell them, that's what the boards tell them, and that's what they read on blogs and everywhere else. However, I try to explain to them who they really need, what they can get off the shelf, and what they actually need to run their business.

That understanding is important and, more often than not, for really small start-ups, bringing in just the right person to put data together, clean it all up, provide the team with a view, and help them to be more self-sufficient can set them up for success later. That's who they need: somebody who's a little bit more mature. They probably need somebody who can look at advanced insights and analytics. So, depending on the maturity of the company, the problem that they're trying to solve, and how well embedded data is in the company's DNA, you need different kinds of people.

In terms of setups, I think that if it's a smaller company, the setup doesn't make a difference. I would probably start with a centralized model. None of this means anything if you're in a 100-person company.

> My measure for how many data people should be in a company is 7 data people for every 100 people in the company.

That's a number that I've worked with, and it's a ratio that's always worked for me.

If there are just seven data people, it doesn't matter where they sit. They just work on solving problems. But if it's a much bigger company – say, with more than 30 or 40 data people – I'm a big fan of a federated model for larger companies, where all of those people belong to a data team. That team should be made responsible for bringing in the right talent, but they are also distributed to work with local teams. They might scrum with local teams, solve problems, understand their needs, and even speak for them. That's a hybrid, two-speed approach that I typically recommend for bigger companies.

AA: If you were to consult for an organization that is new to analytics (that is, they've got some data, they've got some people that understand the data, and that's it) and wants to build up an analytics capability, how would you stage the approach in terms of people? Should they bring people first or technology? Should they create a culture first or a data strategy?

KS: Whether an organization is big or small, I've always taken a *"Why does this team exist?"* approach. You can call it a vision or a mission statement, where I try to understand the team's purpose.

In one of my recent roles, I was in a large company with a large workforce, but the challenge was that the team didn't have a strong purpose. There were a lot of people, but nobody knew exactly how they were adding value. I don't mean it in a bad way, but people knew only about their exact piece of work – but collectively, what was the purpose of the team? Was the team actually helping in making decisions? Was the team just there to just be compliant and do stuff? That clarity was not there.

I've taken this mission-statement approach in quite a few roles, and it really helps the team. We might identify the team as the decision-enablers, for example. Understanding that vision is important. Once it's understood, people on the team can say, *"Yep, this is what I do. This is what I am here for."* Plus, the business knows that's what the team can do and knows what they can go to the team for rather than throwing random things at them.

It's then easy to go into solving specific problems. If a specific problem involves solving key business priorities, and for that, you need the right tooling, then it's tooling that you need to be concerned with. If a specific problem involves taking the business on a journey, then it's a cultural thing. Depending on where the business is, we go down a specific path and take the team along.

AA: So, do you advocate for people first before worrying about what technology stack is needed?

KS: Yes. The organization will typically also need to develop its strategy and have a broader vision. After that, they can worry about how the different things work together.

AA: I'd like to discuss some of the common mistakes and pitfalls that organizations get into when they're quite new to this. What mistakes have you seen some of them make in becoming data-driven, and what should they be doing to get around some of these issues?

KS: I know it's a very cheesy term, but I call it the **constant building paradox** – where the organization is always building. Let me give you an example.

I used to work for a company where the data engineering team was really excited just capturing all this data. Their measure of success was how much data they were capturing and storing. They really didn't care about who used it because nobody had access to it. Analysts and data scientists couldn't use it; nobody could tell whether it was valuable. People having that building mindset is the biggest pitfall I see young organizations falling into. They hire a data engineering team that just keeps capturing more and more data.

I used to work for a media company where their data scientist built a really cool segmentation model. When the marketing team said, "*Can you help me understand it?*" he replied, "*That's not my job. My job is to build models. It's your job to understand it.*" So, that's the biggest challenge: when people only focus on building, the things they build become useless, and projects simply fail. Six or twelve months down the line, when boards or executives look to see whether their investment made sense, they just see that no business impact came of it. Consequently, everyone thinks, "*No point in it – the data doesn't work,*" which becomes a challenge, but it's not true. It's just that we get caught in that building mindset. Somebody needs to balance building with using. If you build something and don't use it, it might deteriorate on its own.

AA: Where do you think that role of balancing "building" with "using" normally falls? Is that right up at the CDO level, or should that be more at the team management level?

KS: I think it's two things. Oversight is definitely needed at the CDO level to shine a light on what's happening and encourage other people to come and ask questions and not build walled gardens around them, but it's also needed at a wider leadership and management level. Management needs to push their teams, ask the data teams tough questions, and give them tough challenges to solve.

> Management and leadership need to continuously probe and push teams to think beyond just building and say, "*Hey, guys. You built this amazing thing. Who's used it?*"

Asking those questions constantly and getting other people to play off them would be great.

AA: Once a team has built a proof of concept for a product, I've often seen organizations struggle when it comes to the productionization of those models. What are some key strategies for having an efficient process for taking models to production? What challenges do you see organizations have in transitioning from dev to prod, and how do they overcome them to develop an end-to-end, repeatable, successful process?

KS: Without a doubt, giving the data scientists and ML engineers in your team the complete ability and independence to build, deploy, and do **continuous integration and continuous deployment (CI/CD)** on their own without having to interface with another engineering or ops team. I saw that shift the needle at one of my older gigs.

> We empowered our data scientists to work on problems from end to end.

There would be someone who was building a recommendation system, someone who was building a delivery estimation algorithm, or someone who was building a **stock-keeping unit (SKU)** allocation: they could do what they wanted, and they didn't have to wait on engineers. They could spin up infrastructure, deploy it, test it, and keep adding features to it. That massively shifts the lead time for a product to be released.

You need to understand the deployment process the engineering team follows in that particular organization, and then get the data scientists trained up on it to a level that means they can play on their own without having to go to an engineering team for every single change request, because then you've got two separate teams and not one.

Apart from all of that, I think that why a lot of organizations fail in the key **proof of concept (POC)** to productionization phase is that people don't realize what can be done with these POCs. We have a functioning **minimum viable product (MVP)** that is an algorithm, but once it is live, most organizations are not sure how to use them. The best examples are Personalization POCs built by so many data science teams that never get deployed because marketing or other teams have no plan in place to use them.

So, somebody has to adapt to that problem and say, "*This personalization model is great. How can we use it to change how we talk to our customers going forward? And how can we start offering better services based on customer tastes across all their interactions with us?*" I think that doesn't happen, and therefore it just becomes a case of the personalization model being seen as not working or not good enough, and it just dies naturally. People knowing what to do with a product is very rare.

AA: Do you think formal processes for the path from POC to production, such as MLOps, are needed to overcome this, or is it a bit more general than that, such as asking the right questions and having the right mindset to manage?

KS: The business needs to be very clear on what they will do with the outcome of whatever it is that is being built. If they can't state what they will do with it, there's no point in building it. Until they know what they'll do with it, there's no point in building something because they're just going to keep asking for more and end up not using it. That'll just turn smart people off, and they'll walk away. People need to know what they will do with something, as in the example mentioned previously. This way, you suddenly supercharge the teams with an ownership mindset. So that comes first: the business side needs to be clear, then MLOps comes second.

AA: We often see statistics showing high failure rates of AI projects, typically in the order of 85%. Do you think this is a reality across most organizations, and if so, what's going wrong? Why does this happen so much?

KS: Interesting point.

> I often joke that data science is 85% change management and 15% technology.

I'll give you an anecdote to explain.

At a place I worked, we had this amazing stock-keeping unit-allocation algorithm that would tell somebody when a product came into the warehouse where it should be put away. Simple. We wrote an allocation algorithm that worked at the SKU level. When a package was scanned, the algorithm would basically say, "*This contains T-shirts. Only one T-shirt sells every six hours. It shouldn't go here. It should go in B6, which is in the sixth row and the top column. It sells very well with pants, which are close by.*"

We did all of that and proved that it could actually bring so much efficiency to the warehouse, so we productionized it. When we ran the tests, the team said, "*No. I've been putting stuff away for the last 25 years. You can't tell me how to do it – how do I know this works?*"

> One of the biggest reasons that things fail is that the people who need to eventually make decisions – whether it's call center folks, operators, or people in sales or marketing – don't believe in it.

So, we said, "*We'll spend one week with you. We will scan for you. We will show you how easy it is and why it will help you.*" We did that for a week. We worked with the team. We opened up the algorithm. We showed them all the component parts. We tried to explain to them in a language they understood, and they gave us feedback: "*Oh, did you consider this?*" We put it back into the algorithm, or at least tried to, and when we presented it back to them, they started calling it their algorithm, and they started using it. It was no longer ours.

I think you need to do a lot of that with any technology, and AI and data are just some of those cool new technologies.

> People resist it because they don't know how to believe it or trust it. If we can bring in believability and credibility, many of these things will actually work.

AA: I've been in that situation many times. On the flip side, while some people become a bit more fearful and concerned about AI, others get really excited. There's all this hype presented through media and vendors, and a lot of senior executives think, "*Well, we need to have this new shiny toy.*" **What warnings can you give senior leaders, in particular, about the reality of AI beyond the hype?**

KS: I think you're right. There are so many plug-and-play tools out there, and everyone thinks, "*Why can't we just use this?*" That's the way these tools are sold, and the go-to-market's very attractive: "*All you need to do is plug it in, and it works.*" I'm not against plugging things in, but I try to provide senior leaders with a clear view of the pros and cons. When a CEO or CTO says, "*Can we do this?*" I try to explain to them the benefits and the shortcomings of using it in their system versus everywhere else so that they know the limitations of their systems. A plug-and-play tool may only work in a banking application but not theirs. Once you explain that, people understand it.

I often try to give a story. I say, "*There's this guy who bought a Ferrari. He loved the red Ferrari. He had it parked in his garage, but he could never take it out. Do you know why? Because he didn't have a road from the house. He got it, he had it delivered by a helicopter, and he can't take it out because there's no road, and so the Ferrari is sitting in the garage. How do you feel about that? Well, that's exactly how this AI is. It's a Ferrari. It's great, but you have to be in the Monza Circuit or somewhere to use it; otherwise, it's completely hopeless!*" People need to make that connection; otherwise, it's very hard for them to understand.

AA: There's a disconnect between how people think something works and how it actually works in reality and what is needed to enable it. When it comes to implementing these solutions, what should organizations do when it comes to measuring impact? More broadly, what metrics do you tend to use?

KS: I'm a big proponent of what I call data **accessibility**, **intelligence**, and **reliability** (**AIR**). It's part of all the strategies that I've done. We come up with scores for all three to benchmark our data capability. Just as you need air to breathe, data tools need AIR to be successful.

Say there's a tool: accessibility is about understanding how many people have access to the tool, how they use it, and how this tool helps them in daily/weekly decision-making. With this input, we come up with an accessibility score.

Intelligence would be about how many intelligent decisions the tool is making. Is it actually making a decision or is it just coming up with insights? Insights are different from intelligence. Intelligence is when you can take action. And thus, you can come up with an intelligence score.

R stands for reliability: how much can I trust this tool, and can I come up with the right metrics to trust it? Is it smart, and can people use it?

The data functions of any organization are measured on those three pillars, so when I report to the board, I report on A, I, and R, and every investment is measured in the A, I, and R lens. Anything that we do in the business – be it an initiative, launching a tool, or creating a metric repository – will flow into that one AIR. It's very simple. I like to have a composite index, and AIR really helps.

AA: We often hear of technical debt in traditional IT projects. There's this notion of data debt in the AI-equivalent space. What do you think organizations can do to limit data debt in terms of being smarter about what they collect; how they use it; the governance around it; and everything related to accumulating, storing, and using data?

KS: One thing is investing more in data observability. It's something that I've been pushing organizations to do. Data observability used to be data engineers writing scripts on key metrics and looking at fluctuations, percentile fluctuations, and averages. But now, there are tools on the market that can monitor trends and see whether there's a drop in data quality, and so on.

Another thing would be coming up with a clear policy, and I think financial institutions do this really well. For example, banks have a rule that you can't store data for more than seven years beyond what is needed. Now that digital is no longer a new thing, we have 20 years' worth of data on things we have no use for. Nobody needs 20 years of data to predict anything – 3–5 years is more than enough. So, having clear cut-offs and definitions for how much data is enough and what data needs to be thrown away is important.

The third and most important thing is this: unless you're regulated, it's the Wild West. In retail, consumer goods, and so on, you have emails, names, and addresses floating around. So, respecting privacy is a key concern.

Back to your question, data debt can only be removed if we think of data tools as software tools. When people build software tools in software engineering, they have the concept of runtime cost. Data tools don't have the concept of runtime cost, so people only cost out what needs to be built. If they're deploying an algorithm, they cost out how much effort is needed to build it on the day, but they don't cost out upkeep for the next 12 or 24 months. They assume that some other data scientists will handle it, and that's not the case. Somebody has to monitor the algorithm for model drift and data drift, so you need to know that data tools are no different from software tools when it comes to budgeting for the long term. The **total cost of ownership** (TCO) needs to be different for data and analytics tools, and without that clear baseline, data debt will be hard to solve.

AA: What does an effective data culture look like? I've been guiding many of my clients lately on how to establish a data culture, and I'd love to know your views. What are the key components? How do you establish one? Have you seen organizations go wrong, or have you had to battle to correct or establish one?

KS: A successful data culture is one where the organization does not think that they have one data team. The organization should feel like they are all empowered to be one big data team. I know it's a very big statement, so let me try to clarify it.

> A successful organization consists of people who implicitly and inherently believe that going down the path of using data is the right thing to do.

They don't do it because someone else is doing it. They don't do it because somebody's forcing it. They believe there is merit in it. It takes a special kind of organization and a special kind of belief system to build that culture. For people to have that culture, they need to see data everywhere. They need to have access to it, and they need to see trust in the numbers. Once they have that, that is the culture.

Another feature would be that people should feel like if they don't use a data recommendation, they're making a mistake and should feel slightly guilty about that. For example, I will only use my digital device to spend money. I would never put money in a parcel and send it in the post. I'd never give it to another person and say, "*Drop it in my bank.*" I do it myself. People need to be able to say, "*If I don't do it this way, it's not the right thing to do.*"

> People should feel excited about the problems they're trying to solve.

They should not feel like they're just pushing paper and doing the same thing again and again. Everybody, whether they're working in the data team or outside of the team, should be thinking, "*This is a new, cool problem, and I'm excited to solve this problem,*" rather than thinking, "*My God, we have no idea what we're going to do. Sales are going down. We have no idea why – somebody's going to ask us, and we have no idea. We'll just have to repeat the same things: it was Christmas six months ago. It is the Queen's birthday next week,*" and so on. Instead of that, they should be thinking, "*Cool. This is a new way to find out why sales are going down.*" People should feel excited about a problem.

I think that when those three things happen, the data culture is definitely better. That's how I feel about it.

Discussing leadership in data culture

AA: An important part of data culture is having senior leadership support. In general, what do you think makes a great data leader, such as at the C-suite or CEO level? What are some of the key attributes that they need to have?

KS: I have a framework for everything; otherwise, I can't remember stuff. This is my AAA framework:

> AAA framework
>
> **Autonomy**: Give people the space to excel and deliver smart things
>
> **Actionability**: Give people the chance to work on actionable, valuable problems
>
> **Appreciation**: Give people the deserved appreciation and recognition

The first thing is autonomy. When you hire people in the data space, you're going to hire really smart, switched-on people: people who can go to a new job and do something cool at the drop of a hat. The thing that they need is autonomy. If you tell them, *"Do this and do that, and write this code, and make sure the p-value is so and so,"* you can turn them off really quickly. So, giving them autonomy is really important. When I say autonomy, I mean giving them the right problem and saying, *"These are the boundaries. These are the parameters. You have to solve this problem. Go."* Then, they'll do amazing things. Autonomy is important. People thrive with it because there's a lot of ambiguity. People thrive in ambiguity.

The second thing is actionability. I mentioned earlier that the reason organizations fail is that they go into an amazing building mindset and just keep building without actioning anything. As an enabler, my job is to shine a light and say, *"Hey, look at this cool new thing this team has built. Who wants to use it?"* and then sell that to everyone. Or, if they're feeling blocked by somebody in a different team, such as a product, engineering, or operations team, as an enabler, you should go to them and use your CDO position to say, *"Hey, guys. Why is this not moving? What can I do? My team's trying to do something important. How can we help?"*

> Providing people with actionable and real-value problems is critical to the data culture.

The third thing is appreciation. If you don't regard and recognize people enough, throwing more and more money or work at them will just turn them off. Teams are doing some really intelligent work, and appreciating that is important. There are five languages of appreciation. Some people like acts of service, some people like acts of kindness, some like acts of time, and so on. Most data people are introverted. I'm not trying to apply a broad brush here, but these are extremely introverted people who don't want anyone to know that they've done the job. They don't want anyone to know their name, but they still need to be recognized and feel like they are being heard and seen.

So, those are the things that we try to do to create a great data culture.

AA: One of the struggles of a data leader is managing competing priorities. Do you have any particular processes or frameworks that you use to manage different demands and the time resources of your team?

KS: People automatically assume that just being in the data team implicitly means they have to work harder, so I have a very clear message.

> I give my team clear instructions that work takes up as much time as you give it, so don't work crazy hours. That's an expectation.

One of my bosses told me, *"Work takes as much as you give it."* If you keep giving, it will keep taking. So, you need to define boundaries so that the team knows when to say *"No."*

When the team keeps getting piled on, I introduce a mechanism where everything shows up in a simple place, such as a **Google sheet**, an **Excel sheet**, or a **Jira board**. It shows a list of all the things that people want to do, and the team and I own that document. I take the effort to clean it up, assign teams, and go through it with the key decision-makers and say, "*These are all the things that have been asked of my team. You need to now go to your teams and help me prioritize.*" People are more than happy to do that as they have visibility – they know what the team is doing because it's a matter of credibility. I used to work in a company where the attitude was, "*You have 50-plus people. Why can't you do it?*" People need to be assured that *50-plus* people are already doing *150* things, and that's why we can't do it. Then, they respond with, "*Yep, that's reasonable. I understand.*" Or, they say, "*Don't do all these 25 other things because they're useless.*" So, that really helps.

AA: One of the most important aspects of being a data leader is being able to manage expectations, as they're the linchpin between the technical and business worlds. What advice can you give people in successfully managing relationships? For example, say someone's a manager of a team and they want to manage relationships with the CDO or someone else in the C-suite – how do they do that?

KS: Instead of saying, "*I want to do this, you want to do this. How can we come together?*" we should be saying, "*What do you want to do? Let me use my team to help you do what you want to do.*" When you do that, people think, "*I really find this team valuable.*" Then, they'll probably be more than happy to help you do other things that you want to do. The position that data teams usually take, though, is, "*This cool new recommendation system needs to be built, but the business is trying to chase customers. That's not our problem.*" That's where the friction happens.

> We should be very clear on why the data team exists: the data team exists to help the business grow.

If you believe in that implicitly, then you go to the business and say, "*My job is to help you grow. Tell me what problem you have. I'll drop everything I have to help you address your priorities.*" I think that works, and nobody's going to say no to it. It's such a win-win situation, instead of people going away and doing what they want on their own.

AA: If you get them too excited, they'll start coming to you with more and more work, or some of their expectations might be a bit unrealistic – maybe you don't have the data to support something that needs to be done. How do you manage the relationship or tensions that can occur? Is it an educational piece that you really need to invest in?

KS: It's an educational piece. I often call it the **freemium model**. The freemium model is that you give them just enough to use something but not use it fully. You can build an algorithm, but it's not in production yet. To put it into production, you then need more resources, whether that is headcount or funding. It was a good test drive, but to buy it, the business now has to invest. If people can see that what they want is so close, they will spend that extra few dollars to buy it.

For ridiculous asks on data, I do this thing called **data access maps** (**DAMs**). These are sessions where we go and tell the business what data we have. You'd be amazed how many people spend 10 years in a business but don't know what data they actually track. They don't care, so it automatically switches lights on when we try to show them. The thing that everyone wants to know is, "*Can I know age and gender?*" We'll respond, "*Yes, but we don't track any of that because we're not allowed to – that's the rule.*" People don't even understand that level of concept, so it's an awareness thing.

Storytelling with data

AA: When teams have gone away and produced fantastic results, we hear about the importance of the data story coming into play – the whole notion of data storytelling to translate those results into what the business understands, especially at senior levels. How important do you think data storytelling is, and what are some key ways to develop the skills necessary to be adept at it?

KS: I cannot stress enough the importance of being able to tell a story so people can understand it, but when I say "data storytelling," it's not just trying to translate math to English; that's obviously the most important part. It involves not only trying to keep giving insights but also connecting the insights to something they can appreciate.

The Ferrari example is an example of storytelling. The other one I often use is this: there is this old mayor of a town, and he grew up in the town. It's a famous beach town. People used to come to the town just to see the beach. But what happened over time was that the beach started getting polluted. It would collect a lot of gravel, and it started filling up with so much gravel that people couldn't walk on the beach anymore. People stopped coming to the town because the beach was inaccessible.

So, the mayor of the town says, "*In my infinite wisdom, I think to make this beach accessible, we need to cover it completely with leather. We should buy a huge piece of leather. We should completely cover the beach so people can walk on it peacefully.*" A young girl raises her hand and says, "*Wait a minute. Why can't we just ask people to wear shoes when walking on the beach instead of covering it with leather?*"

I once connected that story to a decision that a business had to make. The business thought that because their data was not in the cloud, putting everything in the cloud was the answer to all the problems that they had with data. I said, "*Yes, putting it in the cloud is great, but why can't we just put different tools in different clouds? Why can't we ask the people who are bringing this information to wear their own shoes rather than putting the whole mainframe in the cloud, which is like trying to cover a beach with leather?*"

When you try to connect it like that, people understand, and the board understands. If people start using the cloud and things like that, trying to explain the situation in an analogy that they can understand really helps decisions be made.

People often mistakenly think that data storytelling involves using graphs and charts to explain specific events; that again is the limitation of people's understanding of what data actually does. Using graphs and charts to explain specific events is what the **business intelligence** (**BI**) or analytics teams do. But what about data engineering or data infrastructure teams? Don't they have to tell stories, too, to seek millions of dollars in investment? That's why the previous example stands out with boards and leaders.

Getting into the industry now

AA: We've been talking a lot about the leadership part, which is integral. Let's talk about data scientists and data engineers themselves: people doing the work on the ground. For someone starting out in the field (maybe transitioning from another field entirely, which is happening a lot in this industry), what do you think are some of the fundamental skills they need to have as data scientists or data engineers, and how important do you think formal studies are (traditional university studies) versus on-the-job training, doing online courses, and other methods?

KS: Typically, when I interview candidates, there are three things I look for.

First, I like to assess their ability to learn, unlearn, and relearn. Everyone says, *"Yeah, I want to learn a new thing."* That's just the new thing to say in data. I want to see evidence that they want to learn.

Also, it's not just about learning – how willing are they to *unlearn* something? Take a question such as, *"You use Tableau a lot, but what if we don't use Tableau? We use X instead."* How comfortable are they with that?

As for *relearning*, the candidate may already have done stats in the past, and they know algorithms, but how willing are they to learn new, complicated things?

> Their willingness to learn, unlearn, and relearn some things is very important to the culture of our teams.

The second thing that I like to assess is the no-expert mindset. If people spend 5–7 years in a business, they like to think they're an expert. Within our teams, though, we try to continuously tell people, *"There are no experts in our team. I am not an expert. You are not an expert. We are all learning."* Maybe you know a lot more than someone else, but you're not an expert. When you think that way, you stop learning. And so, even for somebody getting into the business, their motive should not be to become an expert; their motive should be finding a habit that means they can keep learning regularly.

The third thing is their ability to be logical – to use common sense. We often give a lot of case study interview questions. How do they decompose it into its component parts, and how do they come up with metrics for it? How do they tie it all up, and how can they solve an eventual business problem? It's not just about how you write a SQL query. It's about, *"You know what happened? Sales went down. What will you do?"* They should respond, *"So, what are the five things we look at when sales go down? Well, we can look at these things, and then I'll look at this metric, and then I'll tie it all together."* If people can think like that, they'll actually be useful. Otherwise, they'll just come in and rerun some code that somebody else had.

I'm a bit biased toward on-the-job training and the school of hard knocks. At one point in time, when I was hiring a lot of interns in a different role, I intentionally made it a rule that they had to have worked for one of the Big Four, with the Big Four being McDonald's, KFC, Hungry Jack's, and Oporto. The reason is that in the place that I worked for, data requests would come thick and fast, and people who worked in one of those fast-food joints knew how to handle "hangry" customers – people who are hungry and angry. In the data world, when you establish that your team can deliver, people become hungry for data. So, you need people who have those customer management skills, and it's really hard. I tried to do that with interns, and it worked quite well because these are people who have taken a year off, done something else, traveled, come back, and then want to pick something up. I'm happy to teach them as long as they have other life skills. I'd take nothing away from people doing courses because any way that people can learn is appreciable and commendable.

> For me personally, it's just that the more practical you get, the more valuable you become.

AA: How important is fundamental knowledge of the mathematics and statistics that underpin ML for a data scientist?

KS: For me, it depends on the person and role we are hiring for. If I'm hiring an ML engineer or somebody who's going to work on advanced insights, I'd really like them to know their math and stats inside and out, to the point where they understand the basics of regression, how it works, and what the fundamental assumptions are. They don't think **heteroskedasticity** is a bad word, and they don't think **normalization** is an abusive term. They don't just know the terms but understand them deeply enough to use them.

However, if I'm hiring an analyst or a BI person, I would be more willing to overlook all of that. If somebody's going to work in the ML space, though, it's absolutely essential. You've seen all of those memes where people come in, import something somewhere, and just write the code. That just doesn't cut it.

> Anybody can write a neural network today. That doesn't make them a data scientist because they don't know how to interpret what any of it means.

AA: What advice can you offer data scientists who are looking to enter or grow a career in the field in such a competitive market? Also, what should they be doing to become a lead data scientist or maybe CDO one day?

KS: There is one thing I tell people.

> Showcase your work constantly in some shape or form.

A lot of people try to write blog posts and stuff like that. For me, it's just showcasing in whatever way works for you. Not everyone has to write blog posts or newsletters. Just continuously do work. If you're really passionate about, say, getting into the retail or finance industry, there are so many open datasets now and so many more competitions to showcase your skills. Pick something, solve the problem, and get a feel for it so you know how that industry works and what's happening. Showcase that work – that will help you transition from whatever role you're doing into becoming a data scientist. Somebody's going to take that bet on you.

Once you get into that role, come up with a career path for yourself. Work with your manager and say, "*Where will I be in 6 months, 12 months, 15 months, 18 months, 24 months?*" Plot that path. It's like a career map that says, "*Do you want to progress from a junior to a senior? You have to pick x skill, y skill, and z skill here.*" It's all about learning alongside a career path. If you can do that, it makes it much easier to progress.

AA: How important do you think mentorship is as they progress through their career?

KS: One part of that career map is finding a mentor.

> To go from being an individual contributor to being a manager, you will need a mentor who has managed people, and you will need to work with them to understand how to do these things.

Looking to the future of AI

AA: I have some questions about generative AI, and particularly tools like ChatGPT. Firstly, how do you think they affect the data science scene?

KS: ChatGPT has brought a lot of attention and interest to data science, getting people outside and inside the area excited about what can happen.

A lot of data scientists themselves are still understanding the various implications of **large language models** (**LLMs**) such as ChatGPT and how they can be used to enhance and expedite their work. While technologists and data leaders are working to understand the implications, non-data and non-technical business users want magical outcomes from ChatGPT instantly. They are asking for artificial general intelligence (AGI)-level solutions to company problems now.

Until a steady state is achieved in terms of capability versus demand, there will be wild swings and a lot of hype on both sides of the spectrum – infinite possibilities on one side and doomsayers predicting the end of days via AI on the other side.

AA: Although, as you point out, understanding all the implications is still a work-in-progress, I'd like to hear your view on what the key ethical implications are for organizations and society more broadly.

KS: Data integrity and security for organizations using LLMs is the first concern.

> Many companies have recently found out the hard way that sharing data with OpenAI (the creators of ChatGPT), or any other organizations providing generative AI tools, is a one-way street. An organization's data and associated IP is no longer theirs to control, once it goes into these systems.

Because these tools are not guarded by the standard data protection and privacy rules, and their workings are mystical to many - the outcome of how these systems would describe and profile someone is quite worrisome. A very pertinent example is an Australian mayor who is suing OpenAI for the wrong characterization of him based on incorrect data that the system picked up somewhere.

> **ChatGPT defamation lawsuit**
>
> *Australian mayor readies world's first defamation lawsuit over ChatGPT content* by Byron Kaye, for Reuters: `https://www.reuters.com/technology/australian-mayor-readies-worlds-first-defamation-lawsuit-over-chatgpt-content-2023-04-05/`.

AI systems already have bias in them due to the systemic bias in our world, which tends to get amplified with ML systems, which are mostly black-box algorithms working to optimize a particular metric or variable.

Few-shot learning is an ML method that enables generalization from a small number of training examples.

> **With few-shot learning being the key way to use LLMs and generative AI applications, the bias of AI systems will only be further amplified, depriving affected groups even further – perpetuating an unending cycle of discrimination and disadvantage.**

The definition of reality (truth) versus imagination (creativity) will become harder for society to understand and delineate.

> **With generative AI's wide-spread availability and the ability of anti-social actors to easily leverage these tools, the ability for citizens to discriminate facts and truths from propaganda will become more and more difficult, amplifying social divisions and unrest.**

AA: In a climate of increasing geopolitical tensions, your point is well made about the potential for actors to leverage generative AI technologies to increase societal division.

Moving back to the infinite possibilities side of ChatGPT, do you have advice for how data scientists/ AI experts can leverage ChatGPT to benefit their organizations and their own careers?

KS: Data scientists can use ChatGPT to stress test their code, as well as check for any security vulnerabilities. A lot of data scientists are already using Copilot and other ChatGPT-like tools to quality check and validate their code (particularly code written using scripting languages).

Summary

I've previously discussed with Kshira various common reasons that data science projects fail to meet their full potential, so it was great to speak in more detail about some of these challenges and pitfalls and to discuss ideas to help overcome them.

Kshira offered some great strategies and advice for anyone facing similar obstacles or having trouble scaling their data science capability. He offered similar advice to Jason Widjaja in that being data-driven ultimately means decision-makers changing their minds in line with what the data is telling them. This often necessitates decision-makers becoming data literate, so they can better understand what data science can and can't do – and, at least conceptually, how it works. According to Kshira, the greatest challenge is asking the right questions – which is a skill. In addition, when it comes to turning that data into scalable and repeatable actionable insights, it's important to have data engineering and DevOps skills in your data science team to help with the transition from development and POC to production.

I agree with Kshira's advice to *"place the data team closest to the decision-maker,"* as this will not only help in gaining much-needed support for staffing and resourcing capability but also in turning the findings into actionable outcomes that are aligned with the strategic goals of the organization. A key part of this is capturing data that's fit for purpose, rather than just accumulating data because you can, and without a clear strategic need. It's also crucial to have a business sponsor who wants a solution to the problem you are solving. As Kshira says, otherwise, you have the issue of people not knowing what to do with a product.

We also discussed the importance of having a successful data culture and some strategies for achieving it. I believe that a data culture can make or break a data and analytics capability within an organization, so it was great to hear Kshira's recommendations on achieving this. We discussed the theme of change management, which Althea Davis also raises in *Chapter 12*, and that it's important to anchor your projects with change management, as well as anchoring your projects with business needs and the creation of business value. An integral part of this is helping people trust the solutions that you're developing for them. One strategy for helping people believe and trust in your solutions is to be able to quantify their value via metrics, and Kshira's AIR paradigm is a fantastic approach.

Kshira also shared pertinent advice on effective data and analytics leadership, which will be helpful for anyone who's new to a senior leadership/management role. His advice for interviewing data and analytics candidates was to focus on three key areas:

- A willingness to learn, unlearn, and relearn things
- A continual learning mindset
- The ability to use logic and first principles to solve problems – especially in new domains

For anyone who's starting or in the process of developing their career in the field of data and analytics, Kshira's advice and recommendations are priceless – especially in relation to helping you stand out and excel in such a competitive field.

5

Consulting Insights
with Charles Martin

Charles Martin is one of the most technical and accomplished consultants I know, with an incredibly impressive career. He is also an active researcher (in collaboration with the University of California, Berkeley) on the foundations of AI, so it was very exciting to have the opportunity to discuss his views on how and why organizations often struggle to achieve success with data science and AI.

I have read about his new WeightWatcher model – an open source diagnostic tool to analyze deep neural networks – and I was interested to hear more details about its development.

I was also keen to discuss his thoughts on the growth of the field of **Natural Language Processing** (**NLP**) and his predictions for what the future holds.

Getting into AI

Alex Antic: How does a chemical physicist end up becoming an expert in AI?

Charles Martin: I think that people may not realize that AI was invented by theoretical chemists. **John Hopfield** was a theoretical chemist. He invented one of the most famous neural networks, called the **Hopfield associative memory network**, and he was a theoretical chemist and sometimes a physicist. My advisor recently worked on something you may know called **AlphaFold**. He and his student **John Jumper** developed AlphaFold. What happened was that Google/DeepMind hired Jumper, and then they took his thesis work and souped it up.

There's a long history of doing AI and theoretical chemistry.

I've been doing this for a very long time, and when I left my postdoc, I went off and worked with my graduate advisor's son-in-law, who had a start-up called E-self. It was a personalized search engine, and that's how I got into NLP, because we applied techniques from AI to try to do searches. I learned about search relevance back then, in the late 1990s and early 2000s. Then, I went to go and work with **George White**, who was the founder of Xerox PARC and was involved with a company many years ago called General Magic. This was in the early 2000s. I worked with him for a while, and he was a very famous guy in AI and the tech industry generally.

I've been doing this for over 20 years. I've been doing AI and NLP on and off, and it's just amazing that technology has gotten better and better; I just try to keep up with it.

AA: How have you seen the field evolve over that time?

CM: The biggest change was when Google and Amazon decided to get involved in AI.

When I was a consultant at eBay over 10 years ago, we were using neural networks, but we were using them for things such as fraud detection. It was the kind of thing where there was a researcher at a university, and we'd hire an intern or a student, and they would deploy to production, but it was very specialized code. I had worked on some of the early forms of RankNet, which were neural networks for the ranking of searches, but this stuff was all custom code.

What happened was that when Google acquired DeepMind and some of the other large firms got involved, they started productionalizing tools such as **TensorFlow** and **PyTorch**, which previously were academic tools. TensorFlow was **Theano**, and PyTorch was a **Lua** program – so, Lua Torch. Once the software engineers got involved at scale, they put hundreds of millions of dollars into it. They stabilized the code and made it totally accessible to everyone. That made things take off.

After that, you could do AI. You could do something complicated, whereas previously, you either had to write the code from scratch or had to go and find some academic code written in C. When I worked at Aardvark, we took academic code written in C and integrated it into Ruby. Nobody does anything crazy like that now. You just download some Python code. Now, you have enormous technical support from the big tech companies. The software engineers have gotten involved in this, and now they think that they do AI because they've written all the software and they have all this technology where you can tune it, install it, run it, and test it. That has been a huge impact on the field.

From a theoretical perspective, there hasn't been much progress. There's been a little bit, but it's more of an applied engineering field at this point. It stopped being an academic curiosity for theoretical physicists and some computer scientists and became applied engineering and applied science, and that's what really changed it.

AA: It reminds me of my early career working as a quant and writing code in C and C++, and using the Numerical Recipes book, which you'll be familiar with, whereas now it's so commoditized. Anyone who's got any vague understanding of Python or ML can just pick a library and without having to hack away at the code at a low level.

CM: When I first started working in Silicon Valley, I would be interviewed for a job, and people would ask me whether I knew algorithms, and I'd go, "*Of course, I know algorithms.*" Then, they'd ask me something such as, "*How do you write a sort algorithm?*" I'd go, "*What's a sort algorithm? What's a binary tree? I have no idea. Why don't you ask me how to solve a differential equation numerically or how to write an eigenvalue solver?*"

I got into an argument years ago. I had an interview at a company, and during the interview, they asked me something about how to solve a problem using a dense matrix. I said I would use a kernel expansion to expand it into a sparse matrix and use a sparse solver. The guy had no idea what I was talking about. Well, I thought, *"I'm not going to work with you. This is ridiculous. You don't understand the question you're asking me. You're in a totally different field."* Now, it's hard to work with some people because they've never seen Numerical Recipes. That's foundational to the whole field. Today, every good engineer coming out of school knows these techniques.

Balancing research and consulting

AA: Speaking of research more broadly, you're an active researcher in this field and you're a consultant, and I find it amazing that you can juggle the two. I'm keen to understand some of the current projects that you're working on and what excites you the most. What areas of your expertise do you love to dabble in and produce actual products and solutions to?

CM: Well, the main thing I've been working on is something I call the WeightWatcher project, which is an AI model monitoring system. What we're able to do is use techniques from theoretical physics and chemistry to analyze the performance of an AI model – such as a deep neural network – without looking at any of the data. I can take a model that's been trained by somebody and feed it into my theory. I can look at some pretty pictures and some metrics, and tell you whether certain layers are overtrained and certain layers are undertrained. I can also estimate the model quality by taking two models that have the same architecture but different hyperparameters or different datasets, and tell you which one I think will perform better, just by looking at the weight matrices in the models. We've developed this theory and a practical tool to try to make AI more like an engineering discipline when you build something and understand why it works.

> Right now, AI is very much engineering, but it's like where steam engines were before thermodynamics. We crossed the great frontier using steam engines, but we didn't have thermodynamics.

We're trying to come up with something just like the thermodynamics of AI so that we can actually understand why things work, and then from that try to build tools that you can offer people. That's the basis of my research.

AA: At a conceptual level, how does the model actually work?

CM: The idea comes from quantum chemistry, and it's the idea that when an AI system is learning, it's trying to learn the correlations in the data that can be transferred to other datasets. You're not just trying to do curve fitting. You're trying to learn some sort of model that you can generalize to other problems. That's very similar to the work we did in quantum chemistry back in the 1970s and 1980s. There were these methods called **semi-empirical quantum chemistry** methods. What you would do is take a model and train it on a bunch of molecules. It was half quantum mechanics, half empirical data fitting. You learn the correlations in the data, but the correlations are somehow transferable to other similar datasets.

What our metric allows us to do is answer questions. How much correlation did you learn? Did you learn a lot of correlation? Did you learn a little correlation? Did you learn too much correlation?

That's all it is, and it's actually very simple, but from a theory perspective, unless you've done quantum chemistry or some areas of quantum nuclear physics, you may have no idea what I'm talking about. It's a very small esoteric little area, but that's what's going on, and that's what we think is going on. We have a new paper coming out where we're going to derive some basic equations that explain not only why these things work but, more importantly, how we can use our theory to make predictions. We can predict which model will perform better. We can predict which layer is not trained correctly. That, to me, is a theory. I'm not trying to explain just why things work. I want to give you a tool with some basic equations behind it, and I want you to use it. That's what we're trying to do.

AA: This is potentially a game-changer in the field, very exciting.

CM: Well, I don't have three billion dollars to invest in writing my own version of TensorFlow. I had to pick something that would have some impact but also would not require an enormous amount of computational resources. So, we fell into this idea of trying to move the needle somewhere where nobody else is looking.

I have an old friend at UC Berkeley – **Michael Mahoney** – who I talk to once a week, and he's like a sounding board. He was also a theoretical chemist, and then he became a machine learning scientist, so he actually has some idea of what I'm talking about. We talk a lot about this and I run experiments, and we publish. It's been good. It gives me something to do between consulting projects, but it comes up because of my consulting.

> The major problems people have in their models are things such as them not knowing why models work, why they break, and how to monitor things. Things break in production – you don't know why. It's all just blind parameter fitting.

A lot of the people in the field are very good technicians, but it's not clear they understand what's really going on. There are a lot of people trying to work in this space, and I just came in through the back door and snuck in. That's how we did it.

AA: That resonates so much with me, especially given my background in applied mathematics where, with traditional modeling, when you've got analytical or semi-analytical solutions, you have an understanding of sensitivity to initial conditions and boundary conditions. You have a much better feel for when things will break and what the constraints are, whereas with deep learning and ML more broadly, you don't always have those insights. Something like this is just so transformational.

CM: What we find is that if you were a quant, you'll know what we're doing. Quants use random matrix theory. Our stuff is based on some form of that. Quants study correlated systems because they study correlations in the markets.

I was a quant at BlackRock, and a lot of the ideas that came into WeightWatcher came from my study of applying theoretical physics to quant problems. A lot of the guys who were quants were applied mathematicians or theoretical physicists, so they had to take a field that was very rigorous and built on the foundations of other experiments, and then apply it to something else entirely. You're pricing options using the Schrödinger equation. You're saying it's a stochastic system. It's obviously not. An option is not the same as a quantum system, but the mathematics is the same. You can apply the mathematics in a way that is not obvious, and I think that's where a lot of computer scientists have been. It's much more about engineering, guessing, fiddling, tweaking, and quibbling. It's not the same as when you're a quant. There, you're saying, "*Look, we have a whole theory of mathematical models. Can we apply these models to some other set of data?*" That's the mentality.

What we've found is that people in the quant world have been much more accepting of the research. As for the AI guys, unless they work in that space, they're not familiar enough with the technology.

AA: That's a very good point.

CM: I interviewed for the CTO position of an AI medical company the other day, and just as with the existing CTO, I could barely talk to the guy. He asked me, "*What is temperature?*" I thought, "*Well, what can I tell you? Temperature is the Legendre transform of the microcanonical ensemble, but it's the intensive parameter that couples…*" I can't say that! He'd have no idea what I'm talking about. He doesn't know physics or chemistry. The thing about this field is that you have so many people from so many different backgrounds. We're coming at it from the theoretical physics/theoretical chemistry background, which to me is traditionally where this stuff came from. But of course, I'm biased because I was a theoretical chemist.

Advising companies on their AI roadmap

AA: In your career as a consultant, you help organizations in many ways, including in their transformational changes around AI/ML. Organizations commencing their journey in this space often don't know where to start. What advice do you normally give them when they look at data, people, process, culture, and technology?

CM: I've been working with some private equity firms doing this: you have to come in with the right executive team that has the vision of becoming a data-first company, and you have to then ensure that you take data seriously. You need to know what data you have and you have to understand at a detailed level whether your data is useful or not.

So, go in and do a data audit. Say to yourself, "*I'm going to audit every single table in every single database and understand what we really have.*" In companies that do that right, they can see a 30, 40, 50, or even 200 or 300 percent increase in revenue, just from you being able to say, "*Here's the data. We've got the data now. We're going to deploy these algorithms.*"

On the flip side, there are other companies that never took the time to ensure their data processes were governed properly.

If you're a manager at a company and you're trying to have an impact, you should try to build something end to end. Try to come up with something that you think will have an impact on your team. It should be a revenue generator for your team, an easy win. Build it end to end – that way, you see every single piece of the data in the pipeline.

I'm doing a project with Walmart. They probably have thousands and thousands of tables. There's no way you could go into a place such as Walmart and do a data audit – it's impossible. But what you can do is say, "*We're going to build a search-relevance system, and when we build it, we're going to see every part of the system, what's in the data and how it's maintained, who the people maintaining it are, what the processes you're using are, and whether we can fix bugs when they appear.*" That's what you have to do, and you have to take it seriously.

One of the problems I see is that a lot of companies bring in guys who are **Facebook, Apple, Amazon, Netflix, and Google (FAANG)** managers, and I use the analogy of Thor. Remember Thor? He's living in Asgard but he gets cast out to Earth, and when he gets to Earth, he can't lift his hammer. He has no power. When you leave Google or eBay or Facebook, you've been living in Asgard. You've been living in a golden city. When you cross the Rocky Mountains, you've fallen back to earth. (Remember: Asgard is flat.) You've fallen back to Earth, and now you're in a company that has no infrastructure, and you can't expect it to be there. You have to build it, and you have to build the processes. You have to understand that you need to have control over it.

It depends on where you are in the organization. If you're coming in and acquiring a company, you've got to do the kind of audit where you really see everything. If you're a manager in a company, you've got to try to build something where you see everything end to end and close the loop. Then, the question becomes the following:

> Are the leaders you work with open to change? Are they willing to do what's necessary, or are they just going to keep doing what they're doing?

AA: That touches on the next question – what are some of the common mistakes and pitfalls you see? What are the challenges that many of these people grapple with in becoming data-driven?

CM: One example I saw was this idea that you can just stuff everything into **Hadoop**, and once everything's in Hadoop, you're good – you're golden.

There is the idea of data democracy with Hadoop. Everything's in Hadoop, and anyone can access the data anytime they want. Great. But that's like just stuffing boxes of junk in your garage. You need to find a screwdriver to fix your lawnmower, but it will take you 6 months to find the screwdriver because you have no idea what box it's in. You've got to unpack every single box to find it.

That's what Hadoop's like. You can't just stuff everything into your garage and then show up 6 months later and expect to know where everything is. You have to have someone in charge who is responsible for the data in the databases. If you have a database, you can't just go in and change the schema anytime you want. If you have data, you can't just go in and change the format of the data in the columns – it has to be stabilized.

What you see is that the companies who do this well are either small enough that they can lock down all the data and have one central database for the whole company, or they're large and distributed but can put in guardrails so that if something changes, the stakeholders for the data get an immediate notification that something has changed. You have to have that kind of change management for your data, you have to monitor your processes, and you have to treat the data with the same level of paranoia that a software engineer looks at code.

You have to have tests: unit tests, regression tests, guardrails, flags, warnings, and signals. You can't just pass data around, and you can't just email people spreadsheets and expect to have a data-driven organization. You can't just have databases running in a foreign country with other people. I've seen cases where we had major problems and bugs. Bugs need to be fixed; otherwise, the system won't work.

You follow a bug. It gets transferred over to a consulting firm in another country, and they take it out of that system and put it in their system. What happens? A year later, it's still not fixed. You can't just hand off problems and ignore them. You have to really go after them, and I think that's where the problems are. People find that they've never built these products, so they just don't understand how one small change in one column in one database can bring down your system for 3 months. I have seen this firsthand; this is where many companies are today.

I read about one incident where a company lost tens of millions of dollars because the training data for their model from one of their clients was corrupted and they didn't know it. When they trained the model, the model was wrong. How could you not know that? It's because you didn't have the guardrail in place to check that when you uploaded the data. How could you not ensure that the data was what you thought it was?

You need guardrails, regression tests, unit tests, and checks. You've got to make sure that things are working correctly. You have to be paranoid about what you're doing.

AA: I like that. You need to have good, strong governance and be committed to governance, not scared of it.

CM: Yes, and to be committed to governance, you have to do the right thing. You can't just have PowerPoint presentations and talk the talk. You actually have to do it.

Understanding why data projects fail

AA: Something I've had discussions with people recently about is the high failure rates of AI projects and data science projects. 85 percent is the number being thrown around.

We're drowning in data. It's easy to get a data scientist to use R and Python and to set up a simple infrastructure. Why are the failure rates so high then?

CM: Well, because data science is not commoditized the way they want it to be.

The IT industry used to be an actual profession. You would go into it with skills and you could do things. After 20 years, the IT industry has been completely commoditized. It's outsourced to other countries. You can hire specialists in different areas. Project managers know how to map things out. It's totally commoditized and optimized.

AI's not like that. I work on search relevance, for example. You can't just bring anyone off the streets – some random data scientist – and have them work in this field. They don't have the experience. It's like saying, "*I'm going to hire an electrical engineer at a school to build a computer chip.*" It's not the same. Companies want data science to be a catch-all so that they can hire people cheaply. They think, "*If we hire five cheap data scientists, that will scale.*" No, what you probably need is one or two really good specialists, maybe one or two data engineers, or maybe an MLOps engineer and a data engineer.

> Companies don't staff things correctly. That's part of it. They want it to be "plug and play".

Another reason is that many projects are just not well formed. Data science is science. You can't just say, "*Hey, wouldn't it be great if we had a pill that could cure COVID-19 and just prevent it? Let's start a company to prevent COVID-19. Let's make a pill.*" That's all they think: "*Let's just make a company where the AI does what we want it to do.*" It doesn't work like that. It's not the same as building a website or a database. You have to understand what you can and can't do. A lot of companies are wishful thinkers about that.

The other problem is that companies still try to apply software engineering practices to what they do. You go into a project and say, "*Well, we're going to be Agile.*" But what is Agile? In most companies, Agile means *waterfall with sprints*. They try to plan out what they're going to do for the year, and then they pretend they're Agile because they have a sprint every 2 weeks. This idea that you have to be able to sit down in an Agile sprint meeting or a scrum call and plan everything out, without any forethought or discovery – it doesn't work.

You can use agile for science, but people don't know how to do it. They try to apply IT and software engineering processes and management to something that is not software engineering.

So, you end up not taking the time to do the discovery, or when you do the discovery, you discover that it's not going to work because it wasn't thought out properly. Other issues come because the data is wrong. Had you been careful about the data and checked it properly at the beginning, you wouldn't be in that position.

Even when you know what to do, there are all sorts of process issues in these organizations. Let's say you build a model and it works, and you know it works, but you can't get it to production because the software engineers have put in so many guardrails themselves for their IT systems that it's become so rigid, and you can't build anything that will fit into this hyper-rigid process.

I'll give you an example. I built a model for a client in 3 months. It took 1 year to convince them to give us a bucket on Amazon so that we could train the model on one machine, put the model file in a bucket, and then have the inference code read it from that bucket. We actually had to hire a guy from another team who had access to their bucket, and the only thing he did that was useful for an entire year was give us access to that bucket.

I'll give you another idiotic example. I built a product for someone using Jupyter Notebooks. In their production environment, the logging system required that you log the standard error, and the standard error then had to be piped to standard output. So, how do you pipe standard output from the notebook? When it runs, where do we store the logs? They had decided that every single system in the world worked like this, and they'd never seen any other piece of software that might be a little bit different. Instead of having a log file dumped to a Splunk log or the standard error, maybe you just make a notebook and stuff it into a bucket. These IT systems are so hyper-optimized that you can't do anything even slightly differently. Even if you build a model, you can't get it to production.

AA: Rigidity of process or thinking will always hold you back, won't it?

CM: What I advise clients is that when you start on a project, you need to carve out something.

> You need to carve out the piece where you're going to do the machine learning and the AI and have as little interaction as possible with IT services.

Another problem that happens is that companies try to hire data scientists or senior people to do machine learning. Well, who does the vetting? The CTO can't vet the data scientist unless they know them. If you know someone who knows them, and you come in through a recommendation or you went to school with them, then yes, you can do it. That's how start-ups work. But if you can't vet the person, who vets them? Well, the director of IT has to vet the chief scientist. OK, but how does the director of IT vet someone who's an expert in AI?

AA: I had a discussion this week with a client where I had to say something that I always tell senior executives – data science is not IT. It's discovery versus delivery. Do not mix the two.

CM: Even if it is IT, you can't do it. I built a model for a major e-commerce company 12 years ago. I still can't get it productionized!

It's not IT, and the problem is that from a hiring perspective, the IT services guy is going to hire the data scientist and vet them. Do they have the ability to vet them? No. And if they vet them, are they going to hire someone who's going to challenge their authority in the organization? No.

I did an interview for the CTO of a medical AI company. The current CTO was a deep-learning guy who was leaving. Why is he interviewing me? He's interviewing me like he's going to vet me. He's leaving! Why is he even involved? You end up just sitting there and thinking, "*I'm going to throw away everything you've done and start over from scratch because you're leaving. I'm not going to maintain your stuff.*" He was mad during the interview because I didn't answer how he wanted me to when he asked me, "*What is the Netflix recommender?*" Really? I'm being interviewed and you're asking me how **singular value decomposition** (**SVD**) works? He was saying, "*Well, we have this LSTM model we're using.*" I replied, "*OK. Well, what if I want to use a wavelet attention model?*" But no – I had to do exactly what he was doing. Even though he was leaving. He was just too junior to vet the next hire properly and asked questions I would ask an intern, not the next CTO.

You have this problem where companies can't hire, and when they do hire, those they hire don't want to do their jobs. I've seen a case where companies bring in a FAANG manager from eBay or Google. There's no infrastructure for them, but also, the process is completely different. They're used to being in the Google process or the Facebook process, and they come into your organization and you're doing Jira sprints. Why are data scientists being forced to use the software engineering process? Why are they using the IT process?

I've seen private equity firms acquire other firms and get rid of all the management, replacing them with people who will actually do the data science for them. That's where I think where the industry is going. If you want to have success, it has to be driven from the top. It has to be driven by where the funding is coming from.

In the past, I did projects with founders. When you do a project with a founder, things work because the founder is determined to do what's needed. What I find now is that I come and work in big places and I'm working with low-level managers. Even if the manager wants to do something, they're stuck within the environment they're in, so you have to make it work in that environment – that's very, very hard. They have people pushing back on them. It's not just because you're not doing science and you're doing exploration. It's because of the internal processes, the politics, and the mentality.

> To be successful, you have to enable the manager to be the champion of the project.

AA: We have already discussed the importance of data quality and data suitability. It goes back to Andrew Ng's theory of data-centric AI versus model-centric AI. Is there anything more you wanted to add to that?

CM: What I assume is that in looking at manufacturing use cases, he realized there are many problems where big data doesn't exist.

AA: Once again, things are different outside of the FAANG world.

CM: Yes. With data-centric AI, what you're saying is that if you're going to make an investment in a company, lead a company, buy a company, or sell a company, you've got to make sure the data's there. It's the idea of doing a data audit.

AA: In your consulting business, over the past couple of years, have you seen COVID-19 have a major impact in terms of trying to rectify some of these issues – people being a bit more cognizant of what they need and what they don't know, hence the need for expertise to be data-driven?

CM: No. I've got to say that I think it's coming from the investor side – the private equity firms that own these companies, or the people who are on the boards and are seeing they're not delivering.

Normally, I would come in as a chief scientist or a data scientist. Now, though, I'm moving toward a more senior level. I'm being brought in by the board and at an executive level to make sure that people are doing what they need to do. We're seeing more of that at the board level – bringing people in to say, "*Look. We need to make sure that you're not flubbing it.*"

That's where I think the industry is moving. People are realizing, "*We acquired a company. It claimed it was a data-driven company, and it turns out that it hasn't deployed a product in 3 years because its data is constantly broken. We need to get rid of the management and put people in who know what they're doing.*"

AA: Some senior executives get very excited by AI. They read an article, they go to a talk, and a vendor sells them some dream idea. Beyond the hype, what warnings can you provide them to make sure that their vision is realistic?

CM: Again, I try to say that you have to build something.

> I'm of the opinion that you've got to build a prototype.

You can't just pretend. When raising funding, as I used to do, you should build a **pretendotype**. You should build something and pretend it's real; fake it to make it and see whether people will buy it. But when it comes to AI, no – you've got to build a prototype. You have to see whether the thing that you think is going to sell actually exists.

What we see is that even if you build a prototype, you have to be cognizant of whether you can get it to the level of accuracy you need. I'm never going to build anything that has 100 percent accuracy. How much give do I have? What happens if I make a mistake? What happens if I build this thing and it doesn't work, or it doesn't work in production the way I thought it was going to?

I had a client come to me once, who said, "*We want to build some sort of document analysis system. We know it'll only work on 80 percent of the documents, but we need to know which 80 percent of the documents it's working on.*" It doesn't *necessarily* work like that. It can't tell you with confidence whether it knows it's correct or not. That's not what statistics is. It's probabilistic. But the clients don't understand this.

You've got to pay attention to what you're doing. The client won't understand something. They won't think carefully about what they're doing. You can't bring in someone with a high-school education – God love them – and have them manage a quantum computer. When you're doing requirement specification, you have to understand the details of what you're doing, and you have to understand whether things are going to work or not. There is this sense that somehow in the business academic world, details don't matter. But in the engineering world, of course, the details matter. Even little things like that can make or break a project.

AA: All too often, when organizations invest in data science, they seem to focus on the data and tech but not the science.

CM: They don't focus on either! They focus on their own opinion!

Say you have to do experiments, so you start doing A/B tests. People are used to doing A/B tests because they're used to doing marketing. For instance, say you want to try five different landing pages and see which landing page attracts the most customers.

Now, let's say I put a learning algorithm into your search engine, and over time, it interacts with your users and figures out what they like. Do you think you're going to A/B test that in a 2-week run? It takes months for the learning algorithm to figure out what's going on. How do you test it?

A company has optimized processes to bring up the margins. You want to have high margins so that you can make money, and you get high margins by optimizing your process. You charge the same price but you make the product cheaper internally. Take the McDonald's milkshake – powdered milkshakes instead of fresh milkshakes. Same price, different margin.

The problem is that those processes that worked and were best practices for things you did in the past are not necessarily best practices for what you're doing now. Companies are very process-oriented, and science is not process-oriented. They want science to be, "*Hypothesis, experiment, test. Hypothesis, experiment, test.*" That's not what it really is! That's what you did in high-school chemistry!

> Real science is messier, and you have to let scientists do it.

When I got out of grad school, I was not ready to go into industry. I went to the University of Chicago. I was ready to win a Nobel Prize. They don't train you to go into industry. You have all of these people coming out of academia who are trained to be researchers, and that is what they want to do. Then, they enter the business world and they don't know how to adapt the academic research process to what businesses need. Bring them into a company and they won't know how to write production code because they've never worked in a production environment. They don't know how to do applied research because they've never really done it. They don't know how to work within the agile software nonsense you're shoving them into. The data's a mess, and you have no one who can lead them.

Maybe 10 years ago in Silicon Valley, what happened was that you could build an ML prototype. It would generate a bunch of revenue. You could put it in production, you could make money, and you were off to the races. Now, companies know that they need to improve those models and do applied research. They need to make the models better. They know they have competitors nipping at them. They need to expand the process. But they don't know how to do it. They don't know how to manage it. They don't have anyone who's ever done it. They're impatient, and they're putting random people in charge. They have this idea that anyone can lead.

You just can't take someone who's been in a graduate program for 5 years working by themselves, throw them into an environment with a bunch of software engineers, expect them to all get along, and then have no one in charge who can manage it.

A lot of companies struggle with this now, and it's like nobody wants to go into these organizations and deal with us. If you bring someone in from a company such as Google or eBay, it's like you're taking their kneecaps out. You're putting them into an organization where they have no infrastructure and no support. They can't do anything. Usually, the solution is to hire people when you acquire a company. A lot of companies will think, "*I don't have a solution other than to hire someone who will come in and do it for me.*" Internally, you just end up seeing managers being kneecapped by senior executives on all sorts of things. You're just kneecapping them. Why are you doing that? I like working with people. I try to support the people I work with. If someone brings me in, I'll go to a meeting and say, "*What just happened here? They totally kneecapped you. No wonder nothing's getting done.*"

AA: One thing I've seen – especially in the public service space, where I do a lot of work – is that traditionally, you'll have a lot of generalists move up to senior ranks and suddenly end up managing technical teams. They'll say, "*We don't know what we're doing. We don't know how to vet these people, how to hire them, or what motivates them.***" I advise them that they need technical people in some capacity, with a background in those roles.**

CM: From talking to very senior people here in the country, such as a fellow who worked in the Bush Sr. administration, in the **National Institutes of Health (NIH)**, my understanding is that there used to be scientists in the NIH, and they ran things. Now, 30 years later, it's all MBA-type consultants who are running the NIH, and they hire scientists. You have no one in the government organization who has any functional understanding of what's really going on. They rely upon other people to tell them. Of course, that's what we do for a living – we tell other people things. But it's a critical problem. You have to be able to make critical decisions and understand things. If you really don't know what's going on, how do you evaluate it? If you go into an organization and there's some cultural friction, how do you determine whether that's really friction because of the person you brought in, a personality issue, or anything else? Is it because there's a functional difference in what two people are saying, and one person is correct and the other person is not? You can't make that judgment call. You're not capable of doing it.

There ends up being stagnation. When you go into an organization, you're trying to help people, work with them, sell to them, and get people to work together. Not everybody in there wants to do that.

AA: In terms of helping people better understand the scientific process, sufficient reproducibility logic, and so on, does it really come down to them having the right people in the right seniority to manage it?

CM: No. You have to have money, for one thing. Science is expensive and slow. You have to understand what it is you're actually trying to do. You have to realize that you can't force it. When I work with clients, I try to give them the simplest model possible that will get them a win. For example, I might give them an SVM. "*Why did you give us an SVM? Why don't we use XGBoost?*" they may ask. I'll reply,

"Well, because the SVM is nonparametric; I don't have to tune it. Because the SVM operates in under 5 milliseconds, I don't have a latency issue. Because I don't have to worry about building feature stores. I can give you the simplest thing, and being able to find the simplest thing possible that will vet the idea is very hard. A junior person cannot do that."

You have to see dozens and dozens of projects and really understand what the operational challenges are. A company is not an academic environment where you're solving a science problem. You're trying to solve it within the operations that they have. What is the operational environment you're working in? Everyone wants to say, *"We have this magic database and we have the data, and you can just run algorithms on the data."* Working at a company, building a product for a company, getting it to production, and running it is not the same as winning a Kaggle contest. I would probably have one of the lowest Kaggle scores. I don't know all those little tricks and nuances. What I know is how to get things into production.

One of my clients is at Anthropocene Institute. It was related to the pagegram. We have a numeracy product. It's an app that tries to help you do numerical calculations about climate change in your head.

When the scientists of the Manhattan Project made the first atomic bomb, they did a back-of-the-envelope calculation to make sure it would work. They didn't start by spending 5 billion dollars and refining uranium. They did simple calculations to convince themselves it would work. It's the same thing when you go into a meeting with a client. You have to think, *"What is the simplest thing I can do to vet the idea? How do I vet it as quickly as possible so if it doesn't work, I don't blow the whole project?"*

You have to be very careful. Consultants, of course, want to bill for a year or two, so they'll just grind away. They don't want the project to die. You have to commit to doing that, and that's part of managing – it's being able to figure out what that is. While you can work with business people and you can use their domain-specific knowledge, you can't let them drive it. If they drive it, they'll drive it into the ground.

As a consultant, one of the analogies I use refers to how, years ago, there was a book called *The New New Thing* by Michael Lewis. It's about Silicon Valley. There's a guy who used to fly sorties in Vietnam, and he runs a helicopter flight business in the Valley here in California. Guys like the super-wealthy would come in and want to learn how to fly a helicopter. This pilot had two sets of controls in the helicopter – one for him and one for his customer. In order for the business to function, the customer has to think that they're flying the helicopter, but if you let them fly the helicopter, they'll kill you. You have to know when to pull the reins in.

> **This is the critical thing in working with clients. You have to let them drive, but you can't let them drive you off the hill.**

You have to be willing to work with them in a way where you can get some sort of prototype into production and tested. You need to do a back-of-the-envelope calculation. It takes a lot of work. That's how I approach working with clients.

Measuring impact

AA: Are there particular metrics you tend to recommend to clients in terms of how they ascertain impact?

CM: I think the question is, *can* you measure your own impact? Do you have the ability to measure your own performance and break it down customer by customer, point by point? Ask yourself, "*What are you actually doing to do your own measurements, and is that relevant to what you're doing?*"

I gave the example of doing an A/B test. An A/B test is not necessarily useful for forecasting machine performance, and a lot of companies simply do not have the ability to ask, "*What is my customer lifetime value?*" They don't know how to measure revenue at a point-by-point level. I think you have to ask, "*What are you going to measure? Can you actually measure what you think you're measuring?*" And you have to work with scientists. You just can't assume "*I'm the business guy. I decide what the metrics are. Here are the KPIs.*" That's not necessarily the best thing. It can be subtle. In a complex environment, having more views doesn't always lead to more clicks, having more clicks doesn't always lead to more purchases, and more purchases in one area may hurt another part of the business. This is something that you have to think through carefully. You're not trying to just say, "*Oh, it's the training accuracy on the model.*"

The main thing you have to do is get the thing into production, and this is something that customers really misunderstand. It's hard to absorb, but attribution to AI models is very hard. When we worked at BlackRock – on Wall Street – we had 250 models running in production. Do you think we could attribute model performance to the trades, especially when there were humans intervening? Attribution is very hard, and that's a scientific problem in itself. You have entire teams who just do attribution, and it really depends on how relevant it is to what you're trying to do.

I've seen cases here in LA at private equity funds where they do this kind of attribution because they're trying to value the company for a purchase. When you start getting to that level, it's a real thing, and if you don't have the data and you can't measure it, you're in trouble. I've been in cases where they've put a model into production and they can't even measure what it's doing. It's just there.

We were going to do a project with a religious organization. Did you know the Catholic Church has a venture capital arm? It turns out that they have all this property. They don't know what they own because all their property holdings are in Excel spreadsheets, spread out all over the world with a load of property managers. There's no central repository. They have no idea what they own. They don't own 10 percent of the world's property, or all of South America, but otherwise they have no idea. They don't actually know at a global level.

Attribution is very hard. You have to take it seriously. There's no one answer. You have to decide whether you can actually do those measurements. In the end, what are you trying to do? Are you trying to reduce risk? Are you trying to increase revenue? Are you trying to increase your margins? Maybe you're trying to decrease your margins. Maybe you have a competitor coming in and you need to get your margins down. Do you know what your customer's lifetime value is? If you put a model in production today and it's performing, do you know what that model does 3 months from now? Especially if you have five other things going into production, it's very hard.

We've talked a lot about internet companies. When you're with them and you do an A/B test, usually, you're trying to make sure you don't break anything. Now, the problem, of course, is if you start deploying lots and lots of stuff, your margins go up because you have to maintain things, but your first step is to ask, "*Did you break anything when you deployed it?*"

We had a case where I was working on an e-commerce system and someone made a change to one of their products internally. They changed the ID and the purchase so that the products that were being carted had a different ID from the products that were being purchased. They rolled this into production, and nobody knew it had happened. They rolled it into production slowly. The monitoring system thought, "*Oh, we're just getting fewer and fewer purchases.*" I'd point out, "*Hey, something's wrong!*" Finally, the whole system started to fail. They rolled it out slowly because they didn't want to cause any damage to the system, but it turned out that rolling out slowly was what prevented them from seeing the bug. The monitoring system itself failed because the monitoring system required doing the join on the carts with the purchases.

So, you can see that this is not a small thing. It's a difficult problem, and you have to take it seriously.

Integrating data

AA: Data integration can often be a big challenge for organizations. What are your ideas around the infrastructure processes that are needed to support this?

CM: When you're going to integrate data from different sources, you have to make sure that that data can be unified. A good example is in the medical industry, where people try to integrate EHR data from different hospitals.

There's been some research that shows that if you have 40 different hospitals, everyone codes things differently. You can't just look at the data and say, "*Well, here are the labels for the data. This is what it is.*" You have to go into the actual details and understand, "*Is this the same data?*" If you're just looking at tabular data such as customer profiles, that's fairly straightforward (but even that has real problems!). But when you start getting into things such as complex legal documents, insurance claims, or medical data – something where there's real complexity as to how the data's represented – you're going to have to make sure that you can map one dataset to another.

One of the problems I see is something I call data-quality mismatch. People try to build data products using data from some other product, but the quality of that data may be very different from what it has to be. You can't just naively say, "*Hey, it's business-school class. We're going to take 10 hospitals, take all their data, and merge it.*" You can't merge it – you just get nonsense. It's no different from saying, "*I'm going to take 10 different soup recipes from 10 different restaurants. I'm going to merge all the soup recipes together to make a new soup.*" Everybody's soup recipe is different, and if you try to merge all the recipes and take an average of the recipes, you end up with something that is not edible.

You have to put that into an analogy that people can resonate with so that they're not stuck in a pseudo-abstraction hierarchy that doesn't make any sense.

What's puzzling is that can you talk to someone who knows what's going on and find that what they're doing is just gibberish. It doesn't make any sense because they haven't thought about the details. The big problem in business is that a lot of business people are trained *not* to think about the details. You have to put something in there to make them understand that "data" doesn't mean the same thing to everybody. If you think to yourself, *"I'm going to make a business that's going to collect all the world's insurance claims and merge them,"* what's the cost of doing that?

We had a client that said, *"We want to go out and crawl all these different websites to extract data from some document that a lawyer produced."* Really? Are you going to hire someone to read every document, or can the AI just extract it? Isn't that what Google does? Google has 10,000 employees. They spend all day dealing with this kind of stuff. There's a whole team that deals with legal documents, a whole team that deals with recipes, and so on.

If you're going to integrate data from different sources, you have to map all that data to some standardized template of information, and that might be an enormous cost for you to do. You may have to hire experts to do it. You have to think carefully in your data ingestion process and ask, *"Does this make sense? Is this cost-effective? What is the margin for doing this?"* You just can't randomly put stuff in. We see a lot of that naivety.

AA: I agree. I've seen similar cases when it comes to entity resolution, where they don't have unique identifiers for people across different datasets or organizations. They say, *"Can't we just bring them together?"* **Well, how do you know Charles M in one bank is the same as the Charles H Martin from another bank?**

CM: Plus, there's entity merging. How do you distinguish Charles M1, Charles Martin, and Charles H Martin in different banks? That's a very common problem. You can't even get your customer database right because you don't know which customer's which.

We had a project with an insurance company where they had a huge problem because if you deny someone insurance or credit because you misidentified them, you can be sued for tens of millions of dollars. You have to bleed tens of millions of dollars every year just to make sure you identify a person correctly. That's a huge challenge. It's not a case of just merging the database tables.

Here's one that comes up all the time. Say you have a website and a tracker; you're using something such as Google Analytics. Google Analytics assigns a unique ID to every user when they log in and it tracks them through a cookie. That's your frontend. Then, you have a backend site. The backend site has user metadata, which is where people log in to your site and you track them. The tables in Google Analytics have different IDs from the table in your backend system. Can you join them? How do we join the data? Did you check to make sure that your data tables can be aligned? That's incredibly common, and even in big sophisticated companies, people make these mistakes because they don't think carefully about what their data actually means.

The problem is, this is very resource-intensive, and you have to have people who can lead the projects and engineers who know enough to do it.

It's particularly hard if you're working in a specific domain – say, accounting or fraud – where you have to find someone who understands data and AI and machine learning but also understands accounting. There are maybe a handful of those guys in the world.

This is the problem, again, with data science being domain-specific – you have to have domain-specific knowledge as well. The average data engineer is not going to understand the nuances of an accounting system when you're trying to do something such as revenue recovery. That's a real challenge, especially in data ingestion. The data ingestion problems that I described earlier, where you can't join the tables and the person – that's naivety. When you start getting into more complex problems – insurance, medical, accounting, and fraud – you need experts who really understand it, and they have to be able to work with data engineers, machine learning people, and software engineers.

AA: I agree, but with some organizations, you'd think they'd bring in a generic data scientist and partner them with a business subject-matter expert, and that would be a solution.

CM: The generic data scientist needs to have domain knowledge. It's like saying, *"I'm going to bring in a generic IT person to build a database system, a website, and a database."* That's fine – that's IT. But this is science. Science, by its very nature, has thousands of specialties. I'm not a specialist in the microbe. There are thousands of different kinds of PhDs. You have to bring people in who have some experience or who can learn on their own quickly. That's an important skill because chances are that what you did on one project is not going to transfer exactly to another. You've got to be able to learn new things, and you've got to give people the time to learn. Also, the other person has to be willing to work with them and say, *"Look. I understand you don't know this."* That's a problem as well.

I'll give you an example. When I worked at BlackRock, I worked with Ryan LaFond, who's one of the managing directors. Before I started working with him, I read his thesis. He has a PhD in accounting. He was an accounting/finance guy. I read his papers. I took the time to read and understand his point of view and try to learn what he knew before I bugged him. You have to do that. That's how you work with anybody. If you're a theoretical physicist and you want to work with a biologist, you've got to take the time to learn some biology; otherwise, they don't want to waste their time with you.

If you're a real scientist and you've worked in an interdisciplinary-type field or you've done interdisciplinary work, you understand this. If you bring someone fresh out of school who's done one simple PhD, where their advisor told them what to do and they published it, they're surely very technical and very good – but they don't have the experience of doing this kind of multidisciplinary scientific work.

A lot of companies have this problem. They say, *"We want to get into this. Let's just do the cheapest thing possible."* Maybe 10 years ago, you could be a general-purpose data scientist. Now, it's starting to fragment and specialize.

AA: I see a lot of organizations think that just because they've hired a data scientist, so they can suddenly have them specialize in NLP, graph analytics, LSTM, recurrent neural networks, or some variation of deep learning. They seem to think that these new people with limited experience, fresh out of university, can do it all.

CM: They want to believe they can just hire the cheapest person possible. An organization like that is not a technical organization. That is an organization whose primary competitiveness is in some other area. Walmart is not a hyper-technical organization. Walmart's competitive because they control the supply chain. Other parts won't be so great, such as their search engine – it's not Google or eBay. eBay has a great search engine. Why? They have no supply chain. Different organizations have different strengths.

They're not going to bring in those kinds of people. That is usually done through an acquisition. If you really want to expand that area, you acquire the company doing it. Maybe they're doing it because their shareholders are pushing them, and they need to somehow show the shareholder or the board that they're doing some AI because they're concerned that Amazon's going to come in and put them out of business.

> Not every project is designed to be successful. Some of them, as we call them in the Valley, are sunshine projects.

They're there to keep the board happy so that they don't come in, fire you, and replace you. A lot of these guys are just trying to hold on long enough that they get their earnout so they can retire.

Finding the limits of NLP

AA: The next question is related to NLP, something that you're quite familiar with. There have been amazing advances. However, one thing I want to ask you is this: what are the real challenges and limits, and will machines ever be able to understand language?

CM: With the advent of NLP, people began to understand how to model sequences and structures. I'm not sure I would call that understanding. On the other hand, does the average undergraduate who takes a psychology class actually understand it?

With what NLP has become, we can do a lot more things product-wise, because there are a lot of interesting things you can do with a product. When you read your Gmail, NLP is quite good at being able to figure out the next sequence of words you want to say. Is that understanding, or is it just surface-level pattern-matching?

What's happened – and this is where a lot of my research with WeightWatcher comes into play – is that we understand much better the fractal nature of the correlations that exist in natural systems, such as images, text, and speech. We understand how to model them because they have a certain type of fractal. We've known they have a fractal nature since Mandelbrot, but we haven't done this for a very long time. What's happening is that these AI systems are able to capture correlations very effectively in the way that these fractal patterns are repeated.

You speak a language, and I speak a language. In some sense, we use the same training data to generate language, because we have to understand each other. When you look at an image, other people have to know what it is. There have to be some common patterns that the brain can pick up, and that's what I think we've figured out how to do.

Now, is that understanding? I don't know. It's certainly not philosophy or religion.

AA: Where do you think the field of NLP is going to go next?

CM: A lot of the work that's being funded is trying to make systems more engineering-friendly. We're seeing larger and larger models. People realize if you just put every single word and every single sentence that's ever been uttered into a model, it can memorize them. It can know every single thing that's ever been said or every single thing that's ever been read. That's where it's going.

Engineering-wise, can you use that in a production environment? That would involve a different kind of thing. It would have to be really fast to be able to work. So, there's a lot of work going on to make these systems accessible to people so that small companies can actually deploy them. There are things such as grammar correctors or things that help you generate fake text. I worked on a product that generates fake text, years before GPT. Or, you might want sentiment analysis. When you read somebody's tweets, you might want to detect whether they're depressed or not. Can you detect when people are becoming depressed? We're going to see more and more product development as these models become more widely available.

Is there going to be any deep level of understanding, though? If you look at how airplanes were developed, we started by first trying to mimic how birds fly. Then, we discovered that that was the wrong thing to do. The physics came in, and we realized that there was lift and drag, and now we have airplanes. We've built machines that fly unlike anything in the natural world. With NLP, it'll be the same thing. You'll have systems that know every single thing that's ever been written, so you're going to have things such as automated translation systems. You and I could speak a different language, and it would just automatically translate for us, just like in *Star Trek* – the universal translator. That's purely statistical – there's no grammar in any of this. The grammar is learned implicitly by the models. That doesn't mean it's not there, but it means you don't have to put it in explicitly as code; the model learns it. That inherent fractal nature of the grammar is there. I think those kinds of products are going to come out, and as consultants, we can help build them.

Companies are trying to make more and more of these kinds of products, and they're trying better search engines. If you want to be able to read every single scientific paper that's ever been written and then ask questions, that's very hard.

A big problem is that when you look at systems such as Google or Apple, people want those systems for other things. I want to have an Apple for medicine where I can just speak to something. I had a conversation with some very senior people in government. They said, "*Why can't the EHR systems in a hospital work the way Apple does, where you speak to it and it translates?*" I said, "*You understand that Apple has a 1 trillion-dollar market cap? That's not a small thing to build.*" They might say, "*I want to have a search engine that lets me search and ask questions the way Google works.*" Well, first of all, Google has thousands of employees doing that, and as well as that, when you list information on Google, you have to make it readable. You have to apply SEO to it and make it readable; otherwise, Google won't process it for you.

There's another aspect that people forget – when you interact with systems, you have to actually *interact* with the system. Google solves the data ingestion problem by making you make your data digestible. But people just want to build things: "*I want to have something that can crawl every PDF document and extract information from it in some way.*" Well, hire one person to read every document, because the NLP systems are not going to do that. You're not just going to magically crawl a website, pull a PDF out, and understand it. People underestimate how many employees Google and Amazon have working on this. Amazon has about 10,000 people working on AI. That's not the same as people at a company doing it with a bunch of junior data scientists.

We'll begin to see more NLP that's product-driven – products that assist people in ways, and NLP assistants. It's the same with self-driving cars. We're not necessarily going to have completely self-driving cars. Although we do have them in a way, they're still a little scary. But, assistance – something that will *assist* you to drive – is probably very reasonable. Consider things that *assist* you to write. I don't know if we're going to get to the stage of *Brave New World*, where they had machines that write books, but we're getting close.

Explainable AI and ethics

AA: How important is explainability in the field of AI, and how do we get there?

CM: It depends on the product. If you work at Wells Fargo and you need to deny someone credit, you'd better be able to explain why you denied them credit. Plus, if you deny them credit today, you shouldn't somehow offer it to them 3 weeks from now or vice versa. There are compliance requirements around what you do with those models, and I think explainability is a larger part of compliance requirements.

When I work with clients, I give them simple models because I know they need to be explainable, as you need to debug them. You can't debug a model if you can't explain what it's doing, and believe me, when you put something into production – such as a search engine – you'd better be able to explain to people.

> The users have to understand why the model did what it did for them. People want to have explainability because they want to accept solutions. If you give someone a suboptimal solution and it seems crazy and psychotic, it will upset them terribly.

You can't just give spam to people. If you start giving people spam, they become very upset: "*Why are you giving this to me? Why are you calling me? Why are you doing this? Why did you contact me? Why did you call me? I don't want to talk to you.*" People hate spam, but they're willing to accept it if they understand why they get it.

Internally, when you build models as products for companies, they're going to break. You have to understand why they work so that you can fix them. That's why we're doing WeightWatcher. If you don't know why something broke, you can't fix it.

I think explainability relies upon the questions, "*Will the customer accept it? Can you fix it? Are there compliance requirements?*" Those are the things that really affect it, and they are critical issues.

AA: That touches on my last question, which is about ethics and the increasing need for ethical, responsible AI. How do the organizations that you consult try and move down that path?

CM: If you're going to deny someone credit because they're a minority, that's a compliance issue. That's not an ethical issue. Medical systems are misaligned because we haven't got enough minorities in the medical databases, and so treatments may end up different based on phenotypes and genotypes. There aren't as many minorities in the population, hence the term "minority." You're not going to be well represented in a dataset because there is a practical issue in collecting data. Moreover, are the phenotypes even distributed by genetics or race, or are they related to something else?

There are a lot of people complaining that datasets are imbalanced. Yes, the datasets are imbalanced. You forget that what you're doing is statistical inference. You're building a dataset, and you're trying to do inference on the data. The amount of data in the classes and what they are is what you're modeling. I don't think people understand from a medical perspective what the differences are. If you're doing medical ethics, are there differences between populations of ethnicities? Are there differences in medical response between men and women or are there not, and does it matter? Those are deep scientific questions, and to say, "*Well, that's just unethical because the data's wrong*" is not right.

It's very challenging, and you have to be very careful that you're designing an optimization algorithm. You can bias it any way you want. If you want to bias it toward certain predictions, add a biased term, and it will predict those terms. If you don't want to add the biased term, don't. It's very problem-specific. From a start-up perspective, you can look at how products are built – people are just trying to get their companies off the ground. They're trying to make some money. They're trying to get off the ground and prove an idea works, and these ethical questions very rarely come up because you're just trying to survive.

People are trying to address some of these biases in things such as credit lending and housing, where it's not really a case of things being unethical but illegal.

AA: Sometimes, it's clearly a legal problem.

CM: If you're working at a bank and you're offering credit, you have to address it. I can tell you this for a fact because I've talked to the guys who do this – for every three modelers you have working on credit models, there's one modeler who's making sure that the models are fair, explainable, and unbiased. That's a compliance issue. If you have 100 researchers working on building the models, you have 30 to 35 researchers working on making sure that they're fair and compliant.

So, companies that have compliance issues are in fact addressing it, but you're not going to have companies that address it blindly without it being a compliance issue. They're not going to do that unless their customers are screaming at them, and usually, it's something else, such as someone trying to extort money out of you.

Summary

I thoroughly enjoyed my discussion with Charles and, in particular, delving into some of the technical nuances of the research and development he is doing on new tools such as WeightWatcher, which help us better understand AI models.

Charles is a straight shooter. He makes the point that it's important to ask yourself, if you're looking at working with or going into a company, whether the leaders are open to change or whether are they just going to keep doing what they've always done. Another key issue that he raises, which I come across often, is a lack of understanding among senior executives and leaders about how data science works. For one, you can't simply apply software engineering practices and management paradigms to data science. They tend to be rigid processes that are not directly "science" projects, and they can hinder the development and productionization of data science models. Data science and AI are not commoditized like IT, which means they can't be managed in the same way.

In addition, the interaction between data science teams and IT teams can be a source of significant friction. Probably slightly controversially, Charles recommends carving out ML/AI capabilities and having as little interaction as possible with IT services.

Charles offers some important advice on data integration and data-quality mismatch. Like Althea Davis, who we'll speak with in *Chapter 12*, he raises the importance of having effective data change management processes.

He also offers some pertinent advice to practitioners and their leaders – senior executives often assume that details don't matter, but in reality, the details can make or break a project. He also suggests focusing on the simplest model possible that will get your customer a win, and I couldn't agree more. His most valuable advice, though, is to understand that measuring model accuracy is really about measuring business value, and how to do that needs to be thought through carefully, as it varies from one business customer to another.

For me, one of the most important points raised by Charles is that organizations need to be committed to investing in data science and AI. As he says, "*You have to have money, for one thing. Science is expensive and slow.*" I also agree with him that not all data scientists are equal – for certain roles, you need specialist skills and domain knowledge – which reinforces the importance of having people with sufficient technical knowledge in leadership roles.

6

Petar Veličković and His Deep Network

I was excited to speak with **Petar Veličković**, one of the leading research scientists at DeepMind – especially given my interest in **graph neural networks**, which is one of Petar's key research areas.

I was eager to learn more about the break-through work that he and his colleagues are doing in using graph neural networks to help solve complex mathematical and scientific problems, and why this approach is so promising for certain problems. I also wanted to learn about the technology's applications in tools we use in our daily lives, such as Google Maps.

Entering the world of AI research

Alex Antic: How did you go from graduating with a PhD in machine learning and bioinformatics at the University of Cambridge to becoming a leading researcher at DeepMind – one of the leading AI research laboratories in the world?

Petar Veličković: I was born and raised in Belgrade in Serbia, where I completed my primary and secondary education. It was a given at the time for Serbian students who were really interested in natural sciences to attend the Mathematical Grammar School in Belgrade. Even if you were more adept at physics or computer science, like myself, it was still kind of a given that you go through the full curriculum of mathematics that's offered there. It offers you the same level of education you would receive in some parts of a first or second year of an undergraduate degree.

It prepared me really well for everything that came after that. Using that knowledge, I was able to attend Trinity College in Cambridge with a full scholarship to study computer science. At the time, I had a very different idea of what I found interesting and what was available in computer science.

I was perhaps partly biased by the culture in the grammar school – which focussed mainly on participating in Olympiads in STEM subjects – but I knew about computer science primarily through classical data structures and algorithms, which is how many people in my generation were first introduced to these concepts.

When I started my degree, I honestly thought that I would probably finish my bachelor's degree, maybe take an extra year to do a master's, and then get a job as a software engineer at a tech company. That's what made the most sense to someone with my understanding of the field at the time. It took a long time, until midway through my third and final undergraduate year, for me to truly discover machine learning and – since my thesis advisor was an expert in computational biology – the transformative power that it can have on bioinformatics. But even then, it was mostly like I was playing with things by myself without realizing how far the picture went and how many amazing things people were already doing with AI.

This was the year 2015, by the way, when the deep learning revolution was already well underway. I still remained completely oblivious to that, but I was so intrigued by the idea of performing something research-like in this area that I was eventually convinced by my advisor to just go straight from finishing my bachelor's to enrolling in a PhD at Cambridge. I started my PhD studies in 2016, which commenced my machine learning journey for real.

I must say, when I started my machine learning journey, I still had no idea about how modern deep learning was properly conducted. I still believed people wrote the backpropagation algorithm from scratch in C++, and it was only many months into my PhD that I learned about the existence of deep learning frameworks, and perhaps most importantly, the existence of online courses where one could learn how to use those frameworks.

I actually learned how to use my first deep learning framework by following an online course on Udacity, six months into my PhD, by which point I would have been expected to already know all these things quite well. What I really found fascinating about deep learning was that in principle, for someone who has the right mathematical foundations, the barrier of entry is remarkably low. I went from completing a beginner's course in deep learning and implementing a few starting-up examples in TensorFlow to, later that same year, being able to write research papers and publish them. It was a pretty hectic ride, and it took me a while to figure out exactly where my interests lay (we will talk later in this conversation about my interest in graph representation learning, where I eventually settled).

That was the seed that started it all. I gradually learned that I'm actually far more interested in research than software engineering, which was my initial aim in life, so to speak.

You asked how I have managed to make my way to a company like DeepMind starting from these foundations. Perhaps the most important and transformative moment in my career, which allowed me to slingshot myself toward a company like this, was my two internships at the Montreal Institute for Learning Algorithms, or, as it's now known, Mila.

I had a chance to do two internships in Yoshua Bengio's group at Mila – Yoshua Bengio is one of the pioneers and "godfathers" of deep learning – and it was there that I had a change of perspective. As for my research group at Cambridge, it was a fantastic group at the time, and it had a lot of people with very different interests that were more or less aligned with computational biology. But, there weren't really deep learning experts in that group. Whenever I had a problem, I did not have people to have deep discussions with and truly understand how research is conducted "in the real world."

When I went to Montreal, it was a proper deep learning lab: big offices packed with researchers focusing on deep learning research. Just being in such an atmosphere led to me absorbing a lot of important knowledge. I actually published some of my top-cited work while I was there.

One last point I would make, which was perhaps very important, is that during my two stints in Montreal, I did not actually learn a lot of new stuff theory-wise. Combining all of my undergraduate foundations with these online courses was, in principle, enough to have the same grounding as maybe even the top experts in the area. What I actually learned at Mila were all of the fancy tricks and interesting ways to engineer machine learning systems that nobody includes in research papers because either they don't have enough space/don't want to talk about it, or their reviewers wanted them to focus on something else.

Ultimately, research papers are very limited, carefully crafted, eight-page versions of research results. They don't tell the story of anything that happened in between an idea and a result. Going to a big deep learning lab such as Mila allowed me to really learn about what's happening in between, and I think that was the spark I needed to supercharge my career onward after that.

Discussing machine learning using graph networks

AA: You touched on one of your areas of research interest being around graph attention and graph representation learning. Can you please explain it in your own words? What does it mean in practice? What are some of the applications and limitations, and where's the research heading?

PV: As I mentioned briefly, graph representation learning is probably my biggest passion in machine learning right now, and I think it's a very important emerging area. Regardless of what area of computer science you choose to specialize in, we're now at a point where you'll probably come into contact with graph representation learning in one way or another.

Essentially, it deals with processing data that lives on graphs. Graphs are interconnected structures of nodes and edges. They're a way to naturally represent networked data.

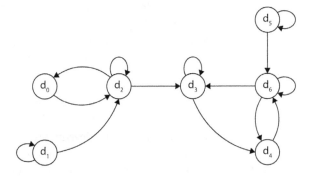

Figure 6.1 – A graph structure with nodes and edges

Now, why is this important? If you follow any of the big-hit news stories about deep learning, you will typically find that most of the successes tend to happen in the domain of images or the domain of text. These kinds of nicely structured data are so easy to attack with deep learning because they're so regularly structured. An image is a two-dimensional grid of pixels, and structurally, every single pixel looks exactly the same, in the sense that every single one of them has four neighbors: up, down, left, right, and that's it. This means we can deploy some very nice, specialized models – such as convolutional neural networks – to process this data in a very idiomatic way.

However, things get a lot more interesting when you want to talk about learning on data which is more irregularly structured. The thing is, most – if not all – of the data that we receive from nature or that we create as a social construct is most naturally represented using graphs.

Think about all the levels of organization of life, starting from molecules, which are interconnected networks of atoms connected together with chemical bonds, all the way to the connectome in the human brain where different neurons are connected to each other with dendritic connections. At all levels of organization of life, from the basic building blocks all the way to the organs that underlie human cognition, everything is best reasoned about as an interconnected network within an irregular structure.

If you look at human constructs, irregular graph structures are also very pervasive there. For example, transportation networks are most naturally reasoned about as graphs of various intersections connected together by routes or roads. Also, the social network, which is one of humanity's most interesting inventions in the past few decades, is best reasoned about as an interconnected network of people connected by friendship links.

Really, when you think about it, even when you have some images of an object, usually those images are not "just" flat, two-dimensional things. They're our projections of something much richer – something, for example, that is three-dimensional with a very interesting, irregular structure. Images are basically our convenient way of flattening things onto a regular grid so that we can process them more easily, but it potentially discards a true wealth of information about the spatial configuration of those objects in three-dimensional space, which you throw away by projecting onto a 2D plane.

Observed from this perspective, what all of us are really doing is graph representation learning – we might be simplifying our inputs somewhat to make the graphs easier to work with. But even then, images are themselves graphs, where pixels have a very simple, rigid, connectivity structure. Maybe we shouldn't relabel all existing research as graph representation learning, but we should all be aware that we might be always doing graph representation learning in disguise.

Just to hammer down the point of how much machine learning on graphs already impacts our daily lives: I mentioned some of these modalities where graphs come about naturally. Molecules, for example, are very standard data modality that every graph representation learning researcher will come into contact with at some point in their career. You can do very simple interventions, such as training a machine learning model to look at the structure of a molecule to predict whether that molecule will be a potent drug or not. This was recently done by a team of scientists from MIT, where they pre-trained a **graph neural network (GNN)** (the standard model in graph representation learning) to take small molecules and try to predict from a given training set whether or not they would be useful antibiotics.

When a model like that is trained, you can apply it to any molecule, including molecules not in your training set. They applied this model to a bunch of drug-candidate molecules that are not necessarily considered antibiotics but treatments for some other diseases. Sometimes, the model can give a really high confidence that a molecule is a possibly potent antibiotic, even if none of the human domain experts ever considered it a potent antibiotic.

They took the molecules with the highest confidence out of the model and sent those molecules to chemists to be investigated more carefully. It actually happened that within that selection, a completely new potent antibiotic was discovered called Halicin. At the time, nobody thought of it as an antibiotic because it has a very unusual structure that nobody would expect for an antibiotic, rather it was considered a potential treatment for diabetes.

In fact, exactly because of that weird structure, Halicin has a very different mechanism of action compared to the antibiotics we might normally synthesize. As a result, it is potent even against the bacteria that have evolved over the years to be resistant to all these antibiotics we keep putting in our bodies. That is, in my opinion, still the most exciting application of graph representation learning to this date.

If you want something more practical, big tech companies often deal with huge quantities of data that is most naturally expressible as graphs, and they realized the utility of graph representation learning as soon as it became more broadly available. The first famous example is Pinterest, which currently uses a graph neural network in production to help decide which pieces of content you should be looking at next, as a recommendation system. Amazon recently reported using GNNs to help you decide which products you should buy next based on the purchases you've made before. Uber Eats has reported using graph neural networks to help you decide which food you should order next.

One application of GNNs that I am personally most attached to, because I worked on it myself in part, is Google Maps. Google Maps is currently using a graph neural network-powered system to tell you how long it will take to travel from point A to point B in the context of the expected time of arrival.

AA: While it's in many ways natural to think about data in a graph representation manner, what are some of the limitations of graph representation learning, relative to other methodologies?

PV: Of course, the field is not without its limitations. You get all that freedom of modeling when you're dealing with irregular structures, but it obviously comes at a cost. If there is some interesting regularity in your data, you now have to work twice as hard to account for it in your model, because it is not guaranteed by the implementation. Recently, there's been a lot of work on topologically inspired graph neural networks that try to include some information about the regularities of the input data that a basic graph neural network might discard. It's a bit of a fun mathematical journey to try to make GNNs more specialized. In their basic form, they're very general. If you were to apply them directly and naïvely to image data, they would probably underperform compared to convolutional neural networks, as convolutional networks are highly specialized for the structure of images. That is maybe one of the main points.

Then, there are other points; namely, that there are problems that can arise when propagating information over a generic graph structure. Specifically, I'll just list a few potentially surprising things that graph neural networks cannot do.

There is a very famous problem known as the **oversmoothing** problem. If you're not careful with how you propagate information across the graph – especially if you make your networks too deep – you might end up with all the features that you compute for the nodes of your graph converging to the same vector. This is useless if you want to classify those nodes in any particular way. It is one of the reasons why, especially in the early days, we didn't really have deep graph neural networks.

In reality, very often, the depth of a typical graph neural network applied to a social network wouldn't be more than two to four layers, whereas most state-of-the-art deep learning systems nowadays have hundreds or thousands of layers.

The second limitation, which is also quite interesting, is called the **oversquashing** phenomenon. Imagine how a node in a graph neural network receives all information from its immediate neighbors, who then receive information from their immediate neighbors, and so on and so forth. If the input graph contains a node that is a very important bridge between two parts of the graph, that node is now put under a lot of pressure, because it has to integrate information from an exponentially growing receptive field of neighbors (that is, the neighbors' neighbors' neighbors' neighbors). Every time you add a new layer of neighbors, you have a multiplicative factor in how many additional nodes you have to work with. This effect is especially pronounced when the input graph is a tree.

It has been shown that if you're not careful to relieve this pressure – maybe by adding more connections so that the node doesn't have to do so much work – you end up in a situation where your model cannot even fit the training data. I'm not talking about generalization – you cannot even fit the training data. So, it's limited in that case.

There's a third limitation, which I think is very important for people trying to run graph neural networks on a new problem. Oftentimes, the interesting labels on a graph will depend on some interesting substructures in that graph. For example, in computational chemistry, a lot of chemical function is driven by aromatic rings. These are a very standard building block of more complex organic chemicals.

However, we now know for a theoretical fact that basic graph neural networks cannot tell apart even the simplest kinds of substructures. To give you an example of how aggressive this is, a graph neural network, when applied naïvely, cannot tell apart a 6-cycle – six nodes connected in a circle – from two triangles – two collections of three nodes that are each connected in a triangle. These two graphs are very different to the naked eye, but if you look at them with a graph neural network, structurally every node looks exactly the same. Every node has two neighbors.

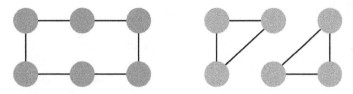

Figure 6.2 – Nodes with neighbors

This is a very important issue, and a good takeaway from this is that if you are an industrial practitioner and you want to apply a graph neural network to a problem where you expect certain structures – say, cliques – you should precompute those structures and add them as extra features.

> I must stress that it's a super fast-evolving field. Every day there are 10 to 15 newly proposed graph neural network architectures that might try to fight these problems in one way or another.

One thing that's maybe important to note in this book is that if we were to have this conversation even one year in the future, let alone five years in the future, my answers on what the main limitations are might be very different. We're always chasing interesting targets and discovering new ways in which these models can subtly fail, so that's very important.

I'll mention just one more limitation. Readers who are well versed in AI literature might be familiar with examples of possible adversarial attacks, which are a known pain point for the deep learning systems of today. For those who are unaware, it means I can make very subtle, imperceptible changes to my input that are sufficient to completely fool a machine learning classifier into being super-confident in an incorrect output. There are famous examples of pictures of pandas where you subtly modify a few pixels in the picture, and suddenly, the classifier thinks that it's a gibbon with 100% confidence.

This is a very important problem to be mindful of, especially if you design something like a self-driving car system. If someone can subtly put a sticker on a stop sign to make the neural network interpret it as a yield sign, that can be a big problem if it's going to be used to power a car in the real world. Graph neural networks, I would say, are – at least in their current form – doubly vulnerable to these kinds of attacks. That's because now, you have two things you can play with. There's the data that lives in the nodes and edges, and there are the nodes and edges themselves: the graph structure. It turns out that in some cases, you can very carefully hack the graph by adding one edge or removing one edge and completely fool some kinds of graph neural network models.

Depending on which model you're using, you may be more or less vulnerable to this, but it's very important. It means that attackers of the future who might want to attack these systems might just carefully create some new accounts and create some new friendship links, which might completely throw the network off. It's something that, once again, people deploying these systems in the real world must be very mindful of and have the right defenses in place for.

AA: That's a great example. People definitely need to be mindful of adversarial attacks on all sorts of architectures.

Applying graph neural networks

AA: One thing we touched on is the growth of industries' interest in graph architectures and the use of graph neural networks, for example, in areas such as law enforcement and national security. You can imagine the applications there in detecting criminal networks. What advice can you give organizations, large or small, looking at leveraging graph analytics for the first time, in terms of the data that they're collecting and storing, the tools they're using, and the problems they are best suited to solve?

PV: One thing that is potentially a bit annoying when applying a graph representation architecture compared to, say, a convolutional neural network for images is that because the problem is not so rigid, we haven't reached the point where we can just give you one architecture and say, "*This is the gold standard for graph-structured data: you should always use this as the first approach.*" Unfortunately, it's a lot less clear-cut than that.

Because of the fact that the data is so irregular and you're passing data over the edges of the particular graph structure you have, the final results you're going to get will depend a lot not only on architecture choice but also on the properties of the data that you have. I'll just mention one example of this to show how rich a problem it is, and then maybe there'll be some hints for interested practitioners to take on.

One of the most popular models in graph representation learning is the graph convolutional network or GCN. This model, in a nutshell, computes a very specific average over all the neighbors at every point. You don't get to control how much you value every single neighbor – it's predetermined by the structure of the graph that you compute some kind of symmetric normalized average over them.

Now, why would this be a good idea? Well, imagine you're in a social network and you want to classify whether or not I'm going to perform a particular action, such as like something or retweet something, or something like this. Usually, it's to be expected that I have very similar views to my friends on the social network, and as a result, if you know to what extent my friends have done something – the feature vectors of my friends – you can make a reasonably robust prediction of what I'm going to do.

Taking an average of all of your neighbors in a social network graph based on friendships is a pretty good way to get a more robust representation of what you are going to do. We call such graphs **homophilous** because nodes that are neighbors tend to share the same labels, and in these kinds of settings, averaging is exactly the right thing to do.

It's also the easiest thing to scale because when you just do simple averaging over different edges of the graph, it amounts to a simple sparse matrix multiplication of your data with the adjacency structure of the graph. This is something that industrially, you can scale up quite well. Many of the impactful, strong applications of graph representation learning in industry tend to be scaled-up versions of these models. This is a fair assumption to make in many cases, simply because if the graph is homophilous, averaging is more or less all you need.

In fact, in some cases, it has been shown that you don't even really need deep learning in such cases. The power of these examples comes not from the machine learning part but from the aggregation: the repeated aggregation of neighbors to get stronger features.

There was a trend of throwing away all the deep learning, all the non-linearities and weight matrices, and so on, and just repeatedly averaging, followed by a simple logistic regression on top of that. This was already performing close to the state of the art on many datasets that we had at the time. Many graphs you will find in the real world are highly homophilous, so edges tell you a lot about which nodes should act in the same way.

However, this is not always the right assumption to make. For example, let's go back to the example of retweeting. Now, say you connect me to someone because I retweeted them. I'm very active on Twitter myself, and I might often retweet people whose work I really like, but I might also retweet someone if I strongly disagree with them. I might enter into very long rants about why I think that some things should have been said or done differently in their work, and so on. In those cases, you should absolutely not give me the same label as the person I've retweeted because I actually have the diametrically opposite opinion to them.

In this setting, your graph is now known as **heterophilous**. No longer does a neighbor have to strongly share the same label as you. In these cases, simple averaging will not work at all. To surpass this limitation, you have to do something more sophisticated, such as use a machine learning model to tell how much you value each one of your neighbors for making your predictions. Those who have worked in natural language processing will probably know that when I have a mechanism that looks at the features of a node and its neighbor and computes such an interaction coefficient, I've just described an **attention mechanism**, which is the cornerstone of Transformers. That's why we sometimes call these models **graph attentional models**. One of my first excursions into the world of graph representation learning was proposing one of these kinds of models, the **graph attention network (GAT)**.

Maybe the main takeaway here is that how powerful a model you need might depend quite a bit on the properties of the data that you have. If your data is homophilous, you can – and should – get away with using a very simple model. However, the moment you need to pass messages between nodes in a more complicated way than averaging all the neighbors in some particular weighted combination (as is common in physics simulations, relational reasoning, and computational chemistry), you might need a more powerful class of model.

There are obviously a lot of trade-offs involved in terms of how scalable the model is.

Graphics Processing Units (GPUs) and **Tensor Processing Units (TPUs)**, which are the workhorses of machine learning nowadays, tend to be great when your data comes in static shapes that are well-defined up front and when you're doing many matrix multiplications. If your graph is small enough to be stored as a dense adjacency matrix, then you can go quite far with GPUs. In fact, Transformers are essentially attentional graph neural networks over a fully connected graph of words – and are commonly used in **Natural Language Processing (NLP)**. You can think of them as a graph neural network that has won the hardware lottery; that is, Transformers are the graph neural networks that are easiest to execute on the hardware of today. But the moment you have to do something more sparse and more carefully pass messages along the edges of the graph, suddenly the operations don't align that well with the GPU and you're not going to be able to scale it up as well on the hardware of today. Of course, once again, in 10 years, this conversation might be very different when new forms of hardware might appear.

Pushing research boundaries with machine learning

AA: You've also played an integral role in helping mathematicians discover new results via ML (by integrating mathematics and AI effectively), and your research has been published in *Nature***, which is exciting. Can you please talk a bit more about this work and the broader implications of using machine learning with other fields to discover new insights?**

PV: It's one of the recent works that I'm really passionate about, and it's part of the reason why, when I told you about my background, I went as far back as my high school. I said that I went into a program that was mainly focused on mathematics. I knew that I wasn't very talented in maths; I was always more of a computer science person. But this inadvertently led to me being surrounded by a lot of very talented mathematicians.

> While I was never that good at doing it myself, I have learned to appreciate the purity of mathematics as an area. Maths is truly, in a sense, the most abstract way to do science. It's a universal language that we all use, no matter what field of science we specialize in, to reason about the interesting phenomena in our fields.

I find our work in this space so exciting because it allowed me to (at least indirectly through my knowledge as a machine learning researcher) contribute to and stimulate the development of mathematics.

I'm also excited about our work because, unlike many commonly advertised applications of machine learning, which might deal with building a machine learning system to potentially automate something that a human annotator would otherwise have to do, the results we present are the mathematical results, which are not derived from the machine learning system. A human mathematician – a professional domain expert – still had to do all the work of formulating a conjecture, explaining the useful structures, and then proving it using the tools of their area. The mathematicians are still in charge of actually executing the creative part of mathematics.

> What does the machine learning system do? In this particular case, the AI system we built is capable of helping the mathematician examine mathematical objects of interest with a sort of "precision microscope," which allows them to zoom in on the parts of the mathematical object that were most important for the particular theory that the mathematician wants to develop.

In some of the areas that we've been able to disrupt, such as representation theory, the objects of interest are super-big before any interesting phenomena even happen. The problem is that even though you're able to generate these big examples with a computer, you won't be able to tell with the naked eye what's going on. There's not a lot of progress you can make on the theory, and indeed, the conjecture we've been trying to settle in our work on representation theory has stood without any attack for 50 years. People were aware of large-scale examples, but they weren't able to make sense of what was going on in them, and that's where machine learning came into the picture.

We trained the machine learning model to take those complicated objects and predict the quantities we were interested in. Sometimes, even though machine learning systems tend to be regarded as black boxes, just the fact that there exists a model with high accuracy on a relevant problem can give a lot of motivation to the mathematician. They might initially only have a faint belief that a relationship exists, but if they're able to train a neural network with high accuracy to predict something they care about, they might already get excited about it and they double down on trying to prove it or make a good conjecture.

But usually, that's not enough. In this particular case, we had a model that was performing well, but without any real explanation of how it did so, or what were the important parts of the structure. That wasn't particularly useful because we already believed the conjecture to be true. Perhaps it gave us more encouragement that it was true, but it didn't really tell us what to do next. This is where saliency methods come into play.

We can generally say that once you train a machine learning model like this, you need to interrogate it somehow and ask it what the important parts of this complicated input were that led to it making a particular prediction.

> **As many of our readers will probably know, explainable AI techniques are still largely in diapers. We don't have sophisticated methods that will reliably tell you why a neural network is doing what it's doing.**

However, in the case of our *Nature* paper, we looked at an example on knot theory and an example on representation theory, and in both of those cases, we were able to successfully use such saliency methods. We were asking what part of the input contributed the most, from a gradient perspective, to the output. It's a very simple intervention you can compute.

In the case of knot theory, it very quickly told us what the most important variables for our problem were, and our collaborators from Oxford – Marc Lackenby and András Juhász – were able to work off that knowledge. They went away and proved an entire theory using those variables. That was quite exciting, and for the first time, it showed a connection between the algebraic and geometric views of knots.

But representation theory needed a different approach. In representation theory, the objects of interest were not only truly huge, they were also graphs. I was actually drawn into this project due to my expertise in graph representation learning.

When the input is a graph, the aforementioned saliency method might tell you about the nodes and edges of the graph that contributed the most to the final prediction. In reality, what we would show to our math collaborator was a subgraph of the original graph. The hope was that even though there was a lot of noise in the subgraph, it was also now substantially smaller, so it should be easier to look at and make sense of.

When we first made these subgraph explanations of our graph neural network, which had reasonably high accuracy, they made very little sense to the DeepMinders on the team. They looked like abstract nonsense. We were working, however, with one of the top experts in representation theory: Geordie Williamson, who is also Australian, like yourself. We sent all those subgraphs to him in the hope that maybe he would be able to see a meaningful pattern. It took a few weeks of very careful staring and analysis, but Geordie is a bit of a magician. He was able to figure out from all those noisy signals that there was a particular diamond-like structure that kept appearing, and that he could relate it to some previous research in that area.

He hypothesized, "*OK. There is a structure. I was able to find it, and maybe it's useful.*" Then, we did the trick that I mentioned earlier, to precompute all the interesting diamond-like structures and add extra features to the graph neural network to tell it, "*Hey, this is where these diamonds are. Can you try to fit the problem again using this information?*"

This is where the most amazing thing happened: empowered with the knowledge of where the diamonds are, the GNN's accuracy shot up from 68% to 99.5%. Now, we had a very strong signal to believe that these structures are going to be very important to the mathematical theory, and after a few more rounds of back and forth and some more discoveries of what the network was seeing as important, Geordie was able to formulate a theorem. He was able to discover a very interesting decomposition of the input problem into two smaller problems, such that when you combine them, you get the correct result. Also, he was able to prove that there exists one way to decompose a problem in this way that will guarantee you the correct answer.

Now, this doesn't completely settle our problem yet because while we've proved there exists a decomposition that works, we don't know how to find the "true" decomposition – there could be many possible ones. We set out to do what a typical machine learning person would do in this case. We build a dataset from every possible decomposition: "*This decomposition works. This decomposition doesn't.*" Just use a binary classifier, right? It sounds like an easy thing.

Well, much to our happiness and surprise, we haven't been able to find a single decomposition that did not work. From the point of view of machine learning, we had a very biased classification problem on our hands, but from the point of view of the mathematics, we could immediately jump to making the conjecture that every possible decomposition would give you the correct answer.

Geordie has left some very interesting clues in our mathematical companion paper on how one might potentially attempt to prove this conjecture. Based on all the computational evidence we've seen, we strongly believe it to be true, and if it is true, then it would settle a 50-year-old problem that is one of the grand challenges in representation theory.

Why do I think this is exciting? Well, as I mentioned, the top experts in the area were able to distinguish the signal from the noise and use it to make very meaningful substantial mathematical results. I am very excited about what's going to happen when we get better explainability methods and model-building capabilities for mathematical objects. At that point, we can truly scale up these kinds of approaches to help virtually any mathematics practitioner, not just the top practitioners who can stare at objects and figure out what's going on, even when there's a lot of noise.

> I'm very excited about how this could transformatively impact all of science, and what makes me very happy about it is that it's fundamentally a model where there's a synergy between the human and the machine.

As Geordie once put it, it's like you went back in time and gave Gauss a pocket calculator to play with. Gauss did a lot of fantastic things, but he did not have access to computation of that kind. You can only imagine what he would have been able to do if he had access to such a powerful tool. You can think of our work – hopefully – as that pocket calculator but for 21st-century mathematicians.

Using graphs for AGI

AA: I'd like to get your thoughts on AGI. Do you think graph representation learning is one of the paths forward? What about algorithmic reasoning, another area that you're working on? How do you see the field of AGI emerging? Do you think it's something that's attainable?

PV: That's a very important question, and one that we must be increasingly mindful of as we become more powerful in building these kinds of systems.

Should we be using graph representation learning as part of AGI? I wholeheartedly think we should be. Think about a truly generally intelligent system: what does such a system (likely) do? It first extracts meaningful concepts out of all of that richness of real-world signals. How does such a model then know how to make conclusions and go forward in their life, so to speak? They do so by taking all the concepts, relating them to past experiences, and using reasoning arguments to derive new pieces of knowledge from the finite means that they have. I believe this is the essence of intelligence, and it's necessary for an AGI because we're not going to be able to expose the AGI to every possible situation it will ever encounter as part of training data. Part of a truly intelligent system is that it's able to adapt to whatever comes in the future, and we don't necessarily know what will come in the future. You need this kind of capability of taking the high-level abstract concepts out of whatever it is that you're seeing, relating it to past experience, and then using rules to compute new pieces of knowledge.

Now, just because I described it in this way, it doesn't mean that it has to be such a clear-cut memory structure with explicit concepts. It might all be captured in high-dimensional real vectors. I'm not saying that a neural network formalism isn't the right way to go, but subtly, even those high-dimensional real vectors must be doing some kind of computation that feels like this. You're extracting meaningful concepts, you're relating them to previous experiences, and you're doing reasoning over those experiences – what did I just describe? I just described a graph of abstract concepts.

> It is my very strong belief (and there exist many famous quotes from pioneers of artificial intelligence over the years that echo this) that if you truly want to mimic what a general intelligence system does, you need to do some kind of reasoning over some kind of graph structure.

To me, it comes as a fairly straightforward no-brainer that any kind of competent AGI we build will have to have some kind of graph representation learning module inside it.

AA: I agree. How far off do you think we are from AGI?

PV: It's really hard to say. If you asked me at the start of my PhD (2016, so not that long ago), I would have probably said we're quite a long way away: several decades, maybe 50 years. But after witnessing the rapid progress in the area firsthand, I honestly am not so pessimistic anymore.

> I feel like it might be within reach in our lifetimes. Probably even in the not-so-far future, we could have the first working example of something that might convince broad groups of people that AGI is coming. Let's say such a proof of concept might even happen in the next decade.

Bridging the gap between academia and industry

AA: I now want to switch gears a little bit: how can we collectively work better to help bridge the gap between academia and industry? I work with both, and I often see a disconnect between incentives and cultures that inhibits the two groups working together. Based on the work that you've done, can you offer any practical advice – for industry and academia to help them collaborate better?

PV: I think my main publicly accessible claim to competence in this particular topic is our work on deploying graph neural networks to predict travel times.

As I mentioned earlier, this is a system that Google Maps has already deployed, together with significant research input from us at DeepMind. Right now, no matter where you are in the world, if you ask Google Maps for the travel time from A to B, it will in part be served by our graph neural network system. It is one example of the sort of sandbox, Lego-blocks research that I'm trying to do in my day-to-day work, actually seeing real-world use. In my interactions with the relevant product teams at Google, I have learned so much about the real-world implications of trying to deploy our fancy Lego bricks in a real-world system.

You've already served me half of the answer, but basically, the incentives are quite misaligned. A researcher usually wants to be able to publish papers, and the reviewers who look at these papers expect to see an accuracy curve that's moving to the top. As time goes on, we want to get better and better at these performance benchmarks, and that is the main thing that a DeepMinder will typically care about, plus potentially some interesting, novel, theoretical insights. Both of these things are very important for a researcher to have a successful career, especially early on, while they're building up a portfolio.

In comparison, what does a typical business client care about? They care about some level of performance, but more importantly, they care about reach. They want to be able to scale up the method to make it easily accessible, potentially even to billions of people who are going to be using that technology. In the Google Maps context, they wouldn't be as incentivized to hunt for the most accurate solution. They might be willing to afford a system that is off by a couple of minutes every now and then when estimating travel times, but if that reduction in performance means they can now reach a billion people, as opposed to far fewer than that, that's a trade-off for them.

We actually put together a research paper on our work together with the folks from Google Maps, where we very carefully detailed the interesting engineering challenges that the product team faced when they had to deploy the system. There are a lot of interesting insights in there. That was my first insight into just how misaligned these two areas are. When you think about these two incentives, it gives you at least a suggestion of how we might be able to bridge the two sides.

It's about understanding what the other side wants. Every researcher wants to promote their methods and the new great theory they have. Of course, if you think you have a model that is highly applicable to a practical problem you're helping solve, by all means, try it. It might work really well. However, you must also be very open-minded about the fact that the optimal engineering solution for your particular problem might not be the great model that you have proposed.

If you look at some of the results in our paper, you'll see that some of the best-performing models are related to research that I have proposed previously, but unfortunately, those particular sophisticated models are not the ones that are currently serving travel-time queries. All that great research gives you a few smaller boosts in performance, but it makes it much slower and harder to deploy. You must be mindful of that.

> I think as a researcher helping out in an industrial problem, you must be willing to accept any solution that works, not just the solutions that you have come up with.

That's one side of it: letting go of ego. It was a liberating experience for me when I first learned how to do that.

While a researcher's state-of-the-art methods might not be applicable to the problem, somebody who's a good researcher is also a fantastic problem-solver – they can deal with the engineering and business constraints you give them. What's necessary, then, is to make constraints clear early on. Make things very clear at the beginning: "*We are trying to build a model that has to be served under these constraints, and therefore you must be mindful in your solution to try to solve these problems.*" That's probably already going to be very fun for the researcher because it gives them a puzzle to play with, rather than telling them, "*Here's a playground in which you can throw around all the models you can think of.*"

There are some projects where publishing a research paper is possible and the engineering team just might not think about it, especially in the case of a public-facing system: it might be a great incentive for the researcher if they were able to publish a paper based on the findings. This is not always directly offered. For the Google Maps project, we only got around to writing a paper when I explicitly asked if it would be OK because I thought it would be cool for the audience to read about it (we had already published a blog post and we weren't publishing confidential information). After a brief discussion, we all agreed to submit a paper together. It got published. It got some recognition, so it gave research points to the researchers as well as interesting exposure to the engineering team.

If you read this as an external observer, you might think of this advice as very straightforward, but I would say that very often, the two sides are not aware that they could be working a lot more efficiently together if they just applied these simple ideas.

Getting into research

AA: What advice can you give to someone wanting to follow in your footsteps and become a researcher in this field?

PV: I think that there's a lot of interesting advice one can give, but I'll try to single out maybe a few pieces of advice that were most helpful for me.

I'll preface this with a disclaimer to say the things that worked well for me might not work well for others, but at least given my temperament and my approach to research, they were quite useful.

One thing that I would highly recommend people do if they want to get started and build a good, solid career in the area, either as a research scientist or a research engineer, is to not just stop at reading papers or blog posts that they find very interesting. There's a lot of interesting content going around nowadays from many different sources, and it's very easy to trap yourself in a cycle of just reading about all of these different things rather than actually going out and doing it. I feel like there's been a lot of great examples of engineers and researchers who just went out and tried to build a GitHub portfolio, thinking to themselves, "*Here are these very interesting papers. I'm going to try to implement them myself and reproduce them.*" That gives you a good set of credentials that tell people you understand how to parse what's going on in those papers and how to convert them into actual efficient implementations.

Moreover, when you look at these papers, you see the authors are always compelled – usually by reviewers – to write about the limitations of their work. You can then think, "*OK. If I make a subtle change to this code base, I can now fix one of those limitations and run some more experiments, and maybe I'll show better performance.*" At this point, you're doing research. A tiny intervention or implementation from you can lead to improved performance.

At that point, you can publish the research, usually at a workshop venue. Workshops are cool, sometimes-undiscovered gems that are attached to large-scale machine learning conferences. They accept a lot more papers, proportionally, than the main conference. They're a lot less about elitism and a lot more about communicating great results as soon as possible.

The great paradox is that because it's a niche workshop attached to a huge machine learning conference, you'll get very particular people in that same area looking at your work, meaning you might even get more attention than if you were to publish it in the main event. You get a lot of feedback, a lot of interesting discussions, and a lot of tips on how to improve your paper when you're ready to submit it to a bigger venue from people who might end up as your reviewers when you submit it to that venue.

I feel like it's a nice, virtuous cycle and a great way to get introduced to the exciting world of research. You start out by tinkering with models and making subtle modifications, but you cannot get there if you don't start with simple things. Even if it's just reproducing existing research, there's a lot of benefit that can be had from these kinds of interventions.

That would be one suggestion in terms of how to get started. I'll also offer a few suggestions when it comes to reading the literature.

The field of GNNs truly started to explode around 2016-2017, and there were a lot of works attacking it from many different perspectives. That was because we did not truly understand what the limitations were and what the expressive power of this new tool (the graph neural network) was.

What we got as a result of that were many different approaches that propose many different ways of processing data on a graph. Many of these approaches turned out to be very similar. There wasn't a big difference in what they were doing besides the language that they were using, depending on the field they were coming from. Was it signal processing? Was it natural language analysis? Was it computer vision? Each field came with its own terminology, and sometimes many papers were proposing the same thing, just under different guises. This can be confusing for people new to the field.

The main takeaway is that when you're reading (especially older) graph representation learning papers, sometimes, the idea that you have and the idea that you publish for everyone to see are two very different things. The ideas in graph representation learning will often be miles ahead of the data in graph representation learning, and you cannot have a good model without good data.

On that note, I will just say it's a very exciting field with a lot of strong activity every single day. Because more and more players are realizing the utility of these things, there's a lot of scope for new contributions and new perspectives, and if we truly want to make deep learning as beneficial and ubiquitous as possible, with high utility for everyone, we need to include everyone in the conversation. I hope this motivates as many people as possible to take part in the action, especially while it's still early days.

Summary

It was an absolute treat to learn more about the cutting-edge application of machine learning to mathematics. Hearing about the work that Petar and his colleagues are doing in solving complex mathematical and scientific problems is very impressive.

I believe that the use of graph representation learning is still in its infancy in industry, and it was great to hear details about its range of applications from one of the leading researchers in the field – including practical tips and insights on implementing it. Notably, Petar believes that AGI could be in reach within our lifetimes.

Petar offered some useful advice in helping bridge the gap between research/academia and industry. The most important point is to clearly understand the interests and incentives of each party and to negotiate a compromise when incentives are misalinged. Finding the right balance can help unlock potential in creating research with a huge impact. For instance, the incentives for researchers to improve model performance by incremental amounts are generally not worth the trade-offs in model complexity, explainability, or scalability for the business. Petar advises researchers to be willing to accept any solution that works, and not just their own, when translating their research to real-world applications.

Petar recommends developing research skills by not just reading but *implementing* techniques in research papers. Authors will note the limitations of their work in papers, and Petar encourages early career and aspiring researchers to see if they can overcome some of the limitations with their implementation. He also adds that trying to implement and reproduce interesting papers and articles helps build a valuable portfolio of research.

As an introduction to the world of research, Petar suggests presenting and publishing your work at workshops associated with the leading ML/AI conferences. Reviewers for workshops accept a high proportion of submitted papers and provide valuable feedback.

7

Kathleen Maley Analyzes the Industry

Kathleen Maley is an experienced and renowned analytics leader with a background in mathematics and statistics. She is currently the Vice President of Analytics Products at Experian and has held many senior analytics roles in the banking sector, including Senior Vice President at the Bank of America, so she is well placed to understand how business and analytics intersect.

I was eager to learn her thoughts on how to establish and manage successful analytics projects at scale, including establishing the right culture and organizational structure.

I was also curious to know whether her background as a teacher helped her prepare for a career in analytics.

Pursuing a career in analytics

Alex Antic: What inspired you to pursue a career in analytics? Were there any pivotal moments or people throughout your career that made you take one path versus another? How have you found the transition to becoming a leader in the field?

Kathleen Maley: I would say, in my case, it was largely accidental.

My background is math; I have an undergraduate degree in pure math. I was teaching, and I stumbled across this opportunity to teach statistics, and it was really fun. I'd always thought about going back to graduate school, but I didn't know what I wanted to do. I ended up going back and getting a master's degree in applied statistics.

What I really wanted to do was join the Census Bureau. I was very interested in population study, but I had to move to Charlotte, North Carolina, and the census work that happens in Charlotte, North Carolina, is not the census work that somebody like me would be interested in doing. It's more about the field collection and less about the analysis, strategy, and sample design. So, I ended up in banking.

I didn't know it at the time, but analytics as a practice was just beginning to take off, so I was hired to build credit risk models. That's where it started, but my career evolved as the practice evolved, so I think I was very lucky to end up in analytics.

I moved around a lot with my first company. I was asked to take on the next role, and I made the decision to say yes for one of two reasons – either I was not afraid to say *yes* because I thought, "*This is an interesting challenge,*" or I was too afraid to say *no*. In many cases, the level of seniority of the person who was asking me to take on the next role, to solve whatever problem it was they were having, was what determined the course of my career.

When I was in college, I didn't take statistics because I was under the impression that statistics was for people who couldn't do math, and when I was asked to move from my risk model development role to a business analytics role, I thought, "*Business analytics is for people who can't build models. Why would I ever want to do that?*" But it turns out that business analytics is exactly the work I was built to do. For me, it is so much more fun. It is so much harder, so much more complex, and in many ways, so much more meaningful.

You asked whether there were any pivotal people along the way. There's one individual I'll mention by name – John Kuntz. He was my manager in my first business analytics role, and he taught me how to be an analyst.

> **Most of us who leave school do not have any idea how to analyze data.**

We know algorithms and we know programming. Hopefully, we still know some statistics, although a lot of programs aren't focused on that anymore, but we don't actually know how to analyze data.

> **Building models is much simpler than addressing a very open-ended business question.**

The first experience that solidified my interest in what I now refer to as investigative analytics was a very simple data request. My executive said, "*Kathleen, I made this change in our contact center. I need you to get me the data that shows that client experience has improved as a result.*" It was an unbelievably important experience for me to have because it's where I learned that my job wasn't just to provide the data. My job was to understand the problem and what the data was telling us about that situation. I was able to go back to him and say, "*Yes, you made a big impact, but it wasn't in the way you thought it was. The impact you made was actually due to this other thing, and so before you spend another $7 million, let's talk about how you can do more of what actually created the goodness and not unintentionally degrade some of the goodness that you created.*"

That, for me, was when the light bulb went on. Doing this role that I've learned how to do and that I fell in love with is first about being able to interact with business leaders – the problem they're really trying to solve. Then, while I'm listening to them talk about their problem, it's about building an analytical plan in my mind. The math behind the majority of the work my team does, the math for any business where analytics has become a business utility, is very light math. It's a lot of summarized data. The hard part is understanding the problem and then making sense of the data. If I hand in 50 pages of data to a business leader, I've not done my job. My job is to answer a question. The 50 pages of data are for me. I analyze it, I summarize it, I answer the question, and I make a recommendation.

AA: You transitioned from being very hands-on and working with data to a senior leadership role. What do you think were some of the key aspects of your personality or skills that helped with that transition? It can be a difficult transition for many technically strong people. What do you think are some of the most important skills, attributes, or mindsets to transition successfully?

KM: Firstly, I was a teacher. I had a lot of practice thinking about breaking down very complex topics in ways that others could understand, but that's only part of it.

The other thing I learned how to do I learned from my first manager, John. He really got it that our job was to answer the business question, and he and I had a lot of conversations about this.

It was also important for me to understand and appreciate what actually needs explaining.

> I don't need to explain everything. I needed to learn how to differentiate the relevant from the irrelevant. When I'm doing an analysis, it's like I've stumbled onto a murder scene, and my job is to figure out what happened.

I get questions all the time: "*Hey, Kathleen, what's going to be the impact of the government shutdown?*" That's a great question, and I'm going to have to do a ton of analysis to answer it. The tricky part is for me to be able to say, "*This is the impact of the government shutdown.*" I have to support my conclusion with a few pieces of relevant data. There will be a ton of data that I look at that ends up being irrelevant. It might be true. It might be interesting. But it is irrelevant to the question at hand.

So, that's part of it. I also think, specific to a centralized analytics function (many of us in the United States will operate under that model), there is a healthy dose of professional humility that has to go along with a role like this because none of the work I'm doing is for myself. It is on behalf of a business leader. I am the caretaker of their analytics resources. I have a very specialized skill set. I have learned how to analyze data.

> I have a responsibility to care for the individuals who work on my team – to make sure that they know how to analyze data, that they have a career path, and that they have a structure in place that makes them feel good about what they're doing and allows them to contribute value to the organization. But they're not doing that work for me. They're doing it for the business leaders.

It's really important for me and my partners to know that I don't prioritize their work for them. I don't prioritize their strategy for them. They have to prioritize the work, and that means I have to be OK with, and even own, the fact that my analytics function is a support function. I may be someone that generates a whole lot of revenue that the business wouldn't generate without me, but at the end of the day, the business strategy belongs to the business. It's not mine.

I think women are often socially engineered to be more comfortable with that setup than men, so I do think there's also something about that typical social construct that makes it easier and more natural for me to take on that position than a male leader might. Very generally speaking, women tend toward collaboration over competition, and that's very important for success in a role like this.

As I was making that transition from being the hands-on analyst to being the leader, I also had to start focusing on things that I was good at but weren't necessarily my interest. I have to make sure the team is always relevant to the business. My job is to make sure they get paid and that the team continues to be funded and relevant across the organization.

> I have to participate in events in very visible ways. I might want to say, "*This is silly. I should be working on analysis,*" but I do have to make sure that the team is visible in meaningful ways. That is my duty to the team.

I've gotten better at that over time. It is still not in my nature. I tend to be back of stage, just doing the work. But if I'm not drawing a solid line between what the team has done and the value they have helped create, nobody else is going to either, and so I owe that to the team. That's probably been one of the most challenging adjustments I've made.

Striving for diversity

AA: You touched on gender, which is something I'd like to look at, along with diversity more broadly.

There's definitely a lack of diversity and inclusion in our field. Have you ever faced any personal challenges with inclusion? What should we be doing as leaders in the field to make positive changes?

KM: When I think about diversity, I think about diversity of thought, diversity of skill, and diversity of experience. But let's be honest about where that diversity comes from. That diversity comes from some much more basic things. It is gender diversity, it is skin color, it is ethnicity – it is all of these things.

I will also say that I look for different experiences for different roles that I hire, but let's just keep it simple and talk about skin tone and gender. Have I been treated differently? Absolutely. I've been in meetings where I've been shushed with a finger. It happens regularly that I will say something that lands with a thud; it is repeated by a man, and it is heard and celebrated.

I can't speak to what it's like to be a person of color; I can only speak to what it's like to be a white woman. Women are told frequently that they need to be more confident. Women are also told, "*Don't be too assertive.*"

> As women, we're walking an invisible line that I've come to understand is like the negative poles of a magnet – as soon as your foot gets close to the line, the line moves. It's actually impossible to walk the line.

We have to decide how we want to be perceived. Do we want to be perceived as nice and pleasant, or do we want to be perceived as competent? I've made a choice for myself, but one of the big differences is that men aren't asked to make that choice. Women are asked to make that choice very often. So much of what happens is invisible to the people who don't experience it, and so if women are not heard, then unfortunately, we need men to point out the fact that we're not heard.

I did have a colleague who would say, "*So-and-so, I just heard Kathleen say the same thing. Are you saying something different, or are you reinforcing what you just heard her say?*" It was amazing, and it was because we're also not allowed to speak up for ourselves in that way. We need that support from our colleagues around us who see it happening. I've had conversations with a lot of individuals who recognize, "*Ah! I didn't realize that was happening, but now that you mention it, I see it everywhere.*" I think men really care about this when they think about their daughters – I don't know about their wives, but I've heard them refer to it a lot more with their daughters.

It is about making a concerted effort.

> It may never be possible for me to see the world through somebody else's eyes, but when they share their experience with me, I can work hard to understand and look for things that I hadn't known to look for previously.

I can trust them when they tell me, "*Actually, this is the experience for us, even if it's not for you.*" I trust them, and I say, "*OK, What can I do differently? How can I help?*"

I think there are also some structural things. I don't shy away from saying, "*All right, look. Right now, I have a four-person leadership team. I have only men. The next role I hire is going to be a woman.*" I don't care if it's harder – I need to do it because I need that balanced set of experiences. Women are given their next job based on what they've demonstrated, whereas men are often given the next job based on their potential.

> It's really hard to demonstrate your effectiveness in a new role if you're not given the chance to demonstrate your effectiveness in a new role.

There's also something about being able to go into a role knowing that people expect you to grow into the role. When I go into a role, I have to already know how to do that role, which is almost impossible the first time you're doing something. When a man goes into a role, there's an expectation that there's a learning curve, but if we're not perfect on day one, then it's a case of, "*Oh, maybe we didn't make the right choice.*"

The other side of this is women not knowing that they're doing it to themselves. Again, take all of this with a pinch of salt; this is not my area of research. This is just my own personal experience. But one of the things I've noticed for myself and for the women who work for me is that they are reluctant to take that risk until they're 100% sure they're going to be successful. I don't think men carry that same attitude into things they're trying.

I think about my daughter learning a new piece on the piano. She's going to be terrible the first 10 times she plays, so get it out of the way. When I'm tackling a new challenge or when I'm operating at a new level for the first time ever, I'm going to make mistakes. Accept it, learn from it, and move on, because there will be a next time. I think there's a different set of pressures that are put on women who have come into a space and are trying to fit into an existing box that doesn't necessarily appreciate how what they do might be different. I think we need to make sure we're deliberate about inviting that difference, recognizing that that difference actually adds value.

It's not that I will hire a woman if I find one or hire a person of color if I find one. It's that I will decide to go to the places where people of color congregate. I will go to the places where women congregate. It is possible to find people. I think it's a cop-out to say, "*Oh, it's so hard to find them.*" If I'm only going to my network and my network is very homogeneous, it's going to be hard. I have to make a concerted effort to break outside of my network.

AA: I believe diversity is an important part of this book. I don't want people picking it up and thinking, "*All the leaders are one particular gender and race, and hence that excludes me.*" **It's something that is very important to me. My fiancée is also a senior leader in this field, and I've heard very similar stories from her.**

KM: It's weird. Part of the challenge is that it's invisible to you because you are not on the receiving end of it.

There are times when I think, "*Did that really just happen?*" In that example I gave where somebody shushed me with a finger, I literally thought, "*Did that really just happen?*" There were close to 30 people, and I was the only woman in the room. One of my employees who was in the room came up to me and said, "*Kathleen, that was the most unbelievable thing I've ever seen, and I wanted to say something so badly but I know that you will handle this in exactly the way you want to handle it.*" He didn't know it, but of course, I needed to hear that from him because I was questioning it: "*Was that real? Did that really just happen? Am I being overly sensitive?*" Sometimes it's obvious, but sometimes it's not, and it's enough to make you crazy.

Becoming data-driven

AA: If you were in a greenfield situation now – say you walked into an organization that had just commenced the journey to becoming data-driven – how would you advise them on how they should prioritize their approach? How should they think about structuring their teams – centralized versus distributed; roles, responsibilities, and reporting lines; and so on? In a general sense, how would you set up an organization that said to you, "*We want to be data-driven. Go for it?*"

KM: Well, if you were to ask me without constraints what would I do, I would answer very differently from how I would answer if I was taking into account the organizational culture and organizational belief about data and analytics. It is remarkable to me how confident individuals who don't know much about this space are in how a solution needs to be constructed. These are people who are saying, "*We are really struggling with data and analytics. We're struggling with getting value out of it, and we need* *you to come in and implement a data strategy, data governance, self-serve, on-demand, and so on. And we also need AI."*

What I want to ask them is, *"Well, are you really interested in a data strategy and data governance and AI? Or are you interested in better business outcomes? If the latter, let's talk about what those better business outcomes need to be and what we need to do to get there. Do you need more of the analytics? Do you need more data to start with? Do you really need self-serve? You might not because you might not be able to analyze this data if you had it, so maybe that's not the number-one priority. Maybe you need some basic reporting and you need to know that you can trust that data, so let's talk about where to start to solve for your specific outcomes, not just a conceptual framework."*

One of the hardest things is getting organizations to let go of their preconceived notions. When an organization is embarking on a data and analytics strategy, I picture them as Vikings in one of those little boats, where you think, *"My God! How did this thing sail as far as it did?"* They are going off in a direction in which they've never gone before, and their only frame of reference is all the other lands to which they've been. They are trying to apply what they learned from those other lands to the land that they haven't yet seen. Well, those lands are completely different! How do you explain what a strawberry tastes like to somebody who's never tasted a strawberry? Things have to tie back to the things they think they know because that is their only point of reference, and if a person has zero points of reference, they're lost.

So, one of the most important things is being able to tap into whatever their existing frame of reference is because there's going to be work across the board. I'll think to myself, *"Where can I start doing work that is ultimately going to move us in the right direction and is going to be the easiest for the business to connect to and hang on to as a point of reference?"* I change what that point of reference is a lot. If they tell me their biggest challenge is too much data, then I know to talk about what a business analytics lead is. I know to talk about what investigative analytics is. If they say, *"We just don't have access to anything we need,"* OK – let's talk about BI. Let's talk about the data infrastructure that we will build over time.

Now, in each case, I'm going to need to rely on the skills and the competencies of the individual that I have in the position as business analytics lead because that individual is responsible for the conversation with the business partner, understanding the problem they're really trying to solve. If we're trying to create a BI report, how do we make sure it's relevant? What is the business going to be doing with it? Who in the business is going to be responsible for logging in to this thing every day, and does that person know what they're looking for? That skill is the same in developing the requirements for a BI tool as it is for saying, *"Hey, Kathleen, what's going to be the impact of the government shutdown?"*

I begin by assessing, fostering, and growing that skill, but I apply it to the thing that the business is anchored to at that moment, because invariably, they'll ask, *"Kathleen, can you get me the data?"* What they're likely thinking is, *"OK, I know this metric is based on these four things and something changed, so I want to see whether it's one of these four things."* But they don't usually explain that context. They simply ask for the data, but in order to help them, I have to know the context of the data request.

We need to get that conversation structure in place: "*Well, what problem are you trying to solve, and what do we need to do?*" That opens the door for the business analyst to do the contextual analysis and discover: "*A-ha! Something changed. Here is your answer in two sentences, and here is the relevant data that supports my conclusion. Now, what do we want to do next? Do we need to fix something? Do we need more granular monitoring of the relevant metrics? Where do we go from here? How do we begin to build a recommendation?*"

That process of gathering context and providing relevant insights always needs to happen, but anchoring it to where the business is focused gives the analytics team the opportunity to build credibility and collaboration with the business. I've got to take them on a journey to see the power of applied analytics that they don't even know they're going to go on, and no matter how much time I spend explaining it conceptually at the outset, they can only connect to it at the theoretical level. That's why I anchor at the places where they do have a more familiar frame of reference and build from there.

AA: In cases where organizations have commenced this journey on their own without the guidance of someone such as yourself, where do you often see them hit pitfalls and challenges and make mistakes? What are the traps they fall into where they really should know better?

KM: I wonder, should they have known better? I'm an analyst, but nobody taught me how to analyze data. Nobody taught me how to define a business question with a partner, and nobody taught them how to engage with an analytics team. We didn't have a blueprint for a high-functioning, centralized analytics function or even a decentralized analytics function. It didn't exist. So, I reject the notion that they should have known better.

Now, at this point, my hope is that there are enough people like me who have developed enough knowledge through trial and error that we can make the recommendations. Once the recommendations are made, if individuals choose not to take them, that's a different question. But what I see most often is, "*Oh, we need automation. We just need as much data as possible as quickly as possible – reports, reports, reports, reports.*"

What happens, invariably, is that there is so much reporting that the executive or the business leader then says, "*Well, this is a lot of detail. I need a report that summarizes the report,*" and we just keep producing more and more and more data. Producing more data without an end is not actually helpful. People think analytics is the production of data – it's not.

Another thing is thinking, "*Well, before we can do anything, we have to get all of our data under control, clean, and streamlined.*" Actually, we can do a lot with imperfect and/or incomplete data. Ideally, as an analyst, I would always have perfect data. But I have seen more data projects fail because the goal became creating a perfect, all-encompassing data asset. Instead, focus on the subset of data that is needed for a specific business initiative. Over time, as various data assets are built with an overarching vision in place, a very robust data asset will emerge that has produced value from its earliest days.

The best work always starts with, "*What's the business initiative? What data do I need to pull together?*" If I don't get into reporting, business analytics, and then the predictive and prescriptive modeling until the data's under control, we will never get there. We're creating new data all the time. New business problems pop up. We acquire new companies. The data journey is never done. That's work that must happen, but taking a historical project management waterfall approach greatly restricts the speed at which we can add value.

There is a Gartner Analytics Maturity curve that says, first, you start with descriptive analytics (what happened?), then you graduate to diagnostic analytics (why did it happen?), then you progress to predictive analytics and modeling (what will happen?), and then finally, to prescriptive analytics (how can we make it happen?).

> The Gartner Analytics Maturity curve
>
> **The Gartner Analytics Maturity curve** is often cited in the field of data science. Here's an example reference: `https://computd.nl/demystification/4-levels-of-data-maturity/`.

Whether they're talking about the graduation of an organization or the graduation of the individual who is doing the job, it's wrong. You will always need reporting; you will always need investigative analytics. Investigative analytics is probably more the starting point from which you understand what you're trying to solve. Analytics is a business utility.

Do you need the same question answered over and over? You need a report. Let's talk about what that needs to be. Do you need to make the same decision over and over and need a best guess at what to do? You need a predictive model. These things work together. It's not a maturity curve – it's an ecosystem.

That's been a difficult challenge to overcome, although I'm seeing fewer and fewer people connect to that maturity curve recently, so that's a nice thing.

In my experience, the majority of our time is going to be spent on investigative analytics because that's where we're trying to figure out what's going on, seize a new opportunity, meet a new business goal, or bridge two separate functions and create a more cohesive strategy. The report is easy – you define it and you build it. The model is easy. But investigative analytics is the foundation of all of it.

So, one issue is thinking you start with the data – definitely not. Another is this maturity curve, which is silly – it's an ecosystem. The third is the idea that analytics exists "over there." Analytics is a business utility – period. There's no such thing as an analytics initiative – there's only a business initiative.

Analytics can't provide value in a vacuum. Generally speaking, there are as many flavors of it as there are of those of us who do the job, but in the sort of work I do, it's more of a support function. Almost never does analytics provide any value in and of itself. The real value comes when it is integrated and a business process is rebuilt around new information or new automation.

The business owner has a problem. Once we answer that question or help solve that problem and have made a recommendation, the business has to ingest it. The business doesn't realize that if it's an hour's worth of work for me, it's 4 hours' worth of work for them, and most analysts don't know how to explain that. They don't even really consciously know that that's what is happening. Maybe because I was a teacher, I think of it as, "*If I'm lecturing for an hour, you're going to have 4 hours of homework.*"

AA: There's so much gold in what you've covered. One of your comments that really stood out is that it's an ecosystem, not a maturity model. I think there's too much dependence on the maturity model, and when I'm asked to come into an organization to help from a consulting perspective, I'm often asked, "*How do we rate on an analytics maturity curve?*"

How important do you think having a data strategy is for an organization? Of course, they need a business strategy first and foremost. Is a separate data strategy integral to that? How do you tend to think about it?

KM: I absolutely think so, but I don't think it is separate from the business strategy. I think they go hand in hand.

One of the other things that has been true for me is that my business partners will talk about the "data folks." They do not differentiate between data, technology, analytics, analytical insights, and predictive models. For them, it's all just the data stuff, and so I don't think of it as two sides of a coin. I think of it as a three-legged stool. Just as we need to have an analytics strategy (how do I make analytics effective at my organization? Where am I spending my time? How am I building and staffing my team to align with what the business is trying to achieve?), we have to do the same thing with data. However, if a business is working with data separately from working with me, or not working with data at all, we're in trouble. We always have to ask, "*What is it the business is trying to achieve?*" Invariably, that is going to require data and technology, and the analytics will inform much of what that is.

I mentioned how I've rarely seen a data project succeed, and that is true when it is about the data project. What I *have* seen succeed is a data strategy that is more in line with my notion for an analytics strategy: "*What is the business trying to achieve? Let's pull the thread all the way back to the data and operating systems that we're going to need now and in the future to support change. How do we start building that so that we immediately start getting a return on investment for that data work that will fund the next round of data work?*" I should be a net positive to the organization, and anytime I produce value through an analytics exercise, I should be able to measure it.

Again, it's tied to a business initiative. We know what a business initiative is worth. We make business assumptions all the time before a business is given funding. My piece of the cost for the analytics, plus the piece of cost for the data, will tell us how much additional incremental value we've created. Just give me 10% of it, and that'll fund my next bit of growth or the next bit of data build.

We've got to have an idea where we're headed, but if we think we can't produce business value until the data work is done, we're shooting ourselves in the foot. The data strategy has to follow the business strategy, and once you're through that initial round of funding, all of the future work should be self-funded.

Dealing with dueling datasets

AA: Speaking of datasets, I've heard you previously refer to this notion of the dispiriting dilemma of dueling datasets. Can you please explain what that is and how to resolve it?

KM: It's one of my favorites.

An organization will say, *"We have two reports with the same metric but we get different numbers all the time, and we're always wasting time rationalizing or trying to figure out why they're different."* It's a problem. It's a lot of wasted effort and energy. My general rule of thumb is that if I've done a piece of analytics work, I want a business to talk about the work I've done for 5 minutes and then talk for 25 minutes about what they can do now that they know something they didn't before. So, if they're spending 30 minutes trying to rationalize two data points on two different reports, that's an unbelievably huge waste of time.

There's a short-term solution, a long-term solution, and a strategic solution, if you will. The short-term solution is, *"OK, everybody stop. We've got two data points. How are you defining your metric? Who's pulling it? At what time is it being pulled?"* That helps me decide whether there really are different metrics, we're using the same name to refer to two different metrics, or it is intended to be the same metric but we're not calculating it equally. I may not need to change anything about the underlying data structures. What I might have to do is some renaming. If it's two metrics, they should have two different names, or I should come up with a consistent calculation for the two.

Very often when this comes up, it's because one leader has asked for a slightly different value to be calculated because they want to use that piece of information to debate with some other leader. That's where I can come in as an objective analyst and say, *"What are we trying to do? Forget about the different metrics. Here's the holistic answer. This is what you want to look at – here's your piece, and here's where it fits in."*

It's partly about reporting strategy. If we're talking about data that doesn't match on reports, there may or may not be something in the underlying data infrastructure that needs to be addressed. More likely than not, instead of a centrally controlled flow of information, what you've got is two individuals building two reports and haphazardly identifying and naming metrics. If a metric is critical and needs to be reported in more than one place, then it is useful to have that metric calculated centrally and then fed into multiple reports. That's a longer-term solution, and one that requires regular attention. There will always be new metrics being developed and we have to always ask, *"When is the right time to bring a new metric into the centralized data warehouse?"*

Overcoming roadblocks

AA: You touched on the question of when to productionize new data metrics. I often see issues occur when some initial development work has been done – say, a proof of concept – which then needs to be productionized. Organizations can hit a bottleneck there. Have you seen that too? If

so, what is a way to make that process – the end-to-end pipeline – a bit more efficient? Is it through MLOps? Is it through processes? Is it through people or technology?

KM: I think it's often a result of not getting aligned at the very start.

For example, if I as an analyst believe in a certain project, I might be tempted to build a proof of concept to demonstrate the value to a business. Obviously, if they see the value, they're going to adopt it, right? No! Absolutely not. There are always more considerations than just potential value to make. Number one – is it relevant to the current business strategy? Number two – has the business resourced and staffed the rollout of a new business process?

> Analysts only see the data piece, but the data piece is only one factor in the adoption process.

I'll give you an example. I was working on a very simple problem for one of my internal partners. It was a contact center problem. He was way over budget, based on his earlier demand forecast – essentially, he thought the call volume would come down, but it didn't, and he needed help reducing cost.

I did the analysis, corrected some misconceptions about who was calling and why, and recommended a simple call-routing algorithm that would save about $6 million annually. He was initially very excited, but when it came time to implement the solution, he was disappointed that there were still so many questions for him: "*Where are we going to install this? How do we get the flag to the reps? Who are the reps going to be? Who's going to train them? Who's going to monitor them during the training period? Who's going to be responsible for a rollout?*" The list of questions went on and on.

Now, here's what was funny about that. This was a guy who ran a contact center. He knew what it took to roll out a new call-routing strategy, but in his mind, this was an analytics project, not a business project. So, the spiritual buy-in was there, but at what point did we ask, "*Does the business have the time and resources to implement this solution?*"

Analysts are often told that a business wants us to be proactive: "*Find the value where we're not seeing it.*" At the very same time, we're bringing solutions to businesses and being told, "*This isn't what I'm working on. I don't have time for this.*" We need to be integrated into the business and say, "*OK, what can I work on that actually matters to you? Where do I see the potential for the value that is aligned with your strategy? Should I work on this proof of concept? Here's my guess as to what it would take if we wanted to fully adopt it.*" As much as we can, we should be working in partnership to avoid any waste.

Me deciding in a vacuum that a project is worth doing and putting a bunch of time into a **proof of concept** (**POC**) is a big risk. I'm just banking on the fact that they're going to fall in love with it and be ready to go. That's unlikely.

And if, by the way, we do the POC but things change and they say, "*You know what? We can't do this right now,*" that's OK. That idea or recommendation sits on a shelf until we're ready to pick it up again.

Analysts are told that if they demonstrate value and focus on value, the partner will act. But that is just a tiny piece of it. The partner actually has to be staffed and funded to act.

AA: That's a great point, and the collaborative aspect, I think, is vital as well. Something I often say to people is, *"You don't want analytics to be isolated. It can't be seen as an isolated function."*

Something we briefly touched on earlier that I'd like to discuss is centralization. I often hear organizations debate about whether or not they should centralize or decentralize, or have a distributed model with matrix structures. Do you have a particular preference? I know it's specific to the organization – their size, maturity, and business goals – but do you have any advice that you can offer?

KM: Number one:

> Organizational structures will not fix your problems – period.

Whatever problem you're having, an organizational structure will not fix it. Now, different organizational structures can be helpful at different times for different purposes, and it matters how big your organization is, how mature the individual analysts are, and how mature the organization is in integrating analytics as a business utility.

I take a couple of things into account. Firstly, I ask, *"Has the organization grown analytics through individual business leaders hiring one or two analysts here or there?"* Generally, you end up with somebody who doesn't know what skills are required for a highly effective analytics partner, hiring people who have fancy degrees but no idea how to interact with a partner or define a business problem. You just get a data merry-go-round.

In those cases, you need to bring people together to form a centralized function. You bring them together for two reasons – a) you need to teach the analysts how to be analysts, and b) you need to teach the business how to work with analysts and teach them that analytics doesn't provide value on its own. It's a business utility. So, how do you incorporate that?

Asking analysts to do things differently is hard. Asking anyone to do something differently is hard! If analysts report to me, however, I have a little bit more organizational influence and authority. A centralized organizational structure isn't a silver bullet, but it makes it easier to influence the deeper changes that need to happen. If a business partner still says, *"No. I just want the data,"* and the analyst still reports to that person, nothing's ever going to change. The organizational authority of a centralized function helps me retrain both the analysts and the business leaders.

Having said that, when it comes to the size of the organization, the amount of analytics resources, and the number of business units that need to be supported, I am in favor of a centralized but fully dedicated support model. Let me explain what I mean by that.

I come from banking. I had a checking partner. I had a credit card partner. I had a savings partner. I had a mortgage partner. I aligned somebody on my team, and probably several analysts, with the head of checking. They were partners. The business context is really important for an analyst to be effective. I don't want a totally fungible set of analysts who never know anything about the businesses they're supporting. I want them to be able to develop that relationship. I want them to develop that business acumen. They're fully dedicated to that group.

Now, some analysts will support more than one business leader, but the business leader has a person who is their point of contact and that is always at their side to manage the portfolio of analytics work being done on behalf of that leader, bringing to bear all of the other resources on the analytics team – be it for data engineering, reporting, investigative analytics, or predictive modeling. That person, for me, is also an analyst. I do not hire "translators." I look for an analyst with the ability to communicate. It can be trained into people, but it's difficult. Not everybody is going to have the natural aptitude.

There's another factor that's important to think about. A lot of organizations will have hybrid or centralized functions, but their risk analytics team will sit separately. So, one of the other factors I usually consider is the career path. Is there anywhere else for this person to go? If this person is an analyst reporting to a business leader, what's their career path? That's a factor as well in the organizational model – we have got to make sure that there's a career path, a way for people to grow, or we'll lose them.

Those are some of the factors I take into account. As I said, generally, I'm in favor of centralization. Most organizations still are at that lower level of maturity, simply because the discipline is still young – 15 or 20 years is not that long. Those of us who figured it out did so through trial and error, and there isn't a whole lot of us. They're still not teaching this stuff in school. Every business analytics program I've come across is focused only on various algorithms and programming, not the core skills of making sense of data and working on behalf of a business objective. And that latter piece makes up about 80% of what we do in most business settings.

Establishing an effective data culture

AA: In your opinion, what does an effective data culture look like, and how do you establish one?

KM: I think an effective data culture is one in which data is so ingrained in a business that it's an integral part of **business as usual** (BAU), where it stops being a surprise that when a business area in the organization asks for data, I'm going to ask them, *"Why? What is it that you're trying to solve?"* and where there is no longer that surprise at the end of a business initiative where people say, *"Oops! We forgot about data. We didn't put any money in to capture the data we're going to need. We didn't think about how we were going to use data through this process. We didn't include anybody from the data and analytics team in the project team."*

I had business leaders when I first joined an organization say to me, *"We don't need you here. We will let you know when we need data."* That is an organization that doesn't yet understand how data and analytics are part of its infrastructure, and I think that's the difference. When analytics stops being just this thing *over there* and starts being just another partner at the table, something has shifted in that organization. Nobody ever questions whether or not you need finance in a staff meeting. Nobody ever questions whether or not you need HR in a staff meeting, but data and analytics? *"We'll let you know when we need data."* If that's the message, then they're not there yet, and I've got to earn credibility for the analytics function.

Learning about analytics

AA: From an educational perspective, in terms of what's being taught at universities, what's missing from the syllabus? You're right – there's a lot of focus on just some niche elements of ML, but as we both know, success in the field is so much broader than that. What do you think we should teach future professionals and leaders in this field?

KM: First, I think there's a little bit about logical thinking.

> I wish everyone was a slightly more logical thinker. I wish people could connect the dots. I wish there was more critical thinking.

That's true across the board.

Specifically in business analytics, modeling, and data science, for those who intend to work in a business, hospital, or any other kind of application where leaders are trying to make better decisions, that is a separate program.

I did my program at the University of Michigan. I had to learn the theory. It was painful. Thank God I had a math degree because it turns out it's all calculus. But I also did applied statistics very intentionally. We talked about the techniques, what they were good for, and when we could and could not use them reliably. I then had to go a lot further and say, "*OK, in the real world, what are the risks I face or don't face when using this technique?*" That was great. We learned programming and algorithms through statistics. Too many programs now (take a Master's in business analytics, for example) have no statistics. It is all algorithms and programming, and that makes me very nervous.

My favorite part of my program was that we had to take a semester of consulting. Now, the University of Michigan had a consulting office, and those statisticians worked for all of the doctors and other researchers at the university. They would bring us a problem. We'd have to say, "*What are you trying to do? Let me see what we can figure out.*" I decided to extend. I did my term, and then I asked to do a second term because the practice was invaluable. I was out there talking to people whose businesses I did not know at all.

I talked to a kidney doctor (I don't remember what they're called!), and she studied the elasticity of small arteries, where there would be higher elasticity and lower elasticity. She gave me all these measurements and said, "*Don't look at the large arteries. We know large arteries are inelastic. Just focus on the small arteries.*" But, as I was doing my work, I thought to myself, "*You know, given the way I've written my code here, all I have to do is replace v1 with v2 throughout, and I can run exactly the same analysis on the large arteries.*" I did, and guess what she discovered? She discovered that actually, in the sickest patients, the elasticity of the large arteries increased. The smaller arteries had completely failed. The large arteries took over. That became the thesis of her next publication. I thought, "*Oh my God! This is it. This is magic. They come in absolutely sure of X, and my job is to test that hypothesis.*"

The experience of doing that consulting is critical. Analysts need to learn how to define a problem. They need to learn how to analyze data and draw a conclusion. They need to learn that when you're presenting your results to an executive, you use one slide. If you want to have 40 slides of what you did, showing it like it's your homework assignment, then that's great for your professor. It might be good for your analytics boss. It is not good for the business leader.

How do you teach your business leader what their role is in the process? How do you know what you should be holding your analytics leaders accountable to? People don't learn any of this stuff in school, and we're teaching it to you now so that you can transform the organizations you join and move into leadership positions more quickly. These are the things that really matter, and there is zero time and attention put into them.

It's like job readiness training for analysts. I did get a really good solid set of core skills from my studies, but it was zero job training. I think if we really want people to be ready to go, we've got to teach them these analyst job skills because they're not getting it on the job. It's too new for there to be on-the-job training.

AA: When you're looking to hire data scientists, analytics professionals, and so on, you're looking for these job-ready skills, as well as technical skills. How much weight do you place on formal university education versus online courses, which are so popular (from Udacity, Coursera, and so on)? How much do you value postgraduate work – honors, Master's, and PhDs?

KM: I have to say, I'm probably stuck in this rut of placing weight on college degrees. At least in the US, that's the door-opener, but it's worth challenging that assumption. That's especially so when what I'm talking about here is that we really need job training; that doesn't necessarily come with a formal 4-year college degree.

So, it's a fair question, and I don't know the answer to that yet. Just from force of habit, I might lean toward a degree. It depends on the role. I may go the other way. I am less likely to hire a PhD than I am somebody with a Master's. I look for somebody with an applied degree or someone who has learned statistics through some other application. It could even be a sociology degree, or a degree in chemistry, physics, or psychology.

I look for conceptual understanding over technical understanding. The technical side is really important, but the conceptual even more so. For example, I built a collections model. There was zero risk. The accounts were already charged off. We'd already lost the money. Who should we call to see whether we can rake any of it back? I got into a fairly heated debate with an individual who was responsible for reviewing my model for independent validation, because I said, "*Look. I don't care about the variance inflation factor.*" She was horrified: "*Any modeler knows you cannot have a variance inflation factor greater than 3.*" I said, "*This is a predictive model, not an explanatory model. I don't care if there's multicollinearity. Doesn't make any difference. I want to get the best prediction possible. It's still really low-cost to run this thing. Why do I care?*" She could not get over this notion, this arbitrary rule. For me, a conceptual understanding is really important.

Another individual – my analytics partner – provided a numerical example explaining why *"medians do not work on non-normal data,"* and my head almost exploded. So, I don't care what your degree is if you can't talk to me about when various measures of center are appropriate and what the downside is of each one. Finding a person who can talk about it conceptually is rarer than finding a person who can give you the equation.

Investigative analytics – that says it all. We walk into a crime scene and we try to solve it. We look for clues. An analyst is a problem-solver. Somebody with some self-propulsion. Somebody who maybe put themselves through school, maybe started at community college before going on to a 4-year degree, maybe had a career change, maybe an out-of-the-box thinker. To me, that is more important.

Now, say I am building very high-risk models, or I'm building a set of models to address data that we've never worked with before, or we're working on a very complex and nuanced problem of forecasting macroeconomic data. Who could do that? Of course, I'm going to go for somebody who's more research-oriented – probably a PhD, probably somebody who's more science-minded. Any tiny bit of incremental improvement is going to be valuable if I'm at a Fortune 100 company. But for a problem that asks the question, *"Hey, I want to know why my customer SAT scores fell three points compared to last month,"* I don't need a PhD-level modeler.

So, it's really important to know what the role is meant to achieve and what skills and experience you need to bring to that.

AA: An issue I've often found, both on the training side at the university level, as well as when employing staff and running teams, is that a lot of those entering the field have less of the fundamental math and stats knowledge – which you alluded to – than they should when running some of these large-scale models. They know how to use the algorithm and the library. They can call a Python library and run something. That's great, but why does it work? Why doesn't it work? What's happening under the hood? These are questions they can't answer. That's something that I think is missing from some programs.

Partly, it's because you don't have the same rigor of assessment that you do at the university level. As a former academic, I'm still biased – I must admit – but I think an understanding of the fundamentals is important.

KM: If I learned R and now the world's moving to Python, I've got to learn Python. But I'm not in there learning the algorithms. I'm just learning the programming.

Here's an analogy – I am great at pressing buttons and flipping switches – as good as any 3-year-old you know. And I don't know if you've ever seen the cockpit of a modern plane, but they are full of buttons and switches. But if you put me in the cockpit of that plane, it is going down – period. Being able to punch buttons and flip switches doesn't make me a pilot. Being able to run some script in Python doesn't make me qualified to do that type of work – period.

That's why when I hear people brag about their automated model development, I think, *"Oh, I wouldn't be bragging about that. I'd be scared of that."*

Looking to the future

AA: How does ChatGPT affect the data science scene, including the field of NLP?

KM: It is a complete and total game-changer. The math and the algorithms are exciting to someone like me, for sure, but what ChatGPT has done is make data science exciting – and personally relatable – to the average person. For example, most people were probably never excited about the data science that underlies their credit scores, even though those scores can have a significant impact on their day-to-day lives and general standard of living. It's a conceptually remote topic, and most people simply can't connect to it in a meaningful way. But with ChatGPT, it is a completely different story! OpenAI has exploded onto the scene with a solution with which the average person can interact, and that has given the average person a new understanding of the field of data science. This is good press for us. Very good press.

AA: What are some of the ethical implications for organizations, and society more broadly?

KM: This is a tough one. It's exciting that ChatGPT is earning so much attention for the field of data science, but most people have no appreciation for how unintelligent these models – like all models – really are. Now, don't get me wrong. There is no question that these models are remarkable, but they are not actually "intelligent" as we typically define intelligence in humans. They cannot fact-check, they cannot question, they cannot produce novel ideas. What they do really well is synthesize and reflect back the information on which they are trained. But because ChatGPT seems smart, the average user will rely too heavily on the information it returns. They will instinctively trust it far more than they should. Also, ChatGPT is trained on the internet and, unfortunately, the internet often produces a concentrated dose of bad behavior. While there are controls in place, it is also true that these models are being trained on the worst of us.

AA: How can data scientists/AI experts leverage ChatGPT to benefit their organizations and their own careers?

KM: ChatGPT can not only debug your code; it can also fix your code and explain those fixes. Imagine how much time and frustration that will save your typical data scientist!

Here's another example: I can ask ChatGPT to translate Python development code into another language that is better suited to a particular production environment. This is a real thing that my team does and, until now, we had to rely on people who are experts in both languages to translate that code from scratch, test it, and debug it. ChatGPT now gives them a huge head-start on these sorts of tasks.

Here's an example of a more research-oriented analysis: *"Hey, ChatGPT… What were the primary drivers of the 2008 crisis? In retrospect, what stocks or hedge funds or positions turned out to be leading indicators of the crisis? Given those leading indicators, are there any parallels or other indicators of a looming crisis in the current market or market segments like subprime mortgage, emerging, student lending? In what markets are the risk concentrations highest?"*

Companies can also train ChatGPT on their own data to answer questions very specific to themselves: *"Here are the timecard entries for all of the people in my company and for all of the projects they have worked. Cross-correlate the projects they have supported with the amount of revenue that we earn from each project/client. Based on this data, tell me which clients earn the lowest returns-on-investment for the company, and tell me what is common among all of them in terms of industry, product, project size, team leaders, and contributors. Now, add in client-support call logs. For clients who have called into tech support, by company, who has called the most and how is that correlated to the revenue we earn from each of those clients?"*

Simply put, ChatGPT can complete in a matter of minutes the same tasks – like the first two examples – that would take a human weeks or more to complete. The last two examples are not dissimilar to the questions that would be asked and answered by a traditional investigative analytics project. What is different is that once these ChatGPT models are trained, weeks and months worth of work can be completed in just minutes. It really is remarkable.

AA: I'd add that the appropriateness of sharing corporate data with ChatGPT also needs to be considered carefully. What are your predictions for the future of LLMs and their broader use?

KM: Honestly, I can't even begin to imagine. Any example I come up with feels small compared to my academic understanding of these models. For now, I think that's where the work lies – coming up with creative applications of this new technology. Additionally, we will look back on this as a period of major societal disruption. For the first time ever, white-collar workers typically considered "skilled" will be displaced by technology. There will be new jobs, of course, but there will be an entire generation of workers who feel the real and very personal pain of progress. It isn't "just" the factory workers anymore.

Summary

It was great to learn about Kathleen's journey to becoming a successful and respected data and analytics leader. As with most of our interviewees, she ended up in the data field via other disciplines and somewhat by accident. She also offers some very important advice for increasing diversity and inclusion, such as searching out women and people of color to ensure diversity in hiring. Publishing a role and assuming they may apply for it is not enough.

I really like Kathleen's replacement of the commonly used Gartner Analytics Maturity curve with the concept of an ecosystem, rather than a linear path. Jason Widjaja raises a similar point, in that the business value that an analytics function drives is the measure of its maturity, not whether it is descriptive or prescriptive. Kathleen's point is that descriptive/investigative analytics is always required – with a direct alignment to business needs – which translates to an ecosystem of different models and not a linear maturity curve.

I appreciate that she refers to analytics as a *"business utility."* She believes there are only *"business initiatives,"* rather than separate analytics initiatives. The standout quote for me is that *"analytics can't provide value in a vacuum!"* Business, technology, and data analytics need to be integrated to deliver value. A business often considers analytics projects to be the sole responsibility of IT/data science, and so isn't prepared to resource the projects with **SMEs** and run appropriate activities to embed a solution. The responsibility for making sure business readiness is in place before starting a project likely lies with IT/data science if we want projects to succeed.

I think these are much-needed mindset shifts that executives need to make to understand how analytics is integral to delivering business value, as opposed to being seen as something that's distinct and isolated from the broader business. However, it does raise the question of whether analytics should be part of their business – that is, integrated directly – or whether it should be a centralized support function of their business. Either way, Kathleen stresses the importance of ensuring the visibility of the analytics team to ensure it remains relevant and funded.

She describes one way she measures the value she brings: *"I want the business to talk about the work I've done for 5 minutes, and then talk for 25 minutes about what they can do now that they know something they didn't before."* I think this is a simple but powerful concept that all data leaders can use.

Kathleen outlines four common pitfalls that organizations can struggle with:

- The business thinks it needs more and more data, and more reports
- The data is not clean and in an appropriate state for use
- A linear waterfall approach restricts the speed at which analytics can add value
- Thinking that analytics is separate from the business

There are so many takeaways from this discussion for all of us to apply to our own organizations!

8

Kirk Borne Sees the Stars

Kirk was a data scientist before the field was even called *data science*. I was eager to learn about his illustrious 40-year career, which has included being an astrophysicist at NASA and the principal data scientist and executive advisor at Booz Allen Hamilton. He is currently the founder of an independent consultancy: the **Data Leadership Group**. It was an absolute privilege to speak with Kirk, learn about his secrets for success, and hear his views on how the field has changed over the past couple of decades.

I was also very interested to hear about the specific challenges he's seen organizations face to become data-driven, and to learn more about his strategies for success.

Getting into the field

Alex Antic: I'd love to hear about your journey from astrophysicist to a global leader in the field of data science and AI. What's the path been like? What are some of the most valuable lessons you've learned along the way? Have there been any particular people that have inspired you? Has there been anything that's happened that's shocked you?

Kirk Borne: It's been a long journey, and I guess you could say I'm near the end of that journey – 40 years in the field. I did start off as an astrophysicist, and I always was in love with understanding how things work, so the universe was my playground. How do stars work? How do galaxies work? How do things work? As a child, I was fascinated by it, but as I grew older and understood there was physics and math involved, I became fascinated with the concept of modeling these things.

That, of course, ties right back to observational data – what does the data tell us about these things that we try to model? That's what I did for the first 20 years of my career. I was modeling colliding galaxies, using observed data from the Hubble Space Telescope and other types of telescopes to build models to understand these things.

As the years went by, I started seeing not only large amounts of data in the science I was working in but also large amounts of data everywhere in the world – in business, government, healthcare – everything. I became quite interested in those broader applications of data. It was natural for me because my career in those first 20 years was working at a space agency – NASA.

> By night, I was an astronomer, but by day, I was managing data systems for NASA's space astronomy missions.

That was my responsibility – managing a team of people who were doing data processing and data management for space astronomy, involving satellites and NASA missions. I was always surrounded by data, and so once I started seeing the application of data for discovery in all the different disciplines, I moved from a focus on astronomy to a broader focus.

It took a number of years to get my focus as broad as it is now, but I was always looking at other disciplines and what they were doing with data, and that grew into sharing my knowledge with the world. When I got on social media, I thought, "*Hey, this is a cool education platform.*" I spent most of my time on social media (I called it my micro-education platform) teaching people about this stuff, and I found I had an audience. People were interested in learning about it, so I kept doing it, and eventually, it grew into a worldwide influence, which took me by surprise. I didn't go into it trying to be an influencer. I didn't even know what that was. I was just doing it because I love sharing information and knowledge with people.

That was an accidental journey, so to speak, to being an influencer, but the data and the modeling were always intentional for me. Ever since I was young, I have wanted to understand how things work. We collect data about things to understand how they work, whether it's healthcare, customers, processes, industries, manufacturing, supply chains, or colliding galaxies. We collect data to understand how it works, and we probe systems by collecting more data and seeing what happens. We ask, "*Does the new data fit our model?*" That's the scientific process of improvement.

So, these things are applicable all across the human-expertise domain. Humans are data-collecting creatures. We have five senses, and ever since we crawled out of the cave, we have been using those senses to make sure that we do the right thing. If you're the first caveman coming out of a cave and you see an animal in front of you, you need to ask, "*Is that animal going to eat me for lunch or can I eat that animal for lunch?*" Either way, if you make the wrong decision, your life is over. It's about object recognition and characterization – not just object detection but *recognition*. Those are very natural characteristics of human beings, and we've now digitized what we do naturally. We collect data from sensors to understand our world.

AA: What are some of the changes you've seen occur throughout the field, especially in AI? I'm thinking of both positives and negatives. From your perspective, what have you seen as the major changes as the field's evolved, especially in the last 10 or 20 years?

KB: In my first encounter with AI, I didn't even know it was AI. This was back in my early days at NASA – more than 35 years ago – and it was what they called in those days **expert systems**. Expert systems were rule-based systems where you defined an if/then rule for every single instance of something: "*If you see this, then do that.*" We had rule-based systems for how to schedule a telescope, how to operate a telescope, how to make decisions in terms of operational facilities in space, and so on.

You had to think of every possible case so that you could write a rule for every case. Imagine doing so for a self-driving car. Imagine if you had to write a case for seeing a child with a blue shirt, seeing a child with a red shirt, seeing a child with a white shirt, seeing a child with a polka-dot shirt, and so on. You would have to write a rule for every single case because every single child is going to look different. Then there are signs along the side of the road, road conditions, and weather conditions. It would be impossible.

In those days, AI was on an impossible journey – there was no way you could write every possible rule. So I went through that AI winter, and I never really paid attention to it. It was just something that was there and people were doing it, but I wasn't involved with it. It seemed crazy to me that you would have to write so many rules.

We have now come to a stage where we have algorithmic rules for pattern recognition. The best machine learning model is a good, *generalized* model – a model that works in general. Any child who walks in front of your car or anytime the road is wet or slippery, it's not safe to drive too fast or to continue driving at all. There are these general rules now. Generalization in machine learning is like fitting a curve through data points. If you fit a curve through every single data point, that's called **overfitting**, but if you follow the general trend in the data, then you have a better predictive model. You're not trying to predict an exact number; you're trying to predict a trend or a general pattern.

We've seen this algorithmic growth in the last decade, so much so that it's empowered all kinds of people to say, "*I want to do AI. Let's do AI.*" It ends up leading to enormous hype, and I've been through lots of hype cycles.

> I've been through the hype cycle with big data: "*Big data's going to cure cancer, end world hunger, solve the world's problems, and end poverty.*" Then, it was, "*Data science is going to cure cancer, end world hunger, and end poverty.*" Then, it was machine learning, and then it was AI. The next thing will be quantum computing or blockchain.

We're crazy about the stuff that we create, and we create these hype cycles.

> Who benefits from the hype cycles? The companies who are trying to sell these products.

I've seen just too much of this. I think that's the negative side of this field.

It's great that there's a lot of enthusiasm, and people are actually catching on to things that I'm passionate about. I'm passionate about the ability to understand your world through data and to be able to make better decisions and take better actions as a consequence of what you learn about the world through data. That's what I think is the positive side of this, and I think AI empowers that so much better.

There are multiple things to avoid, obviously – bias, prejudice, and other things that can get incorporated into our models, but also overexuberance and promising things we can't deliver.

Advising a new organization on becoming data-driven

AA: Say you were going into a new organization that wants to begin their journey towards becoming data-driven – they want to implement data science but they're quite new to it. How would you guide them on beginning that journey, in terms of talking about data, people, processes, and technology? Do you have a framework you use? I know it's specific to every organization, but at a broad level, how do you normally guide them from the beginning?

KB: Well, first I would ask, *"Do you have 3 days? Because I have an 8-hour course for that."*

That's not entirely a facetious answer. There are so many dimensions to that question – it's about talent, technology, tools, data readiness, culture readiness, and more.

I always tell people, there's a challenge at every level. General worker-bee-level people might think, *"Oh, my job's going to go away. AI is going to take my job."* There's a lot of **fear, uncertainty, and doubt** (**FUD**, as they call it). There's also that uncertainty at the senior executive leader level: *"Is this AI going to take my decision-making authority away? Is an algorithm going to make the decisions for the business? Do I even have a role here anymore?"* The middle manager feels like they're just caught in the middle – *"Why do you need me anymore? The data scientists are talking directly to the senior executives. The data scientists are coming up with models for decisions, best actions, optimizations, predictions, and so on. Why do you need me?"*

> Everyone's got this fear that their job's going to go away, but the real thing is that their jobs are going to *change*.

That change often does cause fear in people, but that's the reality. It's an industrial revolution. In every industrial revolution, the type of work that people do changes. That's the nature of it.

I looked back at an old statistic for the first industrial revolution with the introduction of steam-powered engines more than 200 years ago. Back then, in the United States, over 90 percent of the population was involved in agriculture and farming. At the start of the second industrial revolution (of electricity), about 50 percent of the population was working in farming and agriculture. At the start of the third industrial revolution, the introduction of the modern computer 40 or 50 years ago, something like 4 percent of the population was working in farming and agriculture. Now, it's less than a half or quarter percent of the population. But those people who *are* working in farming and agriculture are feeding so many more people. Those people are doing way more effective and productive work for a much larger community of customers. What is everyone else doing? Is everyone just sitting around eating the food that the farmers are producing? No, we're all doing something different.

We have to get used to the fact that this happens in the world – work will change. It doesn't mean the need for people to do work will change. There was a report that summarized that fact in one sentence with two parts, but people only quoted the first part of the sentence. It was a global survey of how AI is affecting employment and the future of work. The first part of the sentence said, *"75 million jobs will be lost as a result of AI and automation."* People just quoted that. The rest of the sentence said, *"But there will be 133 million new jobs created.*" The net gain is 60 million jobs! People just didn't read the second part of the sentence, or they didn't want to focus on it. I don't know what the reason was, but there was all this focus on the negative part without mentioning the second part.

The problem with the second part is, again, the fact that it involves change – changing the work we do. Businesses have to talk about that cultural change. If you have 8 hours, I can tell you all kinds of things about that. One of those things would be that we have a culture of experimentation that enables people to test different things. That's what we do as human beings – we experiment. Just look at any child. Any child will put things in their mouth. They'll test those things, and they'll test their parents. The parent says, "*Don't do that.*" Then, they'll do it again to see what happens. We're natural-born scientists as human beings, and we're natural-born data collectors, as I was talking about earlier.

So, embrace that culture of experimentation, but there's also the culture of fast failing, which some companies get really fearful about. They say, "*Oh, we're not here to fail.*" I say, "*No, you're not here to fail, but you should fail fast to learn fast.*" That's the important thing. You don't just fail because you want to fail. You fail because that's how you're going to learn. Take Thomas Edison. Someone said, "*Hey, you failed to invent the light bulb 1,000 times.*" He said, "*No, I learned 1,000 ways not to invent the light bulb.*" Take that positive attitude of learning from failure. Fail fast to learn fast; foster an experimentation culture.

The most important thing is to think big. What is your big picture? Where are you going as a business? But start small. Don't try to boil the ocean, as they say. Don't try to do everything at once with AI. Build a recommender engine, or a conversational chatbot for your call center when new customers call. Just do things that are straightforward. There literally is an app for a lot of these things. The amazing thing is that a lot of these big cloud service provider companies that provide public cloud services have machine learning and AI services available. Companies don't have to hire data scientists and AI engineers to start doing this stuff. So, think big, but start small.

AA: For organizations that have commenced their journeys, what do you often see as some of the biggest challenges they face when they deploy ML or AI, particularly when they've tried to transition from proof of concept to production?

KB: You need to have the right kind of people. A data engineer or a machine learning engineer is not the same as a data scientist or a machine learning scientist. That scientist is going to build cool tools and models on their laptop. That laptop's not going to deploy in your business. It's not going to deploy at scale for whatever size customer base you have; you really need an engineering team for that.

I think that's what a lot of companies think: "*We've hired some data scientists who are sitting over there in the corner building cool stuff.*" You've got to get beyond that. You still need the cool insight stuff, because you want to get insights from your data. But you can't build and deploy something with the same people and skill set used to do the mathematics and write the Python code. They've got to be people who have other kinds of skills in terms of building and deploying large systems, and those systems engineers may not even understand machine learning in the same way. That's OK, though, because what they're building are things that are going to be scalable, robust, agile, and all the right buzzwords for building sustainable systems and businesses. When it comes to deployment and operationalization, there's another set of hurdles to climb – you're going to have to find someone who can do those things.

I still remember a job description from years ago for just this kind of person. It listed all the different kinds of system skills, web skills, data science skills, and programming skills imaginable, and I said, *"There's not anybody in the world who could do this job with all the skills that were mentioned in the job description."* It was a page and a half of things. It didn't describe a real person. Then, I got to the very last line of the job announcement, which said, *"Salary,"* and next to the word *"Salary,"* it said, *"Anything you want."* Because if there were such a person, they could probably ask for anything they want.

I actually talked with that company because they called me, but I wasn't seeking to work in that kind of job – it sounded like too much. I said, *"There's nobody in the world who could do all that."* They replied, *"Oh, yeah. We don't expect anybody to do that. We expect maybe 10 percent of our list."* They wrote a job description where they knew that they weren't going to ever find anybody, which is scary, because when I worked at the federal agency as a contract manager, I could not write a job description that listed things that were not really required for the job. That's illegal. You can't ask for something and then hire someone who doesn't have those skills. That's just inappropriate. Believe me, I did not want to go down that path.

AA: How important is MLOps? We hear a lot about MLOps as a new progression of DevOps. Is it very much focused on people and processes? How do we effectively use MLOps? I've seen a lot of organizations in my consulting practice stumble here – they don't know what it is or how to implement it. What advice can you offer them regarding that?

KB: I made a mistake myself with this.

The first thing I encountered beyond DevOps was DataOps, and I just assumed DataOps was just DevOps for data. I gave talks on how it was just Agile DevOps principles applied to collecting data, learning from data, and orchestrating the use of data around an organization. It turns out it is some of that, but it's also more than that. It's more about the operations around data. There are DevOps principles involved, but it's beyond that.

MLOps is the same thing. It gives the impression that it's DevOps for machine learning, but it's more than that. People like to use something called observability, which is actually monitoring and observing the performance of your model. This became very important during the COVID-19 pandemic era because, just for example, people's purchase patterns changed dramatically during the pandemic period. I don't know about Australia, but in the US, you couldn't find toilet paper anywhere in the store. You couldn't get certain things; they just sold out. Purchase patterns dramatically changed, so the predictive models of the past no longer worked. For example, if you had seasonal models that predicted certain customer purchase patterns based on the time of year, they just didn't work anymore.

So, MLOps got a very significant increase in terms of the number of companies who were building and selling it and the number of people who were wanting it. It's basically monitoring two things – **data drift** and **concept drift**. Data drift means the input is drifting. The types of data you're collecting now are different, so obviously, the model you're going to build is going to be different. But concept drift means that the output variable you're predicting is different. The data may be the same – for example, you might be looking at customer purchase logs at a store. You're not changing the data. But the thing you're predicting is different because people want different types of things than what you thought they wanted in the past.

MLOps is about monitoring a model for data drift but also for ethical and trusted AI – that is, is there any bias in the model? So, MLOps consists of the following questions: "*Is the machine learning ethical, trusted, responsible, and transparent, and is it consistent with the data and the concepts you're modeling?*"

There's another thing called ModelOps, which is almost the same as MLOps, but it includes all models in a business. Some models in your business are not AI at all. You can have predictive models of customer sales patterns that have nothing to do with AI. They may just be to do with statistics. That broadens the idea beyond just machine learning models to any kind of decision models you use in the business – how do you decide what products to sell and who to sell them to? What should you have in the warehouse, and what should you move out of the warehouse? There are all kinds of questions that come up in business, and they don't always require a machine learning model.

Then, there's another thing called AIOps. I just thought AIOps was the same as all the things I just said applied to AI, and I was also wrong about that one! AIOps is actually used for IT infrastructure monitoring. It's basically a cybersecurity tool used to verify proper network behavior, looking at cyber incidents, weaknesses in the cyber defenses of your organization, or even whether IT infrastructure needs updating. IT monitoring and IT observability are the roles of AIOps. It's watching all those things for emergent problems. There's AI in there because it is doing pattern recognition and anomaly detection. I've said that what we really need is AIOps that involves AI that watches itself – keeping an *eye* on the *AI*!

Anyway, all those things have stolen the Ops part of the DevOps paradigm, but they all have slightly different meanings in actual practice, and it took me a while to figure out that the differences in the meanings were more significant than I thought.

Structuring teams

AA: I often hear a lot of debate internally in organizations on how they should structure their data science capabilities. Some people are huge proponents of centralizing, while others prefer decentralized structures. Obviously, it depends on size, scale, and a number of factors, but in general, do you tend to have a preference on how you guide organizations or structure your own teams, based on reporting lines, skill sets, and so on?

KB: The answer is, it depends. There's no one right answer to that for different organizations.

On the one hand, I'm a big proponent of decentralization – that is, people should be part of business units and not a separate unit. As a data scientist, I can criticize my own group and say that, often, data scientists like to play in their sandbox. They think, "*I've got my cool tools, my cool toys, and my cool Python code on my laptop. I like tweaking my models and tweaking the data. Oh, I've got a new dataset here. Let's try that. Let's see what happens.*" I've got that in my blood because, as an astronomer, I would go and collect my data.

I'm not an active astronomer anymore, but I used to collect data from a telescope and I would play with it forever, just to find a pattern, discover an interesting insight, and try to build a model to explain it. I was just like a lot of other astronomers – we played with our data day and night.

You don't want a team that only plays with data, but you do want an aspect of that because you never know what surprises you're going to discover in your data. There's a whole history of stories in the data science field of surprising discoveries and unexpected things that people have found in their data. You need to have some of that going on, but you also need people to be embedded across the different business units.

Here's an example – I like to think of it as an analogy (although maybe it's a bad analogy). Think about the World Wide Web. Who talks about the World Wide Web anymore? You don't even hear those words anymore. I was at NASA 30 years ago when the first web pages were coming out, and I remember my first reaction: "*That's crazy. Who's ever going to want to use this?*" I just thought it was silly. But once we got over the idea that it was silly and saw that it was actually useful, every organization had a World Wide Web team working on a World Wide Web page for the company. If today you thought that that's the way we should run a business – that we should have a World Wide Web team over in the corner – people would just laugh!

We laugh at that because it's ridiculous. We have a web presence. Our company is present on social media. We have an app. There are all kinds of places where you're present; it's not managed through a central World Wide Web team.

I think at some point, we're going to get to a point where the AI/machine learning/data science part of the business can say, "*This is our presence in this space and how we do business.*"

> Data science is not a thing to do but a way of doing things.

On the other hand, there's still value in people exploring data to find that unusual, unexpected thing in it. I could tell you many stories there. But my favorite one is of a famous retailer in the United States – one of the biggest retailers in the United States. They analyzed their data prior to the arrival of a hurricane/wet season in Florida. There are multiple hurricanes that come up every year – big typhoons, as I guess you would call them in the South Pacific. They're pretty devastating – extremely strong winds, flooding, and all kinds of damage. A lot of times, people lose power, and all kinds of terrible things happen. So, the retailer wanted to discover the main products that people really wanted to purchase prior to a potential natural disaster like that. They found an amazingly huge increase in the purchase of one item in their inventory prior to the arrival of hurricanes. It was a totally unexpected insight discovered in their data. The item was strawberry Pop-Tarts.

Managing data scientists

AA: Something you touched on, which I'd love to dive into a bit deeper, is that data scientists are unique in the spectrum of employees. What do you think is unique about managing people like us, and how do we attract and retain staff? I'm noticing – in Australia, at least – that there is definitely a talent shortage. I'm having conversations with people daily who are asking, "*Where do I find good data scientists and data engineers?*" **A lot of organizations are struggling with attracting and then retaining people with unique skills.**

KB: I'd like to hear *your* thoughts! I don't have a solution!

It is a serious problem, and I think there are a couple of things that I've seen happening. One is, of course, people can "*become a data scientist*" just by taking some kind of training course. You don't need a college degree. It's possible to fill some of these gaps by just training existing people in your organization. They don't need to go off and get a 4-year college degree before they can be hired to do a data science job.

Now, I have also warned young people about this. I've said, "*Yes, you can just take a training course online and get hired by a company with a nice salary, but if you want a lifelong career, you need to have more foundational knowledge and foundational ways of learning and critical thinking, which is what a college degree can help you get.*" It's not the only way, of course, but it provides you with a much broader base on which to build a career. You can't build a career by just learning one specific programming language, because there are so many. There are several programming languages that I used in my early days that are rarely mentioned these days – Fortran, Forth, and VAX VMS, just to name a few.

AA: I started in Pascal.

KB: Yeah. Obviously, Pascal came in after Fortran. I kept using Fortran for a long time.

I was writing a summary of a report recently about all this. A company did a large survey of companies around the world about AI, the AI talent shortage, deploying AI, using AI, operationalizing AI, and more. They asked me if I could just write a summary of the report for them to post online to promote it. One of the findings was that a lot of companies have stopped training people in AI and machine learning, and the reason is that they realize if they do that, most of those people will leave for a better-paying job somewhere else. Why spend money on training people who are going to leave? I don't see why that's a problem because every company I ever worked for always had a clause that said if you leave within 1 year of the training course, you owe the company the cost of the training course. It's written into the contract that if you take that course, you will pay the company back if you leave within *x* amount of time. I don't know if they'd be afraid to do that. But in any case, there's a degree to which companies have stopped training because they realized they're just going to pay for the training, and then the people will leave. And so, we're in a bad situation right now – how do we solve the problem?

That's why I don't have a good answer to your question.

> I think we're stuck here because, unlike other industrial revolutions, which probably took decades to evolve into changing the way business and work are done and the kind of work people do, this one is not giving us any time to retool either ourselves or our businesses to keep up with what's going on. The rapid (and accelerating) pace of new technologies is ridiculous.

AA: I think so too, and there's the boom in data science becoming the sexiest career. There's such an influx of people who are trained in one thing (such as machine learning) but lack the fundamentals in another, so aren't always suitable for particular roles, and that's often not spoken about in my experience.

KB: I don't want to sound like an old curmudgeon here, but through the hindsight of a 40-year career, I can see that aptitudes are more important than specific skills, and aptitudes include *communication*, *critical thinking*, and *collaboration*. I came up with a list of seven Cs, which I called **sailing the seven Cs**, and then it ended up being 12 when I had to add *curious*, *creative*, *computational*, *consultative*, and more (including *continuous* life-long learner). I had all these Cs in there. After a while, I was just forcing the letter C!

One of them was *compassion*. What I meant by *compassion* was having empathy with the person you are working with because, in data science, you're often working with business executives or business leaders. They're not going to know your language and you're not going to know their language, so things can't be one-way there.

Another of the Cs was *consultative*. I did decades of consulting work, and what consultative means is basically being a good listener – that is, finding out what it is your client wants. Don't just try to sell them your fancy, shiny object. Listen to what they want and then deliver what they want, not what you want. Forget, "*Oh, I've got this fancy new deep learning algorithm with this fine, multilevel, convolutional network.*" The customer just wants to optimize sales in Western provinces!

So, I decided that *consultative* needed to be included. Young data scientists especially are very eager and excited to learn the cool stuff – they want to talk about it and want people to use it. But they need to take a step back and make sure they deliver what the customer wants, not what they want the customer to want.

At NASA, and in system engineering more generally, there's a saying: "*Don't just build the system right but build the right system.*"

Why do AI projects fail?

AA: I don't know if you're seeing the same thing, but I hear a lot of statistics around failure rates in AI – something like 80 or 85 percent. Assuming you agree that there's some truth in that, what do you think is the main reason? Technology is cheap and easy to obtain. Data is everywhere – we're drowning in it. Where are organizations going wrong?

KB: Well, I'd like to know the origin of that number, first of all. As I said earlier, I believe strongly in the "*fail fast to learn fast*" principle. If you fail four times out of five in order to get something right one of those five times, then you have an 80 percent failure rate, but I think that's acceptable. If you just deploy the first thing you build, that's probably not going to be a good idea. You probably want to find the failure cases, or what they call edge cases. A perfect example of that is self-driving automobiles. There are edge cases where very unusual things could happen when you're driving that a thinking, logical person would know how to react to, but your algorithm hasn't been offered that training example. Your model's going to fail if it's never seen that training example before, even though you as an intelligent human driver would know what to do.

So, if we measure the failure rate just by the number of times we've built failed models, I think that 80 percent is too small. Maybe we shouldn't be failing as much as Thomas Edison – we shouldn't be failing 99.9 percent of the time. But I sometimes tell young data scientists, just to tweak their minds a little bit, "*When you come for a job interview as a data scientist, do you bring a portfolio of the projects you've completed and your success stories?*" They all say yes. I say the following:

> "*When I interview you, I want to hear about how you failed. Tell me how you failed, what you learned from it, and how you got out of failure mode. That tells me more about you than you showing me all the cool things you've accomplished.*"

It's like *The Beauty Show* – you always see the most beautiful aspects of the individual. There's a lot more to people, as we all know, than the things that we want to show people. There's a lot more to us than that. If you truly love someone, then you love all those parts of that person. That's why I say you should love all your models like you love all your children. They may have their faults, but you love all of them. I love all your models, and I want to hear about those lovable models, even where things fail.

Here is one of my all-time favorite stories – I was working with a team of young data scientists. I was there in an advisory role, so I wasn't actually developing the code. They were just building a predictive model for a client that predicted performance over the course of a year. There were 30 different metrics that they were measuring, and they were predicting what the performance would be of these 30 different business metrics over the course of the next year, based on past performance. They invited me to review the model with them before they showed it to the client.

We were in a business lounge (this was pre-COVID-19 – we could actually get together in person). They were sitting on one side of the table, and I was sitting by myself on the other side of the table – it was a team of three or four people – and behind them was a screen where they were projecting what was on their laptop on the screen so I could see. While they were talking, I was looking at the screen. They had a spreadsheet up there of these 30 variables over the 12 months that they were predicting. Essentially, there was a table of 360 numbers – 30 rows of 30 different business metrics that were predictions for the next 12 months.

It turned out they were all completely maxed out. Each one of the 30 metrics had a specific number – it wasn't 100 percent but rather the equivalent per metric – and it was the exact same number across that metric's entire row. Each row had a different number, of course, because each row had a different metric, but in each row, every number in the row was exactly the same number. That's what I was looking at.

They were explaining their model, talking about Holt-Winters modeling, using seasonal terms and trend terms, talking about autoregression and ARIMA models, and so much other technical stuff. I cut in to say, "*Wait a second. I'm looking at your table. Is that just an example of the type of data that you're modeling, or is that the actual output of your model?*" They said, "*Oh, no. That's the real output.*" I said, "*Well, how come all the numbers are the same in each row? On any given row, the number doesn't change.*" They said, "*Oh, because the metric reaches the maximum value every single month.*"

I said, "*Why are you building a predictive model if you already know that it reaches the maximum possible value every single month?*" They just stopped, looked up, and said, "*Oh, you're right. I guess we didn't need to build this model.*"

I didn't mean it to be offensive! But they got so excited with the tools they were using, the math they were using, and the Python code they were writing. I get it. I used to write Fortran code to do even the silliest things, like getting a list of the square roots of numbers. I could have looked it up in a book, but it was fun to do, so I get it. But on the other hand, you have to think, "*Wait. We're talking about delivering to a client, and there was a very simple answer to the question, which is, the peak value of the metric for each month is the same value as it was last month because it was at the peak last month too and will be going into the future.*"

Obviously, I'm being vague about this story, of the client we were talking with, but it was very clear that there was a little bit too much enthusiasm and too little critiquing of the approach. I haven't talked to those people in years, but I'm hoping that when they go for job interviews and they are asked about how they failed on a project and what they learned from it, they bring up this story, because that to me is more valuable for me to know about that person.

I learned this in my own career. That's why I'm saying this. When I was very young, I was very afraid to admit failure and admit that I didn't know something.

> Believe it or not, most of us have had imposter syndrome at some point in our careers, and I really felt that way.

I really felt like I was surrounded by all these brilliant people when I went to graduate from school at Caltech – Nobel Prize winners all over the place. I was very intimidated. I got to NASA, and all these incredible people were doing all these amazing things, and I was just some young guy. I was afraid to admit the fact that I didn't know something or that I was doing something wrong, but in hindsight, it's more valuable to the team to do that. It also builds trust with people because they realize that if you can admit failure, they can trust you. I want people to feel comfortable admitting that.

It's really the aptitudes of individuals that I care most about because you can be taught skills. You can be taught languages. You can be taught techniques and algorithms, but these aptitudes of who you are, how you work with other people, communicate, critically think, and so on are all far more important.

Building an effective data culture

AA: One thing we touched on earlier is the importance of a data culture. What are your thoughts on the key elements of an effective data culture? I've seen many that are ineffective, and many cases where organizations just don't really have one. Have you seen cultures that have gone awry or ones where they actually have done it the way you'd love it to be done?

KB: I'll tell you two specific stories.

I'm not going to repeat myself, except to say that you need to fail fast to learn faster. That's part of a culture – that culture of experimentation. But I will say two other things. One is related to cultural experimentation.

There's a casino in Las Vegas. Casinos like to keep customers in the house – that's their business. If you can keep the customer in the house, that's good for business. This particular casino – one of the larger ones in Vegas – has an employee culture that can be described in simple words – "*Test, or be fired.*" What that means is that you should test to see what the customer likes. What's going to keep them there? What drinks do they like? Do they like to go to shows? What kind of shows? What kind of games do they like to play? When do they like to play specific games? What is it that they like in terms of food, drink, entertainment, and gaming? Test each customer, find out what they like, and then maximize that customer's enjoyment and experience to keep them there.

That's important in casinos. It's very hard to find an exit there. They also don't have clocks on the walls, so you have no idea what time it is. They have no windows, so you don't know if it's night or day. They have all kinds of tricks up their sleeve, but I really love this example because their employees are empowered to use data to improve the experience of their customers, so much so that if they don't do it, they can be fired. That's really putting your money where your data is, or indeed, putting your data where your money is. It's about empowering your employees to do that experimentation. Give a free drink to someone and find out the results. Did that help that person? Did they stay longer?

We're talking about an industry where the more money you spend, the better it is for the house, of course. But still, it's an idea that can be applied to any customer interaction. Find out what your customer likes. Test and experiment, whether you're in retail or sales – it could be anything. So, that's one thing – "*Test, or be fired.*"

As for my other story, I heard a talk from the CEO and president of a European airline. It was a very small airline – I'd never heard of it before because it's regional within Europe and not one of the big ones. He gave this talk about the culture of his business. He said it was a culture of data democratization.

He described it in very simple words – again, like the "*Test, or be fired*" statement. His statement was, "*If you see something, say something.*" That is to say, if you see something in the data, say something.

I like that. Have you ever been to the New York City metro subway system? There are signs everywhere that say, "*If you see something, say something.*" Now, that's about safety and security. That's a different idea, but it's the same words – "*If you see something, say something.*"

The idea was that a lot of times in businesses, people will say, "*Oh, that's not my job. That's the data scientist's job. That's the data person's job. My job is to do this thing over here.*" But this guy said, "*No. We're all responsible. We're all touching customer data. You may be dealing with customer flight schedules, customer online complaints, the customer call center, customer benefits, or the loyalty program.*" It's like in the casino story – what kind of things can we give our customers to keep them there?

The CEO's story goes like this – there was a database engineer in the company whose job it was to create the flight manifest. The flight manifest is the list of passengers flying on each flight. They would give this list to the flight attendants before the flight so that they'd know who was flying. There was specific information about each person there – whether they were underage, whether they were a member of the loyalty program, their date of birth, and so on. This database engineer's job was to create this flight manifest. It wasn't his job to interact with customers or anything like that. But he was looking at this thing, and he saw one of their frequent-flyer loyalty customers. She was flying with a guest passenger – if you fly enough miles, you can get a free flight for someone to fly with you. It turned out the person she was flying with was her 10-year-old daughter. He spotted this information on the sheet. The thing that stuck out to him, though, was that the girl's birthday was the same day as the flight.

I challenge my students here. I say, "*There's both data and an algorithm to be had here. The data is the date of birth and the date of the flight, and the algorithm should say that if those two things are equal, then…*" That is an algorithm, believe it or not. It doesn't have to be a deep neural network. That's an algorithm – "*If a equals b, then do something.*"

In the past, this database engineer would have done nothing because it was not his job, but because the CEO, the president of the company, said, "*If you see something, say something,*" he was empowered to say something. He brought it to the attention of the flight attendant, and said, "*Hey, look. This little girl who's flying with her mother on this free travel pass has her birthday on the same day as the flight.*" The airline steward and the flight attendants went out and bought a little cake with a fake candle (as you can't light a real fire on planes). I don't know what it cost – probably just a couple of euros. During the flight, they came out with the cake and the little fake candle, and they sang *Happy Birthday* to the girl, and all the people on the plane joined in. They made videos of it on their cell phones, tweeted about it, and posted it on Facebook. They estimated that the positive customer sentiment rating (there's a company that rates the customer sentiment of companies) went up by 1 percent for this airline, just from that one single event, which translates into about 1 million Euros in additional annual revenue – that's the way the metric works.

And what did they spend? They spent 2 euros for a 1-million-Euro ROI, all because a database engineer saw something in the data and said something about it – something trivial, which he normally would have ignored, but he was empowered to say something.

That very simple story illustrates to me the power of empowering your people by saying to them, "*You might see something in our data – it could be customer data, it could be cybersecurity data, some peculiar behavior on your network, or anything else. You might say, 'Oh, that's someone else's responsibility.' But it's your responsibility. If you see something, you should say something,*" and I really love that. It's a very simple way of democratizing data usage across an organization, rewarding people, and giving accolades to people who follow through and do things like that. They're not always going to be major ROIs for the company, but they're going to be enlightening, empowering, and encouraging. It goes back to "*think big, start small.*" If you empower people with these small victories, they'll feel more excited and become advocates for bigger projects. Don't start with the big AI project that just might fail. Start with the small things. And again, just use a very simple algorithm – if *a* equals *b*, then say something. If *a* is greater than *b*, do something.

I teach my students about a lot of examples where companies just set thresholds for a certain number of web clicks or customer purchases, and then do something in terms of interacting with the customer or an offer for the customer, and it returns a tremendous amount of value to the business, just because they have a little alert system set up where you set a threshold. That's a very trivial thing – if *a* is greater than *b*, then do something. It doesn't have to be complicated.

Teaching data science

AA: Speaking of students, one of the reasons you left NASA was to establish the world's first data science program. What do you think are some of the key elements of a modern program these days? What should we really be teaching students that we're not teaching them at the moment?

KB: That's a really good question because I've been involved in a number of panels where they've discussed this, and some of them have gone on for days, and some of the reports have gone on for dozens of pages. You can get really into the weeds with this one.

The key elements are the higher-level things. There are mathematical foundations. There's modeling. People underestimate the power of modeling. A model is a representation of a thing. It's not the thing. We use data as a representation of a customer. The data is not the customer. What you purchase is not you; it's a representation of the things you like or are interested in.

So, there are concepts around modeling and data literacy itself – do you understand data, data management, data formats, and the mathematical concepts that are involved? Then, there are computational languages. At university, we used MATLAB as the language because that was the one common language. It was straightforward for students to learn, and it was used in almost all the sciences, even looser ones such as political science. When people were building political policy models, they were using MATLAB. Students were learning one common language that they could use in all of their classes. It wasn't Python, although our advanced students learned Python.

But the one thing we thought to put in at the very beginning was a course on data ethics, and that was my favorite course to teach. I think we had all the right pieces there at the very beginning, and I really loved teaching that course. I used a little book that was about 70 years old. It's very outdated. It uses ridiculously old, almost offensive language in some places, but if you get around the language they used 70 years ago and focus on the statistical aspects of the book, it's invaluable. The book is called *How to Lie with Statistics* by Darrell Huff. It contains simple examples, the kind of thing you see in the news – "*Four out of five dentists recommend this dental product.*" Well, how many dentists did they ask? Did they ask five dentists?

Students would come to me and say, "*Why should I take your course on data ethics?*" I would say to them. "*Take my course and I'll teach you how to lie.*" They always signed up for my course after I said that.

And I really did teach them that. I said, "*I hate to say this, but the best way to teach AI and data ethics is to give examples of how not to do it.*" Sometimes, I go to conferences and find people talk a lot about these topics and it can be preachy: "*We need to do the right thing. We need to do the right thing.*" Yes, I get it, but when they give me an example of how something failed, I can connect with that. I can say, "*OK. I now know not to do that. This is something that I need to be careful that I don't do.*" Instead of just someone preaching at me, I want to hear the failure examples because I don't want to have to learn from my failure. I want to learn from your failure. We learn from our mistakes. I think a good ethics course goes through those cases, so we can see them and learn without having made those mistakes ourselves, as I think it is a good thing that we don't repeat those mistakes.

So, I taught that course, and I gave all kinds of good examples. I hate to say it, but I really like to play with my students. (Actually, I don't hate to say it.) I gave a great example once, which was a true story – I was at a conference and I went to lunch with two other guys, and the average net worth of the three of us at that lunch table was over 1 billion dollars. In fact, I would guess the average was over 2 billion dollars.

I asked my students, "*What do you think about that?*" Some students said, "*Wow! You university professors sure are paid a lot of money!*"

AA: And I guess you replied, "*You haven't taken your statistics course?*"

KB: Exactly!

Invariably, there is somebody in the class that says, "*Oh, there must be an outlier.*" That is correct.

The outlier was a chief architect in a major tech corporation, which is where he made his billions. He was very interested in astronomy, and after he left his big tech job, he decided he would build observatories around the world for use by students in middle-school and high-school classrooms – robotic observatories that classrooms could access from across the world. He was at this conference where we were talking about using astronomy data as an outreach tool to help students learn math and science. It's basically a STEM outreach tool because most kids love dinosaurs and stars. Well, a major asteroid impact caused the dinosaurs to go extinct, so we had got both bases covered.

Anyway, I went to lunch with him and some other university professor. I have no idea what the other guy was worth – not very much. But I was able to find out the net worth of the billionaire.

So, I took his net worth, added zeros for the other professor and myself – 5 billion plus zero plus zero – and divided by three. The average is over 1 billion dollars. I explained that to my students. You'll see things in the news: "*Hey! The average net worth of the people who went to this event is over a billion dollars.*" Well, yes, but that's because it was Bill Gates and 100 poor people, or Elon Musk and 100 poor people – it's still over a billion dollars. You've got to understand statistical concepts to navigate the modern world, so we threw that into our curriculum.

Predicting the future of AI

AA: Here's a final, high-level question – what are your predictions for the future of AI? What do you think business leaders should be doing to help prepare for it? It could be shifts in employment patterns. It could be the commoditization of technology. Where do you think the industry's heading?

KB: All of those.

In the last few years, when people have asked me about the future of AI, I have always given the same answer, and the answer is convergence.

I see many different technologies. In this past year, we've heard a lot about the metaverse. What is it? It's **augmented reality** (**AR**) and **virtual reality** (**VR**). We've got AR, VR, and quantum computing. We're seeing a lot of growth in the internet of things, and we've seen how 5G and edge computing are going to change how we move intelligence to the point of data collection. AI is moving to the edge. It's no longer necessary to bring all your data back to a cloud service, process it, analyze it, and spend 6 months working on it. You need the answer now. You need the answer at the point of data collection, and that's intelligence at the edge. All these different technologies are converged.

AR and VR in the metaverse require real-time interaction. Now, if you're interacting with things in a virtual environment, you can't just touch something and then wait for an hour for the thing to respond. If you push a virtual button on a virtual machine in a virtual reality environment, you want it to respond now, not an hour from now.

I'm very big on autonomous exploration – mostly in deep space, but it can be deep sea, or it can be what they call robotic journalists who go into war zones or disaster areas and create stories. These are drones or robots that are not just taking pictures but actually creating stories. We have really deep language models now that can generate stories from just a minimal amount of input – they can actually create a story based on an image. Because pattern recognition and object detection are so sophisticated now, you could actually write a story because an algorithm can see what's in the picture and tell its context.

We'll start seeing this kind of convergence of all the technologies – high-speed networks, different immersive realities, the advent of quantum computing, robotic devices everywhere, and even 4D printing. If you haven't heard about 4D printing, you should check that out. We all know what 3D printing is, but what's the fourth dimension? It's time. It's shape-shifting technology. 4D-printed objects can actually change shape in response to signal inputs. A device can actually change shape in real time in response to these inputs.

The example I heard about was heart stents. They put a heart stent in a cardiac patient. Traditionally, it is a static thing – the doctors design it to be the right shape, they insert it into your arteries, and then they seal you up. Well, now they can use carbon nanotubes with piezoelectric sensors that can respond to your body's metrics in real time – that is, your blood pressure, your body temperature, your stress levels, your hormonal levels, and all kinds of stuff that a heart stent needs to respond to in real time. It's obviously not trivial work. It's not trivial technology, but it's something that can be done. The stent can detect the data from sensors, and it knows how to translate that sensor value into an impulse for the piezoelectric device, which sends a signal to the carbon nanotube that changes its shape in real time to the right shape, at the right time, for the right heart conditions. Wow!

So, I say that the convergence of all kinds of crazy technologies is going to be the future of AI. It's about bringing all these things together – a convergence of data insights, knowledge, understanding, optimization, prediction, prescription, and detection.

> Think about all the words you've used to describe your field in the last 10 years. Put them all in one long run-on sentence, and that's the future.

Summary

It was fascinating to learn about Kirk's career, and his transition from astrophysicist at NASA to "accidental" global influencer in data science and AI, who just loves sharing information and knowledge with people.

He shared some pragmatic but valuable advice on success with AI, with the key takeaways being "*think big, start small*"and to focus on putting the right people in the right roles. Remember, as Kirk explains, that the purpose of data science is to understand the world through data so that you can make better decisions and take better actions.

We also discussed the importance of having the right skill sets for moving from research and development to production and operationalization, which can be a challenging transition for some organizations. Paradigms such as MLOps and ModelOps can play an integral role in helping to monitor data and concept drift – and ensuring the models are ethical – at scale. Kirk also advocates for the need to balance pure discovery and experimentation with business-led development to ensure innovation and strategic value.

Kirk also offers some cogent advice about the human attributes needed for a successful long-term career in the field, such as having compassion and being consultative, and understanding that we all have imposter syndrome at some point in our careers – I know I have. As Katherine Maley (*Chapter 7*) also discusses, foundational knowledge is imperative – specifically, mathematics and statistics skills – as are critical thinking skills. Kirk also agrees with Cortnie's views (*Chapter 2*) that skills can be taught but practitioners need to have the basic aptitudes and human skills.

Senior leaders must also redefine what they mean by "failure" in the field. Like Igor Halperin (*Chapter 13*), Kirk stresses that high failure rates are the norm when testing our new ideas and developing data science models, so they aren't the most appropriate metric to measure success or a lack thereof. He also made a point of the importance of empowering all staff to test and make observations from data.

I agree with Kirk's views that the future of AI and technology lies in the convergence of technologies – bringing things together to unlock new opportunities. I think an integral part of this convergence includes humans and machines working together.

9

Nikolaj Van Omme Can Solve Your Problems

Nikolaj Van Omme is a technically skilled and experienced AI leader. He is **chief executive officer (CEO)** and co-founder of Funartech, a company which creates new AI algorithms to solve industrial problems. He has expertise in both the mathematical and computer science foundations of the field and so is very well-placed to offer valuable insights from a technical perspective.

He also happens to be a dancer, singer, and actor!

He's a global pioneer in the emerging field of combining **machine learning (ML)** and **operational research (OR)** – a branch of mathematics focused on solving problems to inform decision-making – and so is in a unique position to discuss the merits and applications of this exciting approach.

I've known Nikolaj for a while, and I really enjoyed finally having an opportunity to get his views on a range of important and timely topics.

Getting started

Alex Antic: How does a singing and dancing mathematician become a game-changer in AI?

Nikolaj Van Omme: Arts and science are closely related. Music, for instance, is really based on mathematics (look at harmony) and in AI, mathematics, and also science in general, imagination and creativity – which are typically associated with the arts – are key.

I guess you can say that I'm a curious person who loves to discover and embrace new fields, not only in AI or mathematics but more generally in all kinds of disciplines such as arts, sports, philosophy, epistemology, or you name it. In every field, there are people who have done remarkable things. I'm always fascinated to see how craftsmen work on their art and how meticulously they solve problems with their ingenuity.

In the past, this was the norm. Nowadays, experts are more and more narrowly specialized, and experts from different fields – and this trend is accelerating – don't talk to each other, don't share their knowledge, and very often reinvent the wheel.

Being so curious and having been trained in both arts – I went to music school as a kid – and science, it was so natural for me to be interested in and even passionate about a wide range of domains. So, the idea of combining different domains and expertise was natural to me.

In a nutshell, this is what I'm doing in AI. I'm taking fundamental ideas from one field and bringing them to another, particularly from ML and OR, but also other mathematical fields.

Pivotal moments? There are so many. Here is a funny anecdote. I never imagined that I would become a mathematician. I met someone in the street, and we spent the whole evening and night in a coffee shop doing mathematics. It was fun, and in the early morning hours, that person told me that I should become a mathematician. I registered at the university to do mathematics. His name was Pierre Bieliavsky, and he is still a mathematician today. He also convinced his father – a biologist at that time – and his girlfriend – a professional pianist – to become mathematicians! A very dangerous guy!

Maybe the second pivotal moment was to understand that mathematics is not what is taught at school or university. Mathematics is about finding relationships between ideas and how they relate to each other to solve real problems. At school/university, you learn about the tools that were developed to do that but rarely about the how and why. Why were graphs invented? Or complexity theory? Or the simplex method? Or the backpropagation algorithm? How are these tools/methods related to the culture/society in which they were invented? What kind of problems were people trying to solve? What were the ideas that people tried, and why and how did they come up with such ideas? Why were some ideas kept or rejected? What were the accepted paradigms?

Assessing the progress of AI

AA: How have you seen AI change from when you began in the field to what AI is now? There's been an increase in the hype, but there have also been more positive elements in terms of how people generally accept and think about AI and ML, which has made our lives much more fun, given that we can play in this field.

NVO: First of all, I should say that for me, AI is probably broader than most people think. I'm more focused on problems than solutions, which means that AI is more like trying to solve something than using the typical methods considered AI.

For instance, for me, OR is really part of AI. Actually, you can throw everything at AI. AI, for me, is not the description of some fields. It's more of a way to try to solve problems.

To answer your question, I saw that the big field that is OR was getting a lot of hype (but not as much as today) in very particular niches in some industries, and then it went away. It faded away, which was really strange for me.

ML was considered a subfield of OR 10 or 15 years ago. There was a consensus about that. Strangely, in the 1990s, you already had wonderful results from ML (in particular, **deep learning (DL)**), but people were not catching on or catching up. They didn't believe in ML, and lots of people went away, particularly from the DL field. We have those three giants in DL – **Yann LeCun, Geoffrey Hinton**, and **Yoshua Bengio** – but there are only three because all the others left.

So, what I saw with my own eyes was something really strange: something that really worked in OR was in use, and then it faded away. After that, you had some ML results, which faded away. It took 20 years before people started to use ML, and now there are these new trends and this hype. Now, the interest is huge – it's probably too much, in the sense that there is huge hype and marketing around AI and ML, and DL in particular. I personally believe – I don't know if you can write this in your book – that it's going to be a little bump in the sense that ML is probably not the best way to do AI. It has its place – it's very nice, and you have incredible results – but still.

> I don't believe that the "intelligence" in "artificial intelligence" is going to be reached by ML alone.

So, if you see AI as a whole as being about ML, I think the bump is going to go down again, and I also believe that people will say, "*Oh, what did we do? How come we were so crazy about that idea of DL?*" That is really nice, but that is so simple. You have hypes: they come, and they go.

ML and OR

AA: ML, and DL specifically, has its limitations, and one way to overcome these is via the integration of ML and OR, allowing us to combine both data and domain knowledge – you're one of the leading advocates of this.

Can you please give a brief overview of OR, its history, how to integrate it with ML, and why this combination has so much potential?

NVO: All approaches have strengths and limitations. Most of the analytical approaches are somewhat related to each other. In theory, you could consider how to transform a problem so that you could obtain the same kind of solutions by one or another approach, but practically, each approach has its pros and cons. Sometimes the differences are huge; for instance, if you try to optimize with ML or OR. Also, the approaches and goals are somewhat different. For instance, ML relies heavily on data and you hope to find patterns through automatic study (optimization of parameters and hyper-parameters), while OR is more focused on gathering domain knowledge and modeling it mathematically.

Operations research is one of, if not *the* best, sciences of analytical optimization. You can retrace optimization problems and the mathematical approaches to solve them since the beginning of mathematics, but most people agree that OR really started with the Second World War and the solving of logistic and operations problems for the military. For instance, a very well-known problem from that time is: what is the optimal number of submarines you need to sink one or several ships?

Too few and they might be destroyed by destroyers; too many and they could hit each other (the answer is three). After the war, OR was still used in the military (which is still to this day probably the biggest employer of OR experts) but also in big corporations to help make business decisions with mathematical insights and to optimize operations.

OR has since become a huge, mature field of applied mathematics. One of the sub-fields of OR was ML. Nobody disputed this 10–15 years ago. ML was basically using OR methods to optimize its predictions. Now, ML has also evolved with the addition of other methods and approaches and has become a detached field from OR. But the basic principles of ML all come from OR and are also used in other OR sub-fields.

ML and OR are really complementary. If we look at the limitations of ML that can be overcome by integrating it with OR, we can certainly mention the following:

- **Optimization**: Some people try to optimize with ML, but that is not always a good idea. OR is usually a way more powerful way to approach optimization. Take the **traveling salesman problem** (**TSP**). ML experts – as far as I know – can solve it to optimality for instances of about 100 cities. OR experts optimize the TSP with about 100,000 cities. And we are talking about a **nondeterministic polynomial time** (**NP**)-hard problem, so a problem that is really hard to solve and for which an algorithm to find a solution with 100 cities probably cannot solve the problem with 200 cities.

> **What is the TSP?**
>
> The TSP seeks to find the shortest and most efficient route between a set of cities or towns that a salesperson must visit exactly once before returning to their starting city. The goal is to minimize the distance traveled by the salesperson in order to reduce time and costs. It is a classic algorithmic problem in the field of computer science and OR.

- **Control**: It is notoriously difficult to control ML algorithm outputs. For instance, use a **neural network** (**NN**) to decide whether a picture contains a cat or not. Most of the time (if the problem has been addressed in a clever way), your system will tell you that, indeed, you have a cat or not with a given probability. But then, once in a while (sometimes very often) you'll give it a giraffe and your system will tell you that it is a cat with a probability of 1. Basically, you don't have very robust systems in ML, while in OR, you can model a problem with constraints that can be mandatory (hard constraints) or that you would like to be obeyed but that the system could bypass (soft constraints). We are currently working on a chatbot where we do have control over its output so that it cannot go berserk, something that is difficult to achieve with ML-only methods.

> Most ML algorithms need a lot of data and a lot of very *clean* data. This is extremely difficult to obtain in general.

Or the real data is slowly but surely drifting away from the kind of data used in training. With OR, you can start right away without data or even with data of bad quality. We once had data of very bad quality, but we could provide solutions of high quality. This is an exception, but it is sometimes achievable depending on the problem at hand.

> The saying "*garbage in, garbage out*" is only valid for ML.

Related to the previous point: ML methods can only reproduce existing data. If your data substantially changed or something new comes along (such as COVID, for instance), your results are meaningless. With OR, there are techniques to be able to deal with such changes or unknowns. Also, if you want to invent new things, ML is not really suited for that. With OR, you really can invent totally new approaches. I'll detail an example in the next point.

With ML, you don't have much insight into a global optimum. Most of the time, you converge towards a local minimum. In OR, you might be able to have insights about the global optimum. This is the reason why OR is so efficient at solving the TSP and ML is not. People have tried for many years to come up with a method to gather local solutions and construct global solutions, and this very rarely works.

Here is an example of the different approaches of ML and OR to tackle the following problem. A company wants to improve its sales by 20%. The ML approach will be along the lines of gathering the data, looking at what the best salespeople do, and somehow mimicking their behavior. This is wrong on so many levels. First of all, you cannot impose techniques coming from good salespeople on your "bad" salespeople. That does not work. Second, the difference between good and bad salespeople is relative. Maybe your good salespeople are good because they deal with the best customers and the best deals, not necessarily because they are good *per se*. Third, salespeople typically try to get the highest bonuses. Most of the time, this is aligned with the interest of the company, but not always. In fact, it can be shown that you can improve the sales for the company globally, but for that, you might need to collectively lower the bonuses of the salespeople! Fourth, new selling techniques might be even better. If you have never experienced those in the past, how can ML find them?

Now, as I said, every field has its shortcomings, and OR is no exception either. ML has strengths that you can use to help OR algorithms perform better.

There are at least four ways to combine ML and OR:

- ML and OR as separate black boxes. This is the most common approach whenever companies propose to combine ML and OR. Often, ML is used to predict, and OR is then used to optimize those predictions.

- Use OR to improve ML. OR can be used to optimize (minimize or maximize) some functions in ML. But wait, isn't this what ML is about? Exactly, ML is strongly based on OR optimization when it optimizes its predictions.

- Use ML to improve OR. The strength of ML is to predict. OR is mainly – but not only – rule-based. When the rules apply, it is hard to beat. Often, these rules don't apply completely and there are some indecisions and some imprecisions. This is where ML can help by reinforcing or avoiding certain rules.

- A complete hybridization of both ML and OR. This is a completely new approach with new algorithms.

The fourth combination is the most interesting one, and we have barely scratched the surface. One example of a new algorithm is **graph neural networks (GNNs)** that combine graphs and NNs, but there are many more ways to create new algorithms based on the fundamentals of both fields.

> No matter what industry you are in or what problem you are trying to solve, *all* problems are best approached with a combination of ML and OR. Most of the time, you also need to add other fields, but for sure, any problem that can be solved analytically can benefit from the hybridization of ML and OR.

Even problems that are said to be in the realm of ML should be solved with a hybrid approach. Our hybridization has been very successful in computer vision and **natural language processing (NLP)**, understanding, and generation.

AA: If ML is fundamentally based on optimization, why is there a slow acceptance of ML combined with OR?

NVO: First, ML can provide incredible results that were not expected, in particular with DL in computer vision and NLP. These successes lead to a massive interest in ML, which is normal. You find something that works; you focus on those methods.

Second, everything is hard. There have been so many new algorithms and approaches in ML, which took lots of people and energy to find.

Third, ML and OR experts are not in the same paradigm. OR is more refined and uses a lot of clever methods. Most of the time, you need to write some solver code to help it find a good solution. This probably stems from the history of OR, where there were no powerful computers at the beginning.

> ML is more of a brute force, where you need to rely on lots of data and big computers. Most of the time, you construct a model and a training/testing method, but once this is done, you use generic libraries without touching their code.

Fourth, you need to understand the fundamentals of both approaches to start to see how to combine them. This is not easy, and there aren't many experts who know both fields. Experts in both fields were certain that their field was the best and didn't need help from the other.

Fifth, to combine both fields, you need to modify them a little bit to make them work together. You need to create new algorithms.

That said, OR is known in a lot of domains, such as manufacturing or supply chain. There is probably not a single airline that is not using OR intensively, for instance. The first combination of ML and OR as two black boxes is becoming more and more known and used with great success. You have hypes and fashions in science too. They come and go. In the 1980s, OR was common in some industries. Now it is ML.

AA: In your area of specialization, what are some of the main challenges you've faced when launching your start-up and trying to drive this field forward? Where's been the main resistance, both with clients, and the field in general?

NVO: It's still hard to advocate for that combination of ML and OR.

Everybody speaks about this first-mover advantage, but actually, it's really hard because you need to convince people. When you start with something really small and unknown, and then you have a huge marketing machine in front of you that tells you that all the biggest guns (Google, IBM, Microsoft, Baidu, and so on) are the best in the world and they use ML, it's really hard. Today, we still need to really convince everybody among our customers. You need to convince the management team, you need to convince the technical teams, and you need to convince the people that will use your tools. You really need to convince them and tell them, "*Yes, DL is nice, but no, this is not what we are doing. We're using a hybrid approach, which is better.*"

So, every time, I have this feeling that we need to start anew. We did eight successful projects with amazing results, but still, every time, we have to start all over again.

It's especially difficult because, inside the internal teams, you have ML specialists. Now, they may start to realize that maybe their field is not the best field to do everything, and they also start to see the results, but at the beginning, it was extremely hard to try to convince those people and say, "*Yes, maybe we should work together; yes, you have something that is fantastic, but look: if you combine it with something else, you can get better results. Your tool is supposed to be so fantastic, yet you're failing. You called us in. Maybe you should listen to what we propose.*"

That's really hard, and it's the same for the research. You would think that people doing academic research would be open-minded. Actually, every time you have a specialization, it means that you're stuck in that specialization. When we started a few years ago to advocate for those ideas, all the doors were closed. No one saw the interest in combining different fields.

> I believe that hybrid methods are here to stay and that they will be the next AI.

People are starting to talk about the next AI winter. I don't think it's coming. Maybe there will be one later on, but first, you'll have those hybrid methods that really let you go further.

AA: It's hard to believe that there's still so much resistance to hybrid methods. I guess part of that is probably that there are so many people who are really invested in their particular area, and they just won't budge, whether they're invested financially or emotionally.

NVO: Exactly.

AA: There's no easy way forward. Unless someone like Google or one of the FAANG companies says, "*We're doing this***," it'll be hard to shift the whole industry to say, "***OK. We should be seriously committing to hybrid approaches.***"**

NVO: If you look at DL, it took 20 years before people realized the potential of this technology on a massive scale, with general adoption. It took 20 years for people to start saying, "*Wow! This is an amazing way of doing things and solving problems.*" I would say hybrid methods only started a few years ago. Maybe six or eight years ago, people started to try to combine these things. Some teams already did this, but not on a large scale or as a new trend or science. Some people say that it's a new science, and it is *relatively* new.

FAANG

FAANG is a term to group the leading tech companies **Facebook, Apple, Amazon, Netflix, and Google**.

AA: Hopefully, this book will help to get the word out – give people something else to think about – because part of the intended audience is senior executives who fund and sponsor data projects in organizations. I'm hoping they'll start thinking and questioning their own team: "*Why are you only investing in ML? What fields beyond ML should we be looking at?***"**

Becoming data-driven

Let's touch on another question. There are high failure rates for data projects in the industry, so we need to think smarter. Say you were to work with an organization just starting its journey. It says it wants to become data-driven: it has data, but it doesn't really understand how to derive insights. In general, how would you advise it to commence its journey in terms of building its internal capability?

No matter which methodology you use, you need to invest in people, culture, technology, and a certain hierarchy and structure. Do you have any advice on how to commence the journey?

NVO: It would depend on a lot of things because if you focus on the problem, then you need to really try to solve it however you can. You should take into consideration the budget you have, the time you have, and what the desired outcomes are, and that's all part of the problem.

So, many companies nowadays try to gather as much data as they can. They have huge data lakes, and the idea is, *"OK, we'll gather the data. We'll figure out later on what we can do with it."* But most of the time, it doesn't work that way in the sense that you cannot do anything with that data because collecting the data is part of the problem.

> Simply having a lot of data does not mean that you will be able to find anything meaningful in that data. You need clean data, and you need the right data.

So, I would not advise starting with data. I would advise starting with the problem, and I would advise starting by looking at what has been done before because there are problems that are probably similar that other teams have solved. You should take the time to look around and have an idea of what exists because you will probably find something in the literature.

My first move would be to look around. Get an idea of what works and what does not because there are things that are working and things that are not working. Only then do you start to gather the people who would be competent for those kinds of problems.

Our approach is more like a street-smart approach. Everything goes. Whatever you need to solve, try to solve it the best you can; just go after it. You don't start to look at what could be done or what type of solution you would need: no. You just start small, and then you grow organically as needed. But be aware of what exists.

Setting your project up to succeed

AA: I've seen in my own experience, and heard from colleagues as well, that many organizations can be OK at doing the initial stage – the development stage, the POC – but when it comes to productionizing those solutions, they can often hit a roadblock. What do you think causes that? How can they overcome that?

NVO: Well, first of all, if I continue on the other question, one thing we see very often is that people try to solve the wrong problem. If I have a customer in front of me that tells me, *"I know exactly what needs to be solved and how to solve it,"* in my head, I'll say, *"Boy, that's a bad start."*

The thing that a lot of companies do is try to do a toy project. They try to see if something could happen, and then they just throw a little bit of data, but that's the wrong way to do it. The problem you need to solve is the problem at scale. It's not to do a POC and see whether it works. This is something that most experts can tell you right away: whether it will work or not. If they cannot, they do a feasibility study, but it's a few weeks or a few months. When you need to do a feasibility study, experts will tell you that, and in that study, you can say, *"This is what works. This is what doesn't work. In our case, it's probably going to work because of this and this and this, and we think that you should probably try this and this and this, or if it doesn't work, here's why that is specifically."*

When we do a project or a POC, it's never a toy project because our customers often call us after all the other teams failed. So, we know it's already hard. We are at the end of the road. So, we know it's not a question of, "*Is it possible to detect things or to be able to optimize something?*" Now, the question is, "*Can you optimize this in reality with the way our company operates, with the data we can collect, with the people we work with, with the reality of the competition, and things like that?*"

So, you need to take that into account from the beginning so that the scaling problem is already in your mind. You start small but have an idea that it needs to be scaled. It needs to be put into production: that's the end goal. The end goal is not to do a POC to see that you can do a POC. There is a huge difference between a POC and an end product that you can use in production. So, of course, if it's not the same problem, you get a different solution.

AA: Why do you think so many people are invested in this notion of a POC? Is it to win support and funding to get buy-in from senior managers? What do you think is the main driver for that?

NVO: Well, there is this notion that you should try something very small to see if it works because if it fails, then it fails rapidly, and you don't spend much money. But the thing is, if you do a POC and succeed, you're going to spend way more money because you have no idea if it's really going to work or not.

Most of the time, you need to have a showcase, and you need to be able to say, "*We did that small POC. It works, so yes, we can go and try to do the real thing now.*" But if you consider the POC as a toy problem without thinking about the reality – the production, the real use cases – that's the wrong way to do a POC.

We never really speak about POCs. We *do* speak about POCs sometimes because our customers like to say "*POC,*" but for us, it's always an MVP at least, in the sense that when we produce a small version of the real problems, we are not very far away from it in that with a few weeks or months of additional work, you can put it in production.

We test that small toy project in real conditions and try to optimize it, which is our specialization. However, optimization is really hard because you don't want to know whether it's possible to optimize. Of course, you can do better than what exists, but that's not the question. The question is, "*Can you really optimize in real conditions? Can you reach 5, 10, 20, or 40 percent optimization?*" That's what we try to do when we do the POC/MVP, but this is because we are already thinking about the end product, and this is the way you should tackle a problem. You have a problem to solve, so try to solve the real problem.

AA: So, your recommendation and advice is to make sure you're solving the real problem and not to do a POC on a toy problem?

NVO: Yes, but with a very good idea of what needs to be done to scale it up and put it in production. This is your main goal. This is the real problem. You should try to solve the real problem and not a toy problem.

Sometimes you cannot solve a problem. In those cases, do that feasibility study and take the time to do it well. Again, it's not a feasibility study on a small toy problem. It's on the real thing.

Then I would say you spend more time and money, but you do a real feasibility study in the sense that that feasibility study gives you a really good idea of what is possible or not. It's similar to the POC.

If your feasibility study tells you that it works, you should also say how and what is needed to reach that because most of the time, you can find a solution that would work, but that actually cannot be implemented in a particular company. So, you always need to take that into account: how people work, how the company works, and things like that. This is part of the problem. So, again, focus on the real problem.

AA: That is fantastic advice and very pragmatic. It's about the solution being scalable within the constraints of the business it operates within. That's like a broader optimization problem.

NVO: It's never happened, of course, but if we had two companies with the exact same problem, we probably would provide two different solutions because of the way they operate and the way the departments work.

Most of the time, you get some data from one department and then some from another one. Most of the time, the one that is creating the tool is the innovation department or the IT department. They're involved deeply in the solution's construction but don't use it. Sometimes they also use it, but most of the time, it's for another department. You have to take that into account. How do these people communicate with each other? How can you gather the data? What is the communication between them? Because sometimes there is secrecy between the departments. You don't have access to the data, or not as you would need to have access to, and things like that. This is part of the problem.

AA: That is actually fundamental advice because for a data scientist commencing their journey in this field, say you were to ask them that question in an interview: "*Two organizations, same problem – would you create the same solution?*" **If they say yes, that obviously means they don't understand the broader business context. That is fundamental.**

Can you offer any insights and tips on how organizations can scale their AI solutions?

NVO: Finding a good algorithmic solution is only half of the solution. The solution needs to be scaled up, put in production, and will need ongoing maintenance and upgrades. The maintenance part often is the costliest part of any project. So, you're in for the long game, and all this needs to be addressed from the start. This means that our POCs are really MVPs that could be put into production right away with only a few modifications. This means much more work, but also, we know that if they work, they will work at scale right away. Scalability should be addressed at the beginning of the project so that this is not a "real" problem later on.

> You construct a product with scalability in mind from the early stages.

Another important piece of the puzzle is understanding how different departments work together. Rarely does an AI project depend on only one department. Often it is initiated in the innovation department or the IT department but will be used by another department or, most likely, you will need some data input from different departments. You need to take that into account right from the beginning. This is why we try to incorporate the way our customers operate because this can be the difference between success and failure. You can construct the best solution, but the project might fail if the people that need to use it are not involved in one way or another.

We had a project where the end user of our system simply did not want to use it. Despite our explanations and games to teach them how the system could help them, they simply refused to use it despite the company trying to force them to. It has been a very painful and slow acceptance. We managed to showcase that the customers of the employees using the system were happy and were paid more. This was the only incentive that worked to motivate the other reluctant employees. In this case, the customer did not want us to collaborate with their employees and told us that they would impose it. The fact is that you cannot truly impose anything: employees will find loopholes to bypass a system they don't see as helping them.

Sometimes, customers have asked us to construct systems against their employees to force them to do things (such as bad surveillance), but we always refuse such projects.

Last but not least, our systems are very complex. We are pioneers in what we do. This means that, no matter how good your engineers or scientists are, they will not be able to maintain our systems. This is why we ask for a long-term collaboration where we do the maintenance. This is a dependency for our customers, but at the same time, they have true experts in the hybridization of ML and OR right at their disposal, and we not only do the maintenance but upgrade their systems constantly.

AA: You've worked with some large organizations around the world and seen how they tend to structure their teams. Sometimes they centralize their data science/ML functionality, and other times they distribute it, depending on the size and scale of the teams and the organization. Has any particular approach that you've seen worked better than others?

NVO: I would say that when people talk to each other, it works better. This is a very simple thought, but the thing is, it's very difficult to construct an AI solution. It's very hard to understand the problem. But then, constructing the solution is also really hard. There is nothing easy.

So, when you have companies where people don't talk to each other, where you have a management team that is imposing things, and don't listen to the people that will use the tools, that's already a bad start. What we see is that when you construct a wonderful solution, it works for IT people because they understand what they're using. But if you try to impose this on people who have no idea about the tools and no idea about the limitations, it's a total failure because they will not use the tools like you want them to.

Most of the time, I meet people telling me, "*Look. We are the boss. We decide, and we'll impose that tool*," or, "*They will use it, or we fire them*," or, "*If they use it well, we'll give them money*." It never works.

> If people don't want to use the tools, they will not. They will always find a loophole.

Then, you see incredible things. I saw people in a huge company using AI tools, knowing perfectly well that they were completely outdated because they didn't correspond to the new situation but had to use them. So, they used them, and then they manually changed the output.

AA: Wow.

NVO: Yeah: *wow*. I think it happens more than you think because most of the time, the people that are making those tools are not really working with the people using them. Once they hand over the tool, there is no follow-up. There is no monitoring.

> If people talk, then they might have a chance to construct something really helpful. If they don't talk, it's going to be a failure.

AA: I completely agree, and I think a key part of that is having the right culture in place, which has to be supported and advocated by senior leadership.

Exploring leadership

What do you think makes a good data leader, such as a CDO or someone who's ultimately responsible for data or AI analytics? What are some of the key attributes that they need to have to make the organization successful and ensure people feel happy and supported?

NVO: I think that one thing they really have to be convinced of, and also convince other people of, is that it's a long and hard journey. If you have a company that is not already using these analytical tools, you don't bring AI in without changing the way the company works and the mindset of the people. It's a huge transformation. It's not only that you're using new tools – you're using a new way of operating.

So, it's really a long-term change. You need to take the time to reassure people, to calm them down. Most of the time, when you have an AI team coming in, people are scared. They think that they're going to lose their jobs. So, there is an emotive aspect to this that is enormous. As a leader, you should really be able to bring all those people together so that they work together to reach the goal of using a new analytical approach to help them.

Also, the tools need to help people because this is not always the case. Sometimes, you bring in tools not to help people but to hinder them. You bring in new tools for surveillance so that you can monitor them better, for instance. Of course, this never works. That's not a good idea. Not in my book, at least.

AA: In the past couple of years, do you think that COVID has impacted how organizations think about investment? Properly investing in AI is not cheap, and some organizations think you can do it cheaply, which is a big mistake. Do you think there's been a change there?

NVO: Oh, yes. So many supply chains have been disrupted. There was false comfort that the world was turning around without a problem: that if something was not happening, it would happen in another way. People were not accustomed to huge disruptions and not having a solution right away. If you look at the semiconductors right now, it's a big deal. It's disrupting the whole process of constructing many things because you don't have semiconductors anymore. Sometimes they will come two years later, and there's no way you can replace them with something else.

So, yes – people are beginning to realize, *"Oops, this is another world. It's a very fast-changing world where you need to make decisions very quickly."* You cannot do this with humans: it's impossible. You also need robust solutions, and again, analytical tools help you to do that. It has had a huge impact on how people perceive AI and the need to start constructing and using AI solutions.

AA: An idea I've had is that when a lot of companies talk about investing in data science and they have people who are working in it, is that they often neglect or simply do not understand the science aspect.

Do you agree with that, and if you do, what should they do to be more scientific in their approach to solving problems in terms of processes, frameworks, ideas, and concepts?

NVO: One thing I see, especially in AI, is that it's a buzzword. We're no longer in a time where managers look after the people or the company. They don't care. They care about their own careers. They just take an AI project to get credit. It's amazing how much public money is literally given away to help companies do R&D projects in AI, which is already a business success in itself. If you do that, you don't need to get results. You just hire a bunch of beginners. They have no clue what they're doing, but they can use some existing tools, and there you have it: you have AI.

AA: Something I see often is they think hiring anyone who's studied data science in some way, such as doing a short course online, is now a data scientist; hence they can help. I think that's to the discredit of those of us who have formal studies and have invested a lot of time studying. Someone who lacks the rigor in math, stats, and coding paradigms can never be that effective, especially in research environments, if they've just done a short online course and have neglected some of the foundational aspects.

NVO: I tend to agree with you, but I would say that it's not because you have a formal diploma or that you worked for years and years. Nothing beats experience. But I'm not drawing a divide between people that are self-taught and people that have lots of diplomas: it really depends on the individual. I agree that you cannot become an expert in a few weeks or even a few months of learning from online courses. Does everybody become an expert in three months? That doesn't make any sense.

Measuring success

We all too often hear of data science/AI projects having high failure rates, approximately 85 percent. Do you think this is the reality, and if so, why is it the case? After all, significant money is spent on people and technology, and we're drowning in data, so what's the problem?

NVO: The reality is that most AI projects fail. There are several reasons, some related to AI, some not. Constructing AI/optimization solutions is hard, and more often than not takes months, if not years, to develop, test, scale, and put into production. And I'm not even talking about the ongoing maintenance and upgrades it needs. So, unless you can do this for a problem where the ROI is provably really big, the chances that a company can accomplish and support an AI project are slim. I've seen sound projects in AI that were torpedoed simply because a director was replaced and his successor wanted to wipe out the past.

> With AI projects, you're in it for the long term.

There are other reasons for failure.

> Most teams focus too much on the solution and not the problem.

For instance, they want to obtain state-of-the-art solutions that are not applicable to the company they are working for. Or they like their new shiny toys (computers, libraries, tools, and so on) too much and want to use all the power of their machines/solvers just to justify a budget planned in advance but without any relation to the real needs of an *optimal* solution. Or they don't talk with the people that will use their tools. These are known problems, and they are not specific to AI, although the hype around it and its complexity makes companies particularly prone to such errors.

But there are at least two main errors that are specific to AI. The first one is the belief that AI can magically solve everything. Companies keep data lakes with the conviction of being able to use this gold mine later on. This is rarely the case because, in most cases, we don't know what to do with all this scattered information. Collecting relevant information is an integral part of the problem to be solved. There is this huge problem that you often try/test a toy problem on data that does not correspond to reality, or if it does, your situation will change, and this will be reflected in new types of datasets for which the models were not trained/don't correspond anymore. Furthermore, if you don't talk with the experts in the domain, AI experts will probably find non-existent correlations that they will interpret as causalities. You need hard work and domain knowledge to interpret results.

The second error is even worse. For most projects, the tools and approaches are simply wrong. In particular, what most companies need are prescriptive approaches, not predictive approaches. It is nice to be able to predict, even essential for many companies, but you want more than that: you want to be able to act on those predictions.

> **You need robust solutions that can adapt to new realities or at the very least can tell you when they stop working and you cannot rely on them anymore. This is difficult to construct with only ML or only OR, for instance.**

I'll give two very simple examples of such failures with ML projects. The first one is about automatic replies to emails. Not so long ago, when you wanted to reply to someone, a model provided an automatic greeting but with the name of the wrong person! The model provided an answer based on the person you are most likely to write to. If you write a lot to your best friend or your wife, that is the most probable person you'll write to. Domain knowledge, however, tells you that if you reply to an email, you're probably writing to the originator of the email.

The other example is companies pushing DL solutions for autonomous and/or self-driving cars. While DL certainly has its place, you cannot rely on it entirely as it is very easily fooled. How many videos exist where you see a car crash without hesitation because this scenario was not in the database/training set?

In some industries, such as marketing, these errors might not be that important, but in other industries, these errors must be avoided at all costs.

> **Important decisions, certainly life and death decisions, should never be taken without some kind of supervision, whether by a human in the loop and/or an automatic verification system based on domain knowledge.**

AA: We've talked a bit about the alignment of the technology and the solution with the business problem. How do we measure success? Have we solved the problem of metrics?

NVO:

The first metric that companies have is, *"How much money can you save or how much money can you bring?"* That's very important to them. But beyond that metric, it should be *"Are people happy? Are your employees happy? Are the customers of your company happy?"*

> **Our metric is, *"Are our customers happy?"* It's as simple as that.**

Of course, it's difficult to define "happy," but at the same time, when people are happy, you see it.

For me, that's the most important metric because the rest are not that important. When it comes to the quality of a solution, if you look at some analytical metrics, it doesn't make any sense. If you can recognize something with a certain probability that is very high, then cool. But if people are not happy, it doesn't mean anything. And on the other hand, you could have something that is really sloppy, but people love it and buy it.

But it's difficult to measure. People tend to like very simple metrics because you can measure them, but we never look at metrics. We don't care because when you have results that make people happy, you see it. It really shines. Of course, you have metrics such as, "*OK, you could optimize by blah blah blah, and we also do this*," but it's useless. It's really about happiness, which is something that is totally out of this world because you never speak about happiness in a business setting. Besides, you can have a beautiful metric that will give you false confidence that your system is robust or does exactly what is needed. It is only the end results that matter, not the metrics.

AA: We need a new happiness metric for AI.

NVO: We've tried to create one several times! It's a difficult problem.

Developing ethical AI in an organization

AA: There's a lot of talk these days about AI ethics and the responsible use of AI. How do you develop it in an organization? What does it look like in terms of the frameworks?

NVO: Again, some people will not like my answer.

Some organizations around the world are trying to define what is ethical or not. They don't quite use letters of intent, but they have some declarations of ethical things, and so on. There is one in Montreal. If you look at the one in Montreal, I don't remember how many people signed it, but it was almost no one. If you look very carefully, you'll find that these companies don't care about ethics. They cannot say it, but the first metric is money. Everything revolves around money and short-term money rather than long-term money. If you account for ethical considerations, you're probably going to earn less.

The other thing is that it's really difficult to respect some ethical considerations because ML, for instance, is based on data. You need data, and it's difficult to get it without doing some ethically bad things. Also, it's difficult not to have biases. Biases belong in data. So, you can say whatever you want about your ethical considerations, but ML alone is not helping. I would say it's an unethical (in the sense of amorality) field from the ground up, and this is one of the strong cases for the use of hybrid methods because with other fields – in particular, OR – you can do things about biases. You can do things about ethical considerations.

AA: I once had a large organization approach me and say, "*We'd like you to help us develop an ethical framework.***" They had a large data science capability, and it would have been a lot of work just to validate, count the processes and frameworks, and try and help them with their journey. I said, "***Sure. Let's have a talk.***"

A week later, I said, "*Let's talk about budget.*"

They turned to me and said, "*What do you mean, budget? We're not going to spend anything on this.*" **I said,** "*Are you serious?*" **They said,** "*Yes. This is not going to generate money for us. Why would we spend a cent?*"

I said to them, "*Why should I work for you?*"

They said, "*Oh, because you'll be associated with us. We're big.*"

I said, "*No, thanks.*"

NVO: I've heard that many, many times. Many very well-known research teams tell us, "*Oh, you should collaborate with so and so.*"

I don't know how many teams we met that told us, "*Look. We are very well known. It's very prestigious to work with us. We're going to work together. Of course, it is prestigious to work with us, so we will not pay you.*" Actually, what they were saying was, "*OK. We're going to take your knowledge and benefit from it, and we won't pay a dime.*"

AA: Final point on ethics: do you think that something will have to change, given that regulation seems to be increasing, especially in Europe? Do you think there will be major changes in how ML, AI, and data science teams are run and how they build their solutions with governing regulations? Will it change auditing and insights?

NVO: It's already changing the game, in the sense that in Europe, they are more and more aware of those kinds of ethical limitations, and they also know that it's going to badly hurt them if they don't pay attention. One solution is to do things in a hidden way, and then you come back with the results and people say, "*Oh, look. We have the results, but we have no idea how.*" But I see more and more European teams really paying attention and trying to find solutions, and at the same time, they're really annoyed that in China in particular, there is no ethical problem. That hinders the process of finding a good solution very quickly and easily. But yes, it's already changing the game.

AA: I've previously heard you speak about the future of AI, and specifically, about algorithms that can morph into other algorithms and adjust dynamically. Can you please share your thoughts on this and the future of AI more broadly?

Is the future of AI, ML, and OR, and can that help us achieve AGI?

NVO: There is a "natural" progression. One big lesson of recent years is that you should leave your algorithms to finetune themselves, and when done well, they will discover some patterns that are hidden or are totally new even to experts in the field. It works pretty well and sometimes returns excellent and surprising results. This is achieved by optimizing parameters and hyper-parameters (what most of the time is called "learning") and can be viewed as **point of function optimization**, or **curve fitting**. The next step is to optimize not points or one (**parametrized**) function but a set of functions or models.

AutoML is a step in that direction, and the results (sometimes) show the effectiveness of this approach. You can think about AutoML and similar approaches as meta-learning: you learn the right model/algorithm to use, then create your model/algorithm from scratch. A kind of meta-meta learning.

The idea is to constantly monitor what your models/algorithms do because most of the time, they are optimized for a given type of input data. It is always a trade-off: the more tailored your models/algorithms, the better they will perform for one type of instance but the less generic it will be. In reality, your instance/input data does change, sometimes in a very significant manner, so you need to readapt/reconstruct your models/algorithms.

Our project about true intelligence allows us to go in that direction, and I have no doubt that if you use such an approach, it will beat any human approach as it will constantly change/morph its algorithms/models to apply the best ones. The industrial applications are numerous, from providing the best solutions in real time to monitoring the way a system is used, and sending an alert if it discovers that it is being used in a sub-optimal way, or tracking bugs and repairing them on the fly.

The future of AI is about hybrid methods. You are better equipped with several toolboxes to solve complex problems. This is obvious from a theoretical point of view: in the worst case, you only use one toolbox. It is also the best approach in practice. One great idea to solve really complex problems on huge instances is to develop several models of the same problem and combine their insights. State-of-the-art approaches cut a huge problem into smaller sub-problems because there is no way to solve one big model that deals with the whole problem.

> **Nikolaj's observations on model size**
>
> *"There is still a tendency from some experts (but not all) to construct bigger and bigger models, especially in ML. In my view, this is futile as you can obtain much better results with refined and much smaller models, especially with hybrid approaches. Not to mention that the statistical approach cannot achieve true intelligence."*

In theory, you should leave as much freedom as possible to your solver/system/algorithms to find the solutions by themselves, but when you are pushing the limits, you are stuck: you don't have the power nor the methods to solve such huge models. Therefore, the problem must be compartmentalized somehow into sub-problems that can be solved satisfactorily. The challenge is that we often lose information or force a certain type of solution in doing this, and it is difficult to reconstruct a global solution for the initial problem from these sub-solutions. Sometimes it is possible, but it is not always the case.

So, a new paradigm emerged to look at a problem from different angles with different and complementary models and approaches. We can even use incomplete models that focus on certain aspects of the problem. When we solve one of them, we have only a partial view of the problem, but we still keep a global appreciation of the problem, and sometimes we can combine these different appreciations to reconstruct a global solution of very good quality despite the size of the instances and/or the complexity of the problem.

I believe AGI is achievable despite some people "proving" the contrary. First, there are plenty of examples of true intelligence on our planet, and we humans are certainly not the only intelligent species. Second, we are really not that clever. Why should we have this anthropomorphic view that we are the pinnacle of evolution and that only we humans can pretend to have true intelligence? Mimicking the human brain might be a good idea, but we know so little about it. I think that a better and probably much easier way to get to AGI is to define a practical definition of machine intelligence and let the machine learn how to become more and more intelligent. I believe it is achievable in part with a combination of ML and OR but probably also with other fields, such as causal theory, for instance. Optimization is a powerful idea to make a system develop itself, and thus, I believe that using OR might be a very good idea. What is also very nice with OR is that you can control (and finetune that control) what your machine is doing, so we could reach a certain level of true intelligence, not AGI, and still be able to control it.

AA: How important is explainability in AI, and how do we achieve it given that we have a limited real understanding of how DL works?

NVO: Again, it depends on the problem at hand. For some, you don't need explainability at all; for others, it is absolutely mandatory. If you need some or total explainability, use the right tools to do so. In particular, OR approaches not only can explain their results (most of the time), but you can also use them to do some sensitivity analysis or control the results. And you should use hybrid approaches. For instance, you could use DL with some OR to explain its decisions during the training or after the NN has been trained to interpret the results. You can also construct a filter to force only acceptable and somewhat interpretable solutions to go through, or you can add a reconstruction procedure to only construct acceptable and interpretable solutions. Of course, you lose some of the divinatory power of DL, but it is a trade-off you must accept. Explainability is a very strong use case for the use of hybrid methods.

Starting out in data

AA: For someone starting out in the field, what are the fundamental skills they need to become a data scientist or data engineer (either in research or industry), and how important are formal university studies and postgraduate degrees relative to on-the-job training and self-paced learning?

NVO: I would definitively put open-mindedness at the top of the required skills. You need to understand that there is a broad set of approaches and fields (ML, OR, causal theory, neuro-symbolic approaches, statistics, and so on) that can help you solve problems. Also, an inclusive approach, where you combine several fields will take you further than any one of them alone, but you need to be able to listen to other experts to combine your strengths.

I'm convinced you need to learn some fundamentals of computer science and mathematics. In particular, be able to analyze the complexity of an algorithm and learn the different algorithms used to do the basic computations such as matrix multiplications (sparse and dense versions).

> You must know about fundamental data structures. It's not only very handy but among the most beautiful results of mankind in recent history, in particular the use and construction of graphs and trees.

As mentioned before, you are at the service of others. Therefore, learn the required soft skills needed to communicate effectively with others.

Are formal university studies needed? Absolutely not. You need strong fundamental bases, but you can acquire them on your own. Curriculums often force you to study some domains you would rather skip, but sometimes they will open up your mind. The big positive point of doing formal studies is that you'll learn in groups with other people. But nothing prevents you from creating a study group with other people or learning on the job if it is possible. Learning with others is easier and much more fun. What is important in the end are the people and their solving skills – soft and hard. There isn't a single path to reach a destination.

Looking to the future

AA: What could LLMs like ChatGPT mean for the future of AI and data, and in particular, how do they affect the data science scene – including the field of NLP?

NVO: ChatGPT is taking the internet and the AI community by storm but, like all previous announcements made with a lot of marketing hype, one should be a little wary of some of the grandiloquent claims. Yes, LLMs and the different versions of ChatGPT (and their siblings) have shown positive and somewhat surprising results, but the reality is much more nuanced than what the official communiqués and the tide of comments would have us believe.

> ChatGPT generates answers that are riddled with nonsense. What some generously call "hallucinations" are – from my perspective – inherent flaws, even though some can and have been fixed.

This is the same "good old" debate that probably can't be settled unless we can show that this approach would lead to true intelligence, and ultimately to AGI, which in my opinion is very unlikely.

At the risk of repeating myself, I'm a proponent of hybrid approaches, in particular the combination of symbolic and statistical approaches. LLMs are so large because in essence they capture (or rather, they memorize) so many combinations. Language in general is so complex that it is probably futile to try to memorise all possible combinations, especially when some combinations are contradictory and/or need lots of external context. So, I – like some prominent scientists – would say that ChatGPT (and transformers) are a nice culmination of the advancements in the current statistical approach, but the approach has fundamental limitations. I believe we should rethink our approach to NLP – using hybrid approaches for instance.

AA: As the popularity and use of these models increases, so do ethical concerns. What do you think are some of the key ethical implications for organizations – and society more broadly?

NVO: This is a huge topic and I don't think we realize how much trouble we are in.

I will only touch on two issues. First, there are lots of implications for privacy and copyright. The way the data was, and is, collected was taken a little lightly. It would seem that entire sections of the internet have been shamelessly plundered. Also, current users are not aware that the use of their data is double-edged: yes, they can improve the generated answers, but by providing access to their data. Employees at Samsung have learned this the hard way.

> **Samsung data breach**
>
> Employees at Samsung used ChatGPT to check their code, not realising that doing so meant the data would be absorbed into ChatGPT, exposing it to the public: `https://gizmodo.com/chatgpt-ai-samsung-employees-leak-data-1850307376`.

Second, we are increasingly inundated with false or approximate information and the ease with which one can now produce texts, images, or sound that are fake, but look truthful or at least plausible, will only be exacerbated by this trend. Not to mention the fact that it may well have a resonance effect: more and more data will be generated by an AI that will learn more and more from analyzing the data it itself generated...

AA: How can data scientists/AI experts leverage ChatGPT to benefit their organizations and their own careers?

NVO: Tricky question. On one hand, ChatGPT can help in finding information quickly, providing code or code snippets, answer some questions about a given topic, write documentation and so on. But, on the other hand, the answers may be incomplete, inaccurate or even completely false.

There is also the risk that ChatGPT might lure you in one direction and you'll have to backtrack at some point – if you are able to! I'm not even sure that I recommend its use to develop solutions or code. Yes, you can use it to generate a first draft but this might push you in the wrong direction. But for more simple uses, like a chatbot that can reply to users about a very particular (and trained) topic, why not? As long as you know that this tool has some limitations and are ready to accept them. To generate some answers that can be quickly analyzed and vetted, it is a fantastic tool, so go for it!

AA: What are your predictions of the future of LLMs and their broader use?

NVO: LLMs will certainly stay in the foreseeable future and be developed with larger and larger models, even if some realize that bigger is not always better.

The general craze for generative AI is such that we are going to see more and more crazy developments, but with risks that we do not understand and that we obviously do not control.

I am also concerned about the expense in terms of resources, both energy and data, and the complexity of these models. Fewer and fewer companies will be able to develop these models and we will be more and more at the mercy of a few giants who will do whatever they want. There is a reason why big corporations are pushing for such developments in AI.

I think that there is a way to significantly reduce the size of current models that essentially use brute force to store some knowledge. As for true intelligence, which is necessary to truly converse and use a language, I don't believe that LLMs will lead to significant advances. At least not without the addition of hybrid methods based on symbolism and logic. At Funartech, we are working on a hybrid ML/OR approach to inject emotion into machine-human interactions, and I don't see how a purely statistical approach could get the results we are getting. Of course, I could be wrong. Time will tell.

Summary

I really enjoyed being able to have an in-depth discussion with Nikolaj and in particular, to discuss the details of his unique approach of combining ML and OR to solve complex problems. ML has become the go-to approach for many problems, so it was great to discuss what other approaches exist that may be better suited to some problems, such as combining it with OR when you don't have lots of clean data available or when you need to control the model outputs. We also discussed the amoral nature of ML – and so it's arguably unethical – and is designed as such from the ground up. The argument for hybrid methods is that they allow solutions to be designed that can handle bias and ethical considerations. Nikolaj firmly believes that the future of AI lies in hybrid approaches, such as in the combination of ML and OR.

It was good to hear Nikolaj's views on investing in POCs, and how effort needs to be put in to ensure that you're solving the "right problem," which can be difficult to define. Also, if people don't know they have a problem – such as an ML solution that is suboptimal for their problem, which Nikolaj often finds – then you need to educate them to understand that they not only have a problem but that it needs to be fixed.

As others raise in this book, such as Kshira Saagar (*Chapter 4*) and Althea Davis (*Chapter 12*), Nikolaj also urges organizations not to just collect data and assume they will figure out how to use it later, but to ensure up front that the data the collect is fit-for-purpose and aligned to clear strategic benefits, and to also think about the required change management. In addition, when commencing a new project, he suggests spending time looking around at what already exists and doing a literature review. In addition, as Kshira also mentions, awareness and consideration also need to be given to the maintenance and upgrade requirements of solutions.

Nikolaj also offers some very helpful advice on how imperative it is to ensure that the solution fits into the way that the business operates rather than just focusing on a brilliant technical solution alone. I've seen many data scientists get caught out by this, not focusing enough on how the business will actually use the solution and the constraints it needs to operate within. This is why understanding the domain and the people are just as important as the problem itself.

Nikolaj also offers some relevant and specific advice to help organizations scale their AI capabilities and draws attention to the maintenance aspect of systems. He strongly advises that scalability needs to be considered from the beginning of a project.

One of the key takeaways for me from our discussion, which resonates with me, is the human element of AI and, in particular, the importance of understanding how the solution will actually be used. This is something that needs to be understood and factored into the solution from the beginning. Nikolaj's measure of success is simple: does the solution make people happy? More broadly, the problem to be solved includes non-technical and cultural factors – the wider context in which a solution needs to operate. My advice is simple: don't forget about the people using the solution and who the solution may ultimately impact.

10

Jason Tamara Widjaja
and the AI People

Not only is **Jason Tamara Widjaja** an experienced leader (and currently the director of AI at MSD) but he also has a background in human resources management, which gives him a unique, and important, human perspective in the field – and one that resonates with my view that a human-centered approach to data science is critical for success.

I was eager to learn about his leadership style for managing a global interdisciplinary team of data specialists, and his strategies for aligning technology with business strategy to achieve successful outcomes.

Getting started in data science

Alex Antic: I'd like to get an idea of your career trajectory, especially with the interesting role you had early on as a police inspector. Did you think of taking another career based on your early studies? What made you shift your focus to AI and data more broadly?

Jason Tamara Widjaja: I think I'm very fortunate to have arrived at data science and AI because, for the earlier part of my career, I was struggling to find an area that intersected with my interests.

My first degree was in computer science, but I always knew I didn't want to be a purely technical "backend" person. I was trying to find something that brought different parts of the discipline together but, unfortunately, here in Singapore where I did my first degree, they labeled you a little bit – you were either a computing person, a business person, or an arts person. Things have improved since then but this was maybe 20 or so years ago, so it's been a while.

I wanted to *work with data on people*. My major in computer science was information systems, databases, and so on. In my idealism and naivety, I said, "*I want to work on people.*" So, I went to study for a second degree in HR, and I think that's how I got involved with definitions of roles, positions, and skills in the data and AI space. And right now, those distinctions are serving me very well.

Later on in my career, I went back to school to do business, and a few years later, I went back to school again to do analytics and machine learning, because I found that the industry had moved on so much from SQL and databases in the intervening decade that I just had such a massive gap in skills. So, I went to school in my mid-30s as one of the older students for the master's in analytics in Melbourne. I think that was a time when lots of fresh graduates were just entering the industry, but there were very few leaders to lead them. This was about 7 years ago. I quite easily moved into a data science lead role, and data science has grown from strength to strength in the past 7 or 8 years.

Recently, I think that the tide is shifting a bit. I hate these buzzwords, but we had a tide of big data, then we had a tide of data science, and now it seems like it is AI. And the three are not the same. My most recent AI role is actually an evolution of data science where we focus more on automated systems, certain AI disciplines, and AI ethics and governance, which I think is a very interesting topic that we might get into later.

As for the police force, I'm sorry to disappoint you: there's nothing interesting there. In our country, there is a mandatory national service system where men need to spend 2 to 2.5 years in the military, and I spent my time in the police force. As for my takeaway from that, it is an environment where you have all the different strata of society working together. You could be sleeping next to the son of a billionaire or the son of a professor – and I think that is a very healthy experience to go through. Having said that, I think data teams are very different from uniformed services, and I certainly wouldn't apply much from managing the police force to managing a data team.

Becoming data-driven

What advice would you offer an organization looking to begin its journey to become data-driven and become better informed through evidence-based decision-making?

JTW: That's a massive question. Whole books have been written on it, and I think you should write one if you haven't done one already.

I'll start with what not to do. I think there is a lot of noise in the industry because there are so many people who want to answer this question. Not all of them are independent, and many of them only see the picture in part. You have a "blind man and elephant" dynamic going on.

I think the first thing would be to choose your source of information. If you choose a client consulting firm, then they'll come with a few chevrons: your current and future state of design, a maturity assessment, and a roadmap. It will cost x dollars, and x will be large.

So, that's consulting. If you go to a product company, almost everyone will promise end-to-end platforms and products, but *end-to-end* is a very nuanced term that means different things in different parts of the data science ecosystem. Often, products will try to define data science as whatever that product can do. I think there are lots of traps there.

The last trap is a particularly important one because of the noise I encounter online: we have something we call the "furniture store problem," which has to do with when we see shiny demos, either of a well-functioning data science team or a well-functioning product or platform. It's like when you visit a furniture store, where the furniture is under these nice bright lights in isolation and it's the only thing in the store. When you look at that, it looks amazing. But when you bring that piece of furniture home, your room is dirty: it's cluttered, there's not enough space, and there are things from 10 years ago.

> The lesson is that when either buying software or importing frameworks from other companies, I find it takes a lot of adaptation and thoughtful working through the realities of what your "enterprise" looks like.

These are some of the problems I see in the industry. As for starting a data-driven approach, it comes down to the value of data science and the way you frame it. Everything cascades from there. I mentioned that I used to lead a data science team and now I lead the AI team, so I will speak from the current AI perspective. The way we think about value is in three buckets.

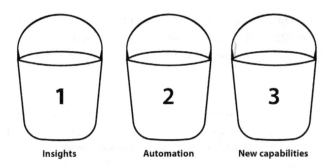

Figure 10.1 – The three buckets of value

The first thing is the one we hear the most about – data-driven decisions and **insights**. If you go down that path, then what you want to do is organize around not the data itself but the *decision*. We start by finding decision-makers who are willing to change their minds when faced with data, because if you don't affect strategic and valuable decisions in the company, then the rest of the activity ultimately will not be too valuable.

So, that is the first bucket. We used to call that team *decision science*, just to emphasize that it's not about just the data or the modeling (although they're both important). It's really about getting to the messiness, incentives, psychology, and environment of the decision-maker. The KPI would be, "*Can you change their mind?*"

The second bucket is about AI for automation. That's a very different beast from data science because many of these things are closer to an Excel macro on steroids than a statistical forecasting model. This one is very system-oriented, and all about the process and workflow. It's a very different source of value from insights. That's the second bucket of value – achieving productivity using AI for automation and actions.

Then there's a third bucket, which is about **new capabilities**. With recent advances in technology, we have many domain-specific or state-of-the-art models coming out that don't fit nicely into either automation or insights. Examples of these can be found in healthcare, where we might look at new methods of drug discovery. We might look into new generative models for language. We might automate tasks such as searching through images and video.

Those are our three buckets of value: **Insights**, **Automation**, and **New capabilities**. Each one of them requires a different team, a different skill set, and a different paradigm of management. When we want to start on that long journey to being data-driven, it's often nice to box up what exactly we mean so we can attack the problem in a more coherent way rather than trying to do everything all at once. If you try to apply a single paradigm to manage all three buckets, that's not often very compatible.

AA: Why shouldn't organizations get caught up in the often-quoted maturity models, such as Gartner? I know you've written a piece recently about this, so I thought we could talk a bit about that. I think that with some of the insights you provided, you hit the nail on the head.

> Jason's article
>
> *How analytics maturity models are stunting data science teams*, by Jason Tamara Widjaja on *Towards Data Science*: `https://towardsdatascience.com/how-analytics-maturity-models-are-stunting-data-science-teams-962e3c62d749`.

JTW: When we think about maturity models, a couple of them come to mind, and for one of them, the narrative goes like this. You get all your data in shape, then you get all the insights from it, and then you get all the value from the insights. That's one very simplified version of the narrative. But it's also almost impossible for a business. If you start with, "*Let's get all our data in place*," that's an impossible task because the amount of data increases exponentially. Data also comes in many different types, so even if you wrangle every single database in your business, chances are that you'll probably get to less than half of your data because lots of it exists either locally, such as on people's laptops, or realistically, in people's brains – and then there's unstructured data.

So, the idea of, "*Let's sort all the data, then let's start data science*," is a race that will never end. I think it's an endless road where value is far, far off. The winner, in this case, is the provider who sorts out your data management, as opposed to the company that is trying to create value.

That's one pothole. A different one is, "*Oh, let's do descriptive analytics, then let's do predictive analytics, and then let's do prescriptive analytics.*" This is one of the famous Gartner models. It sounds very appealing, but it is not so good in practice because when we think about descriptive analytics, we think of teams that do data warehousing and dashboarding in practice, whereas in predictive and prescriptive analytics, you have use cases where data scientists are modeling, and so on. These two are very different practically. One is very IT- and data-heavy. The other is very uncertain and experiment-heavy. If you treat a data science team the same way you treat a reporting team, that is a recipe for bad news.

There is uncertainty in the data science world, which is very different and does not exist in the data management world. If you try to be deterministic, promise outcomes before you see the data, and treat data science projects with schedules and milestones in the same way you treat an IT project, that is a route to project failures. That's a different trap.

AA: In addition to the advice you've just given to organizations broadly, is there any specific advice you'd offer start-ups?

JTW: That's an interesting one because my brother and I run a start-up of about 20 people, and I also had this question from my friend who has a start-up of about 40 people.

This is not a very popular point of view, but I tend to ask, "*Would your start-up actually benefit from data analytics at this point of size and maturity?*" There are, for instance, many B2B start-ups in very closed, relationship-driven industries where the incremental value of analytics is probably not very high. Sometimes, you have an enthusiastic **chief experience officer (CXO)** or founder who has gone to a conference and comes back saying, "*Oh, we need data science!*" I'd reply, "*No you don't,*" or rather "*No, not yet.*" I think that we should be realistic about whether the company at its current stage actually requires that capability.

On the other hand, if you are a start-up with lots of customer and operations data, you would probably benefit from analytics. In that case, I'm of the view that you should do lots of small experiments before you try to draw a big architecture diagram and start buying proprietary tools and black boxes, and so on. I think you need to find a pattern that works for your start-up.

Often, that means trying open source tools. Try to monetize off-the-shelf services and models rather than trying to train your own. When it comes to many tech start-ups, they have software engineers, and what is often underappreciated is that there are a lot of machine learning services that you can call as APIs from cloud providers, rather than training your own models from scratch. Try to push these as far as you can before you hire a data scientist. If you know exactly what you want from a data team, then sometimes you don't need a data scientist. You just need a data warehousing person or a data engineer to get your data in place and do reporting on top.

If you have software engineers, you might want to consider the commoditized AI/ML services from cloud providers. Only if you have real clarity, a use case, and a pipeline of work would I invest in more sophisticated capabilities.

It's not the best thing to say from the perspective of a data scientist but, at the same time, I've met many data scientists who are in start-ups, and because they were mis-hired, they end up doing work that is not data science at all, and I think that's a lose-lose all round.

Managing data science projects

AA: Something we touched on a moment ago was this notion that data science is not IT – they have to be managed differently. It's about discovery versus delivery. What's your advice on successfully managing data science teams, projects, and initiatives?

JTW: I completely agree with what you said. Data science is not IT. I've worked on both the *business* side and the *IT* side, and they are very different.

One mental picture I have is that if you're trying to get value from data science, with data pipelines running efficiently, it's like water from a drinking tap. IT is very interested in the tap, the pipe, and the water, but not very interested in the drinking, whereas the business is all about the drinking and doesn't really care about the pipe, water, and faucet, although it's all one picture. The two play complementary roles; they are very different mindsets and very different concerns.

To add to what I said previously, I find it's important to find a pattern that works, and that is the discovery portion of it. What this looks like is that you have teams with good business acumen, seniority, a good understanding of modeling, and a good understanding of data, all trying different use cases and seeing what is truly valuable.

> It's very common for use cases to fail but, for me, there's no shame in failure.

It's just inherent to the space due to the data you have and the complexity of the use case. There are ceilings to prediction and different models.

I have nothing to fear from failure. I think the art of failing well is key before you go from the discovery portion to the delivery portion.

Another way to frame it is that first, you do the science, and then you do the engineering. You don't engineer things when you have not done the science, and when you're in full-fledged engineering work, you don't staff engineering teams with scientists.

So, I think there are different aspects of it, but two-speed science/engineering is the way I think about it.

AA: I've often seen organizations struggle to go from science to engineering. They may begin with a proof of concept or a small prototype, but when it comes to productionizing it, they tend to hit a roadblock or struggle with that transition from end-to-end data discovery, model development, and productionizing to iteration, maintenance, and so on. How do you think organizations should overcome these challenges?

JTW: I'll share an illustrative case of how these things happen.

The most basic way it happens is that the same people doing the science are also asked to do the engineering. Often, the skill sets are so different that it takes a heroic effort from the scientists trying to take on skills that are wholly not their own to try to get engineering to work, and that's not a sustainable pattern.

To give you an example, there's a whole spectrum of what putting a model into production means. It could just be, "*I'm going to run it on my laptop and hit "Run" every morning when I come in,*" or it could be full-fledged systems with a frontend and a backend in the cloud, using model operations and MLOps. For that second format, you're actually no longer thinking about data scientists but maybe about an MLOps person. But then, if the model powers something more like a full-stack application, what you want is a software engineer at the backend, as well as a frontend engineer, maybe a designer for the frontend, and maybe a product manager as well. These things should not be tasks filled by data scientists. The skill mismatch would be too large.

I have to add: practically, I have seen many data teams heroically power through and try to take on these things. What happens is you have backend frameworks, such as R Shiny and Dash for Python, and these become the entirety of your application landscape, as opposed to a nice JavaScript frontend and a proper, more efficient backend, and so on. In these cases, there's another failure mode where, without the operations layer, data scientists sink into quicksand. Say I have one app in production and I spend half a day a week doing refreshes and managing the app. Say I then succeed, and I have four apps in production. Now, a quarter of my team is doing operations work. At some point, the team members say, "*This is not what I signed up for,*" and they leave.

The fundamental issue is that the skills for the proof of concept are very different from the skills needed for production. That was actually one of the things that drove my recent career change. The company realized that a product team and a data science team look quite different. So, they sent the data scientists closer to the decision-maker, and the product teams closer to IT, where they could borrow resources such as software engineers, frontend designers, and so on. Having that mental model where there are different teams for the proof of concept and production is useful.

It also comes down to early collaboration. You don't want to say, "*Ah, here is my model. Please help me: what do I do with it?*" You want to have the production process at the back of your mind from the start. What that means in practice might be that instead of doing things on a Jupyter notebook on your laptop, you might consider doing them in a form or environment, which you can eventually containerize, because it's easy to put something like that into production. Alternatively, your company may have invested in an MLOps product or platform. There are so many of them around now, from massive companies with very loud voices in marketing. I'm sure your company will have at least investigated some of them. Starting with those platforms in mind, or even starting work on the platform itself, is not a bad idea.

AA: How important do you think the role of project managers, delivery managers, and the like is in this broader field? How important is it to have dedicated people for that?

JTW: It's a really interesting one.

I think there is the full-time role of a project manager, and then there's doing it as part of being the leader of a data science team. I've seen both, and I think they are a little bit different. I think it's always helpful for a project manager to have a little bit of a data science background if they're managing a data science project. If not, there are all these potholes that we could mention that many project managers fall into, and it's no fault of theirs. In their own certification and qualifications, they learned how to manage uncertainty, time, cost, quality, and scope, which are very useful, but they may not know things such as modeling uncertainty, data uncertainty, and some of the other nuances of data science projects.

They may also not intuitively fall into that discovery/delivery mode. They're almost perpetually, in my experience, in delivery mode. They see themselves as a delivery specialist: *"I know that's my job. You entrust delivery to me."* With that golden delivery hammer in mind, the POCs and the science also become delivery.

That's my main hesitation with project management taking leadership. I think once they understand some of the fundamental differences, they are great. But if not, then I would prefer a data science person to lead. Having said that, there is some confusion now in the market. I think once you get into the more sophisticated space of AI applications (recommendation engines, chatbots, or anything else that's product-oriented), *product* managers are extremely useful. That's more than just delivery. It's also setting the vision and roadmaps for a product beyond the scope of the project, and I think those skills are distinct from the project manager skill set and a useful thing to have.

> **Note**
>
> Jason expands on this point in his article *What every project manager should know about managing data science and AI projects*: `https://medium.com/towards-data-science/what-every-project-manager-should-know-about-managing-data-science-and-ai-projects-d13f3f8f62a`.

Why AI projects fail

AA: We've talked about developing these products, but we also should mention that, often, we read articles that cite a Gartner report in which it states that supposedly 85 percent of projects fail in the broader data science and AI space. Do you agree with that, and if so, what is happening?

JTW: That's a great statistic that you can just riff on for quite a while. I believe Gartner has one of those statistics, but also I've seen a statistic in the context of MLOps platforms, basically complaining that models don't make it into production. To me, those are different things because just because a model doesn't make it into production doesn't mean the project is a failure.

As an analogy, I work for a drug company. Over 80 percent of vaccines fail, but I don't hear people complaining about how 80 percent of vaccines fail, because that's the nature of science. You do your best with the data. You do your best with the model and the distributions that you have. When you train the model, you see whether the intervention works. I think data science is very similar to that: there is uncertainty inherent in the data that you have and the model that you want to train. You want to get into a place where you are running through lots of models as fast as you can so you can explore use cases quickly.

So, I have no trouble at all with an 80 percent failure rate.

> **A model not going to production is not a failure.**

I think a second thing I should mention here is the distinction between a data science *project* and a data science *model*. If you are a project manager and you think, "*Ah, this is my delivery scope and that's a project,*" that's a very different mindset from a data scientist saying, "*OK. I'll try method one, method two, and then method three.*" These are different ways to approach a problem. Sometimes, different data scientists with different backgrounds will come in with different methods, and where one method may not quite work, little tweaks at a method level may give you success.

To me, some project failures are premature, because you might just need to reframe the project. Other times, project failures are actually not failures because it was a successful experiment that showed that a dataset was not right for the use case.

> **Projects that don't enter production**
>
> Jason has written a longer article on this topic: `https://jasontwidjaja.medium.com/8e245a03bd53`.

Communicating a realistic expectation to clients and partners

AA: I'm so happy you went into those details. Sometimes, things are labeled as failures because senior executives don't really understand what the reality is or what to expect. Often, they can be sold hype by a vendor through a talk they go to. They get caught up in an ideal world of AI. How can you give them a more realistic picture of what AI is and isn't and what they should expect? I think one important part of that is discovery. You don't know what you'll find, sometimes.

JTW: I think there are a couple of lenses that I've tried that I think executives find useful.

The first one is we generally reach an agreement on what I call these three buckets of value. In some circles, people use data science and AI analytics very interchangeably, but if you are doing it in your company, you might want to be a bit more thoughtful and nuanced about it. I go back to those sources of value: insights, automation, and new capabilities, and the different ways to approach each one.

Besides that, one thing pertinent to AI but not not data science is the idea of autonomous execution. AI systems are sometimes just data science systems with a human out of the loop; or, to flip the same thing the other way, the data scientist is the one that disintermediates the model from the consumer because they're the one who interprets it and presents it to the consumer. But for AI systems such as a recommendation engine, there is no data scientist there monitoring the fact that I got recommended coffee because I went to Starbucks yesterday. These are automated systems, and so the element of automated decision-making is one of the roots of why AI governance is so important. You don't have that disintermediation.

Sometimes, these automated decisions are applied to use cases that are risky enough to cause harm, and because they are scalable, the harm can multiply very quickly. That's how we enter all these discussions on biased systems, unfairness, and so on, but I think the more insidious things include industrialized distraction – the way data scientists are so successful that we end up being addicted to our news feeds.

There are a whole bunch of things in there that stem from the differences between data science and AI. As a starting point, one useful way to think about it is to say, "*Ah! I need to think about what happens when I'm not just a data scientist but there's an autonomous system working for me.*"

Establishing a data culture

AA: How important is it for an organization to have a data culture, and what does a successful one look like?

JTW: I think the data culture question is really hard. It's easier to engineer software than to engineer data culture. Inherently, some people are just not built this way, and I think I need to acknowledge that. In the distribution of people in the company, there are large populations who are very number- and data-averse, and we need to live with that.

I also think data culture itself has different aspects. Sometimes, we think about it as the ability to interpret the output of a data science project because that seems the most applicable to a data team. But it is a lot of other things, and I think the way we approach it must have different facets.

For instance, someone might never want to see the output of an R model, but they might be OK to understand their own cognitive biases and how their mind plays tricks on them. We have this idea of data literacy as more than just being able to read dashboards; it's also thinking about ideas such as cognitive bias and the way our mind makes decisions.

There is also the discipline of being suspicious of data. You mentioned you work in government. I actually had one government example I came across just earlier this year, where they took a survey of whether people were satisfied with the train system – and they took a survey *at the train station*. Of course, people who didn't like the train system would never even take the train. There are many of these very relatable examples that just get people suspicious and thinking about data. Long before you reach an understanding of the output of a machine learning model, you have all these ways our minds think, which I think are a lot more accessible. When I think about data culture, I tend to go for that a lot more than going straight for data interpretability.

AA: How important do you think training for enterprise-wide data literacy is to try and overcome some of these issues?

JTW: I think there are two separate questions in there.

I have a strong belief that companies benefit from data literacy because not all important decisions are made at the level of a data science project. Improving the general conditions can be very useful, but when it comes to high-value use cases, to be very frank, I think it often depends on the person who is interpreting and making the decision. We don't want to fall into decision-driven data making, where a senior person gets a data science team to validate what they already think. That's not very useful, but it's quite common.

The other thing is that it's really important to have senior data-literate people as sponsors for long-term data programs. I sometimes see data programs being sponsored by people who want to be associated with data analytics but, actually, they don't want to be the consumer themselves. To me, that is very dangerous. It's like saying, "*I want to be fit so I buy my weights, but I don't want to exercise.*"

To me, organization-wide data literacy is great, but you need to pick out some sponsors who have an inclination to not just pay the money but also get into it themselves.

AA: **Like data literacy champions who champion the cause.**

The importance of data governance

Apart from the importance of data culture, how important is data governance? What are some of the key elements to making it successful?

JTW: I think it's certainly important. There's another statistic that's floating around that says that data scientists spend 80 percent of their time cleaning data. I'm not sure whether it's as high as 80 percent because data scientists also spend lots of time trying to *get access to data*, and that is a data governance issue. I think it is a tricky thing for a stranger to come up to you and ask for a company asset that you are not very comfortable with yourself. To me, that's the problem that data governance tries to solve. In terms of greasing the wheels to make analytics efficient, data governance is very important, but also, in terms of managing data risk, data governance is very important.

I'll also make a distinction between *data governance* and *AI governance*. When it comes to elements of data governance, I think about things such as security and access and cataloging and visibility. For AI governance, I'm thinking about the considerations when you're building autonomous systems: is it a high risk? Is it a low risk? Is it human-in-the-loop? Out-of-the-loop? Are you perpetuating some discrimination because of the algorithm? Are you putting controls to mitigate against the effects of black-box models?

So, we do have both these teams – a data governance team and an AI governance team – and they are not the same, and both are important. They both scale according to the importance of data use versus the amount of AI that's in a company.

AA: **Related to that is the work you've done in Singapore around the AI ethics body of knowledge. There's a lot of interest among organizations globally in adopting various principles, and establishing ethical AI frameworks. How can we help organizations take the policies and frameworks, and all this discussion, and put it into practice?**

JTW: This one is definitely worth another book. It's a great topic.

Just to show a statistic out there that there's a lot more policy than practice at the moment: the most recent meta-analysis (2019) showed there were 80 policy documents around the world. Earlier this year, it was well over 120-130, but unfortunately, it's quite patchy.

> Some countries have a lot, some companies have their own, and some countries have nothing at all. If you are, say, headquartered in Australia but with overseas operations, it gets very complicated because of the disparity in maturity in this space.

The way we approach it is to first distinguish between *policy* and *operational* processes. We created our global AI policy, which happens to be going live this month. We operate in over 140 countries, and so we looked at AI documents across different continents. We eventually arrived at the IEEE, the OECD, and whatever makes sense locally. I know Australia has a pretty good document. Singapore has one as well. I can take a local document, but can I join it with globally recognized standards?

We made a superset of those and then adapted it to make a company policy. The reason we did that is that if you jump straight to processes without policy coverage, governance becomes very hard because governance needs to tie into policy.

> We do have an internal policy as a first layer, but lots of these policies, when you look at the language, are really vague. They look amazing to a policy expert. They look terrible to a data scientist or MLOps person.

They say, "*What is 'human-centric'? Tell me. Show me the code for human-centric.*" It just doesn't quite work.

So, what we did was an AI impact assessment at the operational level for projects that use AI. What constitutes AI is a whole question in itself, but we tried to make it not stand alone; we tried to hook it to existing risk management processes. I think this is a very interesting area because the skill set for it is all over the map. We work with policy people, we work with cybersecurity people, and we work with data science people, and I've not been able to find all three in one person. You need lots of relationships and handshaking.

> We have a saying that the most difficult language to translate is English into English.

We have policy people, compliance people, ethics people, cybersecurity people, and data science people, from very disparate domains, speaking in what's supposed to be English, but are actually very different languages. We're trying to bring worlds together. It is quite a difficult space, and I think it's a space of great potential, great growth, and great importance.

I will add one last comment here: I'm very passionate about this area. I think ethical AI is only the midpoint. We've gone from exploitative to ethical, but the attitude is, *"I'm going to do business as usual, but I'm going to try to minimize the harm that I do as I scale up my business."* I think the opposite is basically to think of AI in a more redemptive way: to think of AI applications that can really help the world. You can appoint AI systems to address child trafficking, security, poverty, and hunger. So, I think *ethical* is good, but there's a lot of other good you can do in AI by pointing AI at the right problem.

Discussing leadership

AA: This all involves strong leadership. What do you think makes a great data leader? What do you think you offer as a leader, and what inspires you in other people who you see as leaders in this space? What advice can you give someone looking to move from the technical space into a leadership role?

JTW: That's a really tricky one. I think you have lots of experience in this space as well. I think you've had a tremendous journey yourself.

> Sometimes when I hear this question, I almost feel like I want to apologize to the technical person asking it because there is this notion that leadership is not technical: either you're a data leader or you're a technical person. To me, that in itself is quite a dysfunction.

In an ideal world, you would have a track where a leader would say, *"What exactly are you managing? Are you managing business strategy, vision, budgets, and projects? Or technical strategy, methods, and research?"* I think there's a space for an equally senior technical subject matter expert who's extremely deep in a domain. That's a different flavor of leadership that people will look to for specific problems, guidance on methods, and technical leadership. These work side by side.

The way I hoped to build my team, which is a bit of a struggle, was to have myself and someone who is less keen on people management but who is a technical specialist in our key AI areas. We would lead as equals and model two different flavors of leadership. That was the way I thought about it.

In the absence of that, I think we can apply our data instincts to this. I think the most difficult thing to do, if you're put in a place of data leadership, is to hire someone that is very different from you, and to me, that is a very human and difficult thing. For instance, I'm not amazing at working with project manager schedules, so I would choose someone who is. They may happen to be an ex-data statistician who is really good at Scrum and daily Agile things. That picture looks different for every person. I think to have someone that complements you is the most useful thing to do, but that's also the most difficult person to hire because you have things within you that are screaming out to say that this person is not like you and you can't work with them. But actually, that person might bring exactly the skills you need to excel as a team. I think building the leaders of a team and the members of a team is probably the most important thing there is.

AA: Are there specific attributes that you think a data leader should have in general, either from a technical or non-technical background, depending on their remit?

JTW: I think that's a very tricky one. This is actually a debate that I sometimes lose. I'm not sure how common this is, but I have the notion that if you're going to lead a data team, you should have a fair grounding in the technical world itself, and if you don't, you should learn about it. We live in a world where you no longer need to take a Ph.D. to get that depth of knowledge. There are lots of other offerings that can allow you to upskill in these areas.

I personally find it very hard to lead a team if I have not experienced at least a little bit of what they've done: the mindsets and the realities of it. That's my point of view.

> I often get overruled by people who think that leadership is a generic role in itself – that it's agnostic of the team. I'm very much of the camp that you want to have some grounding in data, and if you don't, please go and get some.

AA: I agree completely. I see so many roles where people are appointed to a particular senior role who don't have the background where the organization thinks, "*We can put a generalist in the role, and they'll just learn all the data and tech stuff.*" **It's often not as simple as this.**

As for data science leaders, what advice can you give them on managing senior execs from the business who come to them with sometimes unrealistic demands or multiple priorities?

JTW: That's quite a tricky question because you're talking about managing the expectations of your boss, over whom you have no control.

So, what exactly are you managing? The answer is *expectations*, and that's probably one of the most important things to manage.

What I've found useful is to give an analogy. Ask your senior leader, "*If you were to think of data science as a department, what department would you think it's closest to? Are you thinking about it as a manufacturing department, where you're expecting things to go in and outputs to come out? Do you think of it as a shared services department, where you drop requests in, we have SLAs, and things come out?*" I often challenge them to think about it as an R&D department, where you need to fund it and not have very tight, granular deadlines for deliverables. We give you visibility of effort and experiments.

R&D are not the same. There's the *R* portion and there's the *D* portion, which is basically the production portion. I always try to get them to use that analogy of R&D, where you need to fund the *R*. Once you have the output of the *R*, that's when you can go into *D* mode, where it's a lot more similar to project delivery. Every leader can say, "*Yes, I know that bit.*" Use the analogy of asking them what department they frame data science as.

AA: As leaders, how can we better mentor, support, and develop our staff? I think that that is one of the key parts here. I think – and you probably agree – it's about giving people more exposure to how the value of their work directly impacts the business and decisions.

JTW: Absolutely, thank you for bringing it up. That is a very unappreciated area. If a data scientist doesn't know the context of what they are working for, they lose motivation very quickly.

Advising new entrants to the field

AA: **For aspiring data scientists, the field is continually growing, and there's so much demand and competition. What advice can you offer those people? Do you have any particular views when hiring as to formal university studies versus being self-taught via online courses?**

JTW: If you asked the question 10 years ago and then asked it now, I think the answers might be quite different because the education landscape has changed so much. I got into data science at a time when Drew Conway's Venn diagram was very popular, and I thought, "*Oh, that's really cool,*" because here in Singapore, we are quite qualification-minded. I said, "*Oh, you want programming, business, and statistics? Sure. I will get three degrees.*" I went ahead and got those.

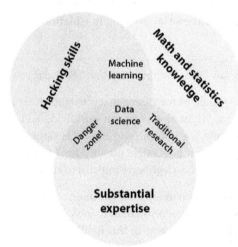

Figure 10.2 – Drew Conway's Venn diagram of data science.
Reproduced with the kind permission of Drew Conway

Firstly, I would say that you can get the same thing in a lot less time today because there are targeted, granular, on-demand courses. You can probably get the content and – to a lesser extent – some experience in a lot more of a concentrated way and in less time today. So, that's the first thing.

The second thing is that I think there's also a danger in going completely online because with everyone being able to publish, you lose the filter of quality. There are many courses out there that strike me as scams. They overpromise and charge you lots of money, and you don't get very much out of it. If a hiring manager were to see that on your resume, it would actually be a negative signal as opposed to a positive signal. So, I think there's a bit of a balance needed there.

To take a different lens to it, I think getting into data is also not a homogeneous single job; there are so many different nuances – whether you want to be a researcher, a data scientist, an AI/ML engineer with a bit more of a software engineering background, or a reporting BI professional. All of these are well-established, well-trodden career paths, and if you are able to try a few paths and then pick a path that you want and are interested in, that removes three-quarters of the work you need to do. I would advise people to be more granular in getting into a data career rather than just saying, "*I want to work in data.*"

AA: You're right: the education space has changed. What are your thoughts on what universities should be doing to adapt these days? Are there elements they're not teaching that should be integral to courses, such as ethics or governance? Should they be part of the curriculum for a data scientist?

JTW: Absolutely. I think this is a loaded question because you have a lot more time in academia than myself, but I certainly think that the interface between academia and industry could be a lot closer.

> There could be a lot more industry consultation based on what we think are the skills you need next year, as opposed to universities working in isolation and coming out with their curriculum.

As hiring managers, we'll often say, "*Why did you study that? Why did you not study this?*"

I think, as you mentioned, ethics should be something that is taught widely in any data program. In the past 2 or 3 years, we've started to see legislation come out worldwide, and data teams themselves may soon be liable for these things. To me, it is not a nice-to-have: it is almost existential because, one day, you might be doing something that your boss might get sued for. I think we definitely want to do the right thing, but also, there is starting to be a legislative impetus to study proper governance and ethics.

AA: When you're hiring – just to pick two of the most common roles in the field – a data engineer or a data scientist, what skills and attributes do you tend to look for? What should they be cognizant of when they're trying to develop their careers in this field?

JTW: Whether they are a data scientist or an engineer, or any of the other data roles, I look for the soft elements of collaboration, teaming, and the intuition to not always do things in their own team and silo.

> If you're a data scientist, you need to get into the mind of a decision-maker. You almost need to have a second major in *empathy*. You need to understand how a person is thinking and collaborate with the teams that will provide you with your data and MLOps. So, teaming is definitely a big must.

The same goes for the engineer. If you're an engineer, you're probably building a pipeline to offer a full-stack product or a data science product, so you have all these very different roles working with you. I have observed many interesting interactions between engineering teams and user experience teams. They're extremely different personas, but they need to work together. The ability to collaborate with teams that are very different from yours is the key to getting the larger team to work.

There is another thing. Much of the HR work goes, *"OK. Let's define what an engineer is. Let's define what an analyst is."* But at the same time, when it comes to the individual, I have a very strong desire to not fit people into boxes but rather to draw the box around the person because there's so much nuance in the data and AI space. Beyond the level of the "data scientist" headline, there could be half a dozen sub-flavors of data scientists: NLP specialists, optimization specialists, statistical time-series people, and so on. You want to be thoughtful about what exactly a person brings and what exactly a person wants to do and be very granular about career paths at an individual level. To me, being a people manager means considering, for every person in the team, how people want to develop and trying to create that future-state picture. It's a jigsaw puzzle where everyone fits together. Everyone has something to aspire to.

Generative AI and ChatGPT

AA: What does ChatGPT and generative AI mean for AI and data?

JTW: The architecture of transformers is 1) is adaptable to multiple purposes, and 2) can be multi-modal. This marks a drastic move away from the era of single purpose models and early steps away from what everyday language calls "narrow" AI. This also unlocks multi-modal uses cases and I expect a sharp increase in applications that incorporate and integrate multiple data types (for example, text and images)

The way that running generative models like ChatGPT multiple times with the same input yields different results, together with the multi-purpose nature of such models, means that governance and testing will have to undergo a paradigm shift. Governance on the underlying models of ChatGPT is quite meaningless as it can be used to generate both pizza or poison. Governance will then have to shift to the point where the model is fully formed into a use case and /or to the point of usage.

The impact of ChatGPT on "citizen" AI is similar to the impact of the pandemic on remote work – It took a gradual trend and accelerated it manyfold, cementing it as a permanent feature of work life. Going forward, any AI strategy would require citizen AI as a central pillar, as opposed to a peripheral consideration.

With models like GPT-3 being both available for fine-tuning and provisioned via an external service, organizations will have to navigate an increasingly complex "model zoo."

In terms of AI implementations, early research outlines the footprint of ChatGPT and similar technologies to be:

- A productivity increase, especially for **No One Right Answer** (**NORA**) tasks rather than historical, mathematical or scientific queries.

- An overall quality floor and uplift.

- A change in the structure of work towards less time taken to draft and more time needed to check.

AA: How can data scientists/AI experts leverage ChatGPT to benefit their organisations and their own careers?

JTW: For data scientists and AI teams, it is fascinating that English (and other spoken or written human languages) are now *programming languages*. **Prompt engineering** is now a new skill to learn and a new job role. For data science and AI teams it is also an opportunity, giving:

- A method to shape model behaviour by simply appending text invisible to the user (for example, "limit the output to 200 words")

- An enabler for mass personalization, generating dynamic descriptions of users based on other features, and appending those descriptions into a prompt input

- A basis for competition, with different companies chaining different combinations of models together to create new applications

AA: What are your predictions of the future of Generative AI and LLMs, and their broader use?

JTW: I predict that advertising business models will consist of training language models to respond to Generative AI prompts with product placements

The arms race of generative models and detector models is on. However, detectors will always be a step behind, and generative work will always sit on a spectrum (is this generative if a human checked it and edited it by 20%? How about 10% or 1%?)

This will lead to a digital environment of post-trust, where citizens will have lingering doubts about whether a piece of content is human authored.

> Doubts over authenticity could lead to trust becoming a premium commodity, and likewise humans will be a premium commodity.

Possible ways this will play out is a swing back towards trustworthy sources of information, and new markets for assurance that a piece of work was done by a human.

Predicting the future

AA: Final question, and it's a big one: what are your predictions on the future of AI, attaining AGI and in particular, humans and machines working together. I think it's an area that we both strongly believe in, and I'd love to hear your thoughts on it.

JTW: That would be another book!

Implicit in thinking about AGI is thinking of AGI as human-human intelligence: "*When will AI be as smart as I am?*" To me, that is sometimes not very helpful framing. It carries a bit of arrogance in that it implies that a human is the gold standard in intelligence, but where the most sophisticated models are already heading today means that AI systems will far outperform humans in a few things and greatly underperform humans in other things.

When it comes to AGI, for me, it's broadening things out so that those peaks become broader and so the AI becomes more general. But without fundamental scientific breakthroughs, the troughs that underperform will not go away. General reasoning, causal reasoning, common sense: all these things are very intuitive to us but totally unintuitive to machines. So, I think rather than using the term *artificial general intelligence*, we should track the shape and capabilities of AI as we progress, pay particular attention to the places where AI will still underperform human beings, and wrap risk management, ethics, and so on around that. That's my general framing of it.

There's also a much longer-term view, where there is superintelligence as well as concepts such as safety engineering. That is a very important area of research because already – actually, this year – we are seeing, not exactly evil AI systems, but AI systems run by evil people. We have seen lots of growth in the autonomous weapons space, which, to me, is a really serious threat. You can see autonomous systems that are able to kill humans without a human in the loop. These are areas where, if you scale them up, they can cause imaginable harm. Safety engineering is a concept that means we want to design systems that don't even fail once.

Those things are probably a lot closer than most people realize. I believe the median prediction for AGI is 15 or 20 years, but autonomous weapons exist today, and the need for these things is quite important. The current ethical AI landscape of trust and human centricity is a very immature landscape because, when we think of trust, we almost snap to, "*Oh, let's do explainability.*" But there are so many things beyond that that would garner trust. And even explainability itself right now is one-way explainability – "*Can the machine explain its decision to me?*" One of the research projects I was working on last year looked at whether we could do the opposite. Instead of us saying, "*Hey, Alexa, these are the five features behind why I chose this prediction,*" could it be the machine saying, "*You've not labeled this yet, but I'm going to guess you would label it in this way, and these are the reasons why I think that.*" I think two-way interaction between humans and machines is lacking as a foundation for trust. We are still in a quite nascent stage in terms of humans and machines working together. It's a big topic, but I think I'll leave it at that.

Summary

I really enjoyed my discussion with Jason. He provided great insight into the human side of making an analytics capability successful. I particularly liked his concept of categorising the value that can be gained from analytics into three buckets: insights, automation, and new capabilities.

One of the themes we discussed was that for an organisation to be data-driven, decision-makers need to be prepared to change their minds if the data suggests it. This means the change to become data-driven involves not just data and modeling but also the psychology and environment of the decision-maker.

Jason raised the a heavy reliance on data and analytics maturity models without understanding their context and the nuances of doing data science in practice can be a recipe for failure. Business needs and value ultimately drive what type/mix of descriptive/predictive modeling needs to be done over time, rather than some notion of which type is more "mature." Maturity is measured by the business value that an analytics function is driving.

What I've learned throughout my career is that building a high-performing team is like a jigsaw puzzle – you need to focus on people and how they all fit together as a team, -complementing one another's skills and personalities, rather than having preconceived ideas of fixed roles. Jason's view is that if you're going to lead a data team, you should have a fair grounding in the technical aspects of what the data team does. I share his view, but Jason mentioned this isn't always the case, and he's lost some debates on this point. One important topic we discussed was leadership, and in particular, that leadership is not only for those with direct reports. Senior technical **subject matter experts (SMEs)** also play an important leadership role in an organization. Jason advocates for leaders to build teams that have a diversity of personality types, strengths, weaknesses, likes and dislikes, and skill sets, in order to build the most productive team possible.

In regard to the tricky task of managing the expectations that senior executives have of your data science team, Jason shared a pearl of wisdom: have them think of it as a research and development function that needs to be funded and supported, rather than a delivery function such as IT which comes with rigid and specific deadlines and expectations.

Jason also shared some valuable insights on managing data science versus IT projects, and how to think about the failure and success of projects. He sees it as two-speed - first you do the science, and then you do the engineering. By contrast, my view is that data science and engineering need to be thought of as a continuum throughout a project's life cycle, and not two distinct steps. At the beginning of a project – and especially during the development stage – data science plays a leading role, but as the project progresses and nears production, engineering will then play the leading role.

As to the future of AI, and in particular Generative AI and tools like ChatGPT, Jason believes they will lead to a digital environment of post-trust, where citizens will never be completely sure whether a piece of content is human authored. Trust and the human touch will likely become a premium commodity.

11
Jon Whittle Turns Research into Action

It was an honor to chat with **Jon Whittle**, the director of Australia's national center for R&D in data science, **Data61**. Prior to this, Jon was the Dean for the Information Technology faculty at Monash University, and early in his career he was a senior research scientist as NASA. Jon is an experienced research leader and academic who is well-placed to speak about innovation in the field of AI and strategies for success. I was particularly interested in learning about Jon's views on how to overcome challenges in translating research into actionable outcomes – especially at scale.

Building a career

Alex Antic: First question – I'd love to hear about your career broadly – your transition and trajectory up to this point. In particular, what inspired you to take on the (what I imagine is challenging yet exciting) role as director of Data61?

Jon Whittle: I've actually been in AI for a long time. I did my PhD in AI at the University of Edinburgh over 20 years ago. That was in automated theorem proving – so, getting AI to do mathematics – back when AI wasn't quite such a hot topic as it is now. In fact, it was on a bit of a downward curve at the time.

From there, I moved to the States, and I went to work for NASA, where there was a lot of early AI in space exploration going on. So, there was the Deep Space 1 probe, which actually took full autonomous control of a deep space spacecraft. There was the Mars Rover stuff going on there as well, but I actually left AI thinking that all this stuff will never work and went more into core software engineering until more recently, when I moved to Australia in 2017 and found that AI was on an upward curve again.

At Monash University, I repositioned what we were doing in AI, and I created the Monash Data Futures Institute, which was a multidisciplinary, university-wide AI data science institute. Then, this opportunity came up at Data61, which was a very interesting opportunity because Data61 is part of CSIRO and has a national impact. I love Australia's universities – I think they're fantastic universities. But, at the end of the day, you work for one university, whereas if you work for Data61, you've got an opportunity to work with all universities and industry and government, and you can really have an impact at a national scale.

CSIRO and Data61

Commonwealth Scientific and Industrial Research Organisation (CSIRO) is Australia's national science agency and is responsible for scientific research.

Data61 is CSIRO's data and digital specialist data science arm. It is Australia's leading digital research network, offering advisory services, applied research to solve business problems, and product and platform development.

AA: I guess that's probably the best part of your role, having the broader remit rather than just being confined to one institution.

JW: It is – you can make stuff happen when you've got that national footprint that would otherwise be very difficult to make happen. I think that Australia's new National AI Centre is a very good example of that. We were asked to host the AI Centre and really work on bringing the ecosystem together, which a single university would never have been able to do by itself.

AA: What would you say would be the key skills and attributes that helped you make the transition from researcher to leader? Did you feel like it was an easy progression, or are there certain transferable skills you had to utilize to make that jump?

JW: I suppose, in some ways, the Data61 role is a leadership position, so it wasn't that much different from any leadership position. I always think one of the key things that any leader needs to be good at is articulating a very clear vision. I did that in all my leadership roles, but it's about being able to take a look at what the strengths are of the institution that you're going into, but also being able to imagine a future, set a North Star, and then put a plan in place.

Monash is an interesting example. I was head of the faculty of IT there, and when I joined – according to *Times Higher Education* – the IT faculty was ranked around 300 in the world, but Monash is now ranked 57. We shifted the dial quite significantly in a very short space of time – about 3 or 4 years. When I left, somebody said to me, "*When you first came in and you spent the first 6 months figuring out what the vision was going to be, and then you stood up and told us about it, we all sat there listening and thinking you were a little bit crazy because it was so ambitious.*" But actually, you need to be ambitious, you need to set goals, and it's OK for those goals to seem unachievable, as long as you put the scaffolding in place so that people can see that there is actually a pathway.

I think it's a similar thing coming into Data61. One of the challenges that Data61 has, just like any institution working in AI or digital, is that digital really is everywhere and can be applied everywhere. I often say it's a blessing and a curse to be working in digital. It's a blessing because everybody wants to work with you, but it's a curse because everybody wants to work with you. If you're not careful, you can end up spreading yourself too thin.

So, coming into Data61, it was again about how we could actually articulate a clear vision and a strategy.

> **Moreover, that strategy needed to talk about things that we *wouldn't* do as well as things that we would do.**

We've got a strategy in place now that's got three elements to it. The first of those is to support the AI ecosystem nationally and bring it together so that Australia can take its rightful place on the world stage. The second part is using digital technologies to support recovery and resilience. So, that's both recovery from pandemics and resilience against the next pandemic, natural hazards, or cybersecurity. The third one is the one that people often find very interesting, which is what we call reinventing science. There's a huge opportunity, I think, to reimagine the very way that we do scientific discovery and digitize that, not just to increase the productivity of scientists but also to enable science to take place that just couldn't take place without those digital tools.

So, we've got those three things now. They're still quite broad, but they're three North Stars that people can get behind and that inspire people. As with any leadership role, it's the ability to come up with that very clear vision, articulate it, and then keep reinforcing and repeating it at any given opportunity.

Translating research into real-world impact

AA: Following on from that, what do you think are the key challenges in translating research into real-world impact, particularly within the AI space? How do we address some of these challenges?

JW: There's a common narrative within Australia when it comes to AI, and that's that we're very good at R&D but pretty bad at commercialization and translation. We certainly are absolutely excellent in terms of R&D. We appear in any top 10 list of AI nations globally, but it's probably true as well that we don't necessarily implement the commercialization of those technologies as much as some other countries.

In Data61, I think we've been relatively successful at this. Since Data61 was formed, there have been almost 20 spinouts, many of them extremely successful – Coviu, for example, which is one of the biggest telehealth providers in Australia with a Data61 spinout. More recently, there's been Emesent, which is using some of our 3D mapping technology in all kinds of different applications and is going brilliantly. There are lots of great successes.

We are actually, though, in the middle of rethinking our approach to commercialization. In the past, it's largely been a case of trying to develop technologies in-house and then supporting them with resources to become sufficiently mature so that they can stand on their own and spin out, and that can be very successful, but I think it's not the only way that things can be done. The big missing piece for me is actually focusing on the people first rather than the technology, because one thing I've learned about commercialization and entrepreneurs is that it's very much a mindset thing.

> **You've got to have a founder or a CEO who is driving things and is passionate about that particular technology and the problems that it can solve; without that, you've got nothing.**

They've also got to have some business sense or partner with people that have got business sense. I think we miss a trick in Australia more broadly, because if you look at our PhD programs or even our postdoc programs, they still operate on a very traditional academic model. People come in and they start a postdoc. They're told just to do some good research, publish some papers, and then it might be that 2 or 2-and-a-half years into a 3-year postdoc, there's actually something with real commercial potential there. By then, though, it's too late to go and commercialize that because that person's already thinking about where their next job is going to come from.

So, I'm a firm believer that we actually need to be training people in what it means to be an entrepreneur from day one so that every PhD student gets exposure to that and they know that it's a possibility. Every postdoc gets exposure to that. That doesn't mean that they'll all come in and then go on to be entrepreneurs, but at least they've all been given the opportunity. I think now we have PhD students finishing and postdocs finishing, and they don't even know the first thing about what it means to commercialize technology, so it's just not a pathway that they're exploring.

> I think we need to focus on training up a whole generation of research entrepreneurs.

AA: That leads to one question I was just about to ask – what should universities and research organizations be doing to educate, train, and support future AI experts? Entrepreneurship is one aspect, especially with all this appetite for founding a start-up these days. They need support. What else do you think we should be doing, more broadly?

JW: I think probably the other great opportunity is looking across the disciplines.

We launched a new program called **Next-Generation AI Graduates** recently, with some federal government funding – about 25 million dollars – and that is going to support up to 250 students at all levels working in AI. We designed that program in a very particular way, because I actually don't think that we have so much of a gap in terms of AI experts or computer science experts. Australian universities are very large – we churn out many graduates every year that have IT/computer science skills, including AI.

Where the gap is, I think, is in other disciplines. We know that AI is affecting just about every industry, and if we really want those industries to fully adopt AI, then it's actually the engineers, medics, and lawyers who need to have some level of basic understanding of AI. They don't necessarily need to be experts at it, but they need to have an understanding beyond just the broad definition of what AI is so that they know how to apply it in their industry.

With the Next-Generation AI Graduates program, we're doing a couple of things. First of all, it's focused more on people from other disciplines and training them in AI, rather than focusing on computer science. Secondly, we're taking very much a team-oriented approach to it. We're having students work in cohorts – for instance, 10 to 20 students – and this allows you to tackle larger-scale problems in a particular area. Rather than every student working on their own project and everything being disconnected, you have teams of 10 or 20 working on the same problem but from different disciplinary backgrounds and perspectives, all with the aim of looking at how AI can be applied to address the problem.

It's a bit of an experiment and we'll see how it goes, but I think more broadly, there's an opportunity within universities not just to train computer scientists but also to have AI as a core module across every program in the university, so to speak.

Developing AI that is ethical, inclusive, and trustworthy

AA: With all these people developing AI-capability products and so on, there also comes risk. How do we instill knowledge and educate these people to make sure that they're developing responsible AI from the get-go with ethical notions in mind, rather than it being an afterthought?

JW: You just hit my favorite topic. This might be a long answer, but we'll see.

First of all, the good news is that there is a lot of interest globally in so-called ethical AI or responsible AI right now. How do we develop AI systems in a responsible way? I think if you turn the clock back even 5, 10, or 15 years, there was very little exposure to that going on, although it's worth pointing out that even from the very early days of AI, people were talking about the ethics of it.

If you go back and look at the early proceedings of the **International Joint Conference on Artificial Intelligence (IJCAI) conference** (which is the big AI conference that started in 1969), even in 1972, there was a paper that talked about the ethics of AI and run all these thought experiments about what AI might become and how we can make sure it is a force for good and not for evil.

So, it's got a long history, just as AI has itself, but it's probably only in the last 10 years that it's really gained a lot of traction.

Since then, we've had lots of ethical AI frameworks that have been developed. Data61 led in producing Australia's AI ethics principles – eight high-level principles, such as transparency, human-centered values, and fairness. Now, pretty much every country's got a similar set, and there are also organizations such as the OECD that have got a set.

They're all a little bit different, but the core of them is essentially the same. There's a big missing piece, though. We've got a general consensus that these principles are important, but the question is, how do you actually implement them? How do you operationalize them? That's the area that we're really working on, because you have to think of it from the point of view of a software engineer or an AI technologist who's building these systems. They're used to having quite well-defined specifications or requirements of what it is that they're supposed to build.

Even if you're doing agile software engineering, you have conversations with your customers, and it's pretty clear what the next feature of your product is that you're going to build. Imagine saying to them, "*OK, now we want you to build that feature, but we want you to do it in a way that is fair, or in a way that satisfies human-centered values.*" They, of course, have no idea what that means. We actually need to produce methods, tools, and techniques that can operationalize those principles, and we need to come up with things such as design patterns, guidelines, or assessment frameworks so that the technologists who are actually building these systems have something a bit more concrete to work with. So, I think that's one piece of it – how do we provide the tools, techniques, and methods?

Education is certainly another piece of it. I think we're missing a trick when it comes to computer science education, for example, because your average IT or computer science undergraduate gets very little exposure to the ethics of technology. They will probably take a computing ethics course, but that will typically be one unit – one module that they take in isolation from everything else, and as soon as they've taken it, they've forgotten about it.

What I think needs to be done is embedding that ethical thinking across the whole of the curriculum, and it's actually not that hard to do. If you're teaching a course in machine learning, you can introduce some machine learning ethics. If you're doing cybersecurity, you can introduce some case studies that talk about the ethics of cybersecurity, and so forth. It can be done, but it requires universities to take a lead on that. That's the second element of it.

The third element of it is probably related to standards and codes of ethics. There is a very well-established ACM code of ethics that has been around for a long time. **ACM, the Association for Computing Machinery** – the big global professional body for computing – created its first code of ethics in the 1970s, but most software practitioners who you talk to will have no idea that it even exists. So, there's a bit of work to be done in terms of making sure that people know about the ACM code of ethics and how to actually implement it in practice.

I have a strong view – and this is perhaps a little bit controversial – that it's actually the engineers and the technologists that need to own this problem, and I don't think that's happening right now. It's often social scientists, philosophers, and lawyers that are out there talking about the ethics of AI. There's almost a sense that if you're a technologist or an engineer, it's not your job to think about ethics. That's why we have social scientists. That's why we have philosophers. People will often say that the solution to the problem is just to have multidisciplinary teams, and so you bring the social scientists into the team. Yes, that can work, but I don't think that's a practical solution for the most part.

> What we really need to do is to get the people who are actually building the systems to be thinking about the ethics of what they're doing and making informed choices about what they're doing on a day-to-day basis.

They need to be trained and educated in the ethics of AI, and they need to be given the tools, methods, and techniques that mean they can implement it in practice. It can't be delegated to other people from the social sciences.

AA: I agree. I think there's been a lot of pushback from technologists who say, "*This is not part of my role – not my remit.*"

JW: We did a study a couple of years ago, when I was at Monash, where we went into companies and tried to understand how they were implementing technology ethically. Now, the good news is that there are some examples of companies out there that are doing absolutely the right things. They're embedding it in their hiring practices and their culture, and so forth. But the bad news is that a lot of the engineers that we talked to on the ground, when we'd talk to them about things such as values and ethics, would say, "*Oh, that's not my role. That's for the CEO to worry about.*" Of course, at a high level, it is, but pretty much every software decision we make can have an ethical implication. It's not enough to just say it's the CEO or the board that needs to think about that.

AA: Even more broadly, anyone who deals with data within an organization should be empowered and take responsibility and ownership for its ethical use.

When we speak of responsible AI and technology more broadly, it goes beyond just ethical implications. It's also around having greater diversity and inclusion in this field, which is something that people are becoming a lot more vocal about. What do you think is being done to address this, either broadly or through Data61 and CSIRO? What do you think we should be doing to try and increase diversity and inclusion?

JW: First of all, it is absolutely still a problem. It's a bit of a shame that despite many decades of trying to address diversity issues in the computing field, we still have these problems. If you look at the history of this, it's quite interesting, because a lot of the early computer scientists were women. Margaret Hamilton was a NASA engineer who coined the term "*software engineering*" 50 or 60 years ago.

So, a lot of the early pioneers were women, but if you look at why that changed, it was largely with the introduction of the personal computer and the way it was marketed. If you look at enrollments in computer science from a gender perspective, for example, there was a sudden drop off the cliff in terms of female enrollments into computer science in the mid to late 1980s, and that was largely because when the personal computer was brought out, it was marketed as a "toys for boys" thing, and that hasn't changed since.

Without wishing to divert things too much, I'll tell you a story about my own daughters. I've got twin daughters, 7 years old. When they were 4 and transitioning into primary school, the school assigned a buddy that they would have at school to help them settle in. A few months before that, they got a letter from their buddy, introducing themselves. I was sitting with my 4-year-olds reading this letter, and the letter talked about all the great things about the school. Then, at one point, it said, "*I particularly enjoy the robotics club at school,*" at which point one of my twins burst into tears. I couldn't understand what was going on, so I said to her, "*Why are you crying?*" We didn't know the gender of the buddy at this point, and she said to me, "*I don't want my buddy to be a boy.*" Even at the age of 4, she was associating robots and technology with the male gender.

That, I think, illustrates the core of the problem. Even today, if you go to North America and look at AI professors, 80% of them are male.

We definitely have a problem, so what do we do about it? Well, we're doing a couple of things. Just coming back to our National AI Centre, we've identified three core themes. One of them is, "*How do we build AI systems at scale?*" The second one is, "*How do we build AI systems responsibly?*" But the third one is, "*How do we support inclusion and diversity in AI?*" We think of that from two perspectives.

First of all, it's about getting more underrepresented sections of the population into AI and working in the profession, but we also think about it from a design perspective – how can we design systems that are more inclusive and that respect diversity? If you think about the fact that 80% of the software practitioners building the systems are male, then they're going to build systems that are designed for that majority. We have to change that.

There's some very interesting work, for example, by Margaret Burnett at Oregon State University. She's come up with a tool called **GenderMag**, which is actually a way of going through the source code of a software system and identifying whether there are any gender biases in the way that that code is written. We think we need more things like that so that we're designing things in the correct way.

There are very simple examples of this that have got nothing to do with AI. Think of trying to hire people. When we form interview panels, we are now very used to making sure that we think about diversity, take unconscious bias training, and so forth. But what about the actual HR systems that those panels are using to select and shortlist candidates? If someone's taken maternity leave, for example, that's typically not very prominent in the user interface of those systems. If you want to find out whether someone's taking maternity leave and you want to take that into account because they might have had a gap in their career, you have to hunt for that information in their CV somewhere. If you've got 300 CVs that you're trying to go through, that's very difficult.

You could easily design that system in a different way, where you make it so that you ask candidates to make that information prominent up front when they're applying, and then you highlight it as part of your user interface. Nothing to do with AI, but it illustrates that these ethical and diversity issues are not just to do with AI – they're to do with software more broadly.

As for another thing we're doing within Data61, we've just appointed somebody who's now our lead for AI and diversity and inclusion, Didar Zowghi, who was a professor at the **University of Technology Sydney** (**UTS**). She's now leading a team that is doing R&D in exactly those areas.

I think another example of great success is the **Women in AI Awards**, which started 2 years ago and has become the awards ceremony of the year that everybody wants to be at. I was there in Melbourne just a couple of months ago. There are hundreds of amazing women working in the AI field. Whenever I go to that, it just illustrates to me that we're not short on talent when it comes to diversity, but we need to do more, and not just on gender diversity but other forms of diversity as well.

There's some very interesting work going on at the CSIRO around working with Indigenous communities and AI. One of my colleagues, Cathy Robinson, is leading some very interesting work, trying to work with Indigenous rangers to help them manage invasive weed species in the country, and they're essentially doing this by sending out drones and then using computer vision to track where these weeds are, and then they feed that back to the Indigenous rangers so they can do land management much more effectively.

> The really interesting thing about this – and it's a lesson, I think, for us all – is that it starts with the people first and the technology second.

They're co-designing the user interfaces. They're co-designing the AI algorithms with the people who live in those communities. They're not coming up with some great technology and then going in and saying, "*Hey, use our technology. It'll solve all your problems.*" It starts with the people.

AA: Another thing that complements a number of topics we just discussed is trust in AI. We're hearing more in the mainstream media about abuse and misuse – facial recognition systems, data profiling, and so on. What do you think is needed to build trust in AI?

JW: I might give a bit of a leftfield answer to that one. I could talk about regulation and standards, and there are definitely lots of people working on that, but I think as much as anything we need a public-awareness-raising campaign about AI.

AI is one of these things that are just overly hyped to death, and I think it erodes trust in the technology. A classic example of this was the recent story that hit the media across the world about the Google engineer that claimed that he'd invented a sentient machine. Now, anybody working in AI knows that that was a ridiculous claim to make, but it was a great story for the media. So, they went crazy with that one.

> **The "sentient" Google chatbot case**
>
> A Google software engineer (who was subsequently dismissed for violating data security policies) claimed that its AI chatbot is sentient – that is, it has the ability to experience feelings. Google itself does not support such claims. Even though current AI systems are not sentient, it does, however, raise important social questions about the field of AI. You can read one report on the event here: `https://www.theguardian.com/technology/2022/jun/12/google-engineer-ai-bot-sentient-blake-lemoine`.

Unfortunately, if you ask most members of the public what AI is, you'll find they have a very limited understanding. They're probably thinking of things that they've seen in the media or in the movies, where some robot has gone off-piste and is taking over the world. I think the more we can educate, the better.

There are some great initiatives. There's the **Day of AI** coming up, which is all about going into schools and telling kids about what AI is. We've got some plans in that space as well to do more public education around AI, and I think that the more we can educate people about what AI is, the more people will be able to cut through that hype. I think the hype is not good for anybody, at the end of the day.

AA: With increasing scrutiny and regulation, it seems to me that the case for explainability in AI is also rising. Given that we have limited understanding when it comes to the application of deep learning in particular, relative to more traditional methods, how important do you think explainability is, and if you agree that it is important, how do you think we should go about trying to achieve that? What's realistic?

JW: I think the answer is that it depends. Take the regulations going through the European Parliament that basically classify AI applications according to risk. If something's high-risk, it's got more regulations around it. I think of explainability in that sense.

If I'm sitting at home on a Friday night and Netflix is using AI to recommend movies that I should watch, it's probably not that important that it explains exactly why it's recommending one movie over another. Maybe it'd be useful to get some hint of why it's chosen a particular thing, but it's probably not that important. But if it's in a safety-critical domain, then it's 100% important.

I think that example supports taking this risk-based approach to AI and the adoption of AI. It also suggests that there are implications for the types of AI that are being used in different applications. We only ever seem to talk about deep learning nowadays, and that seems to be synonymous with AI, but we shouldn't forget that there's rule-based or symbolic AI, which has been around for a lot longer, is still used in many applications, and is a lot easier to explain.

In fact, going back to my NASA days, more than 20 years ago, I worked on an explainable rule-based AI system. Even that's not trivial because typically, within a rule-based AI system, there are hundreds of thousands of computations that go on. You can easily spit out a list of all the rules that are triggered, but that's not going to be very intelligible to a human being. So, your challenge there is, how do you actually raise the level of abstraction in a way that's intelligible to a human being? But it's at least feasible. Once you come to deep learning, it's not feasible anymore.

There is quite a lot of work going on now in the research community about how you can explain deep learning systems, but it's still very emergent. I think that where we're at now is about understanding what system you're applying AI to, what the risk level is, and then the implication for that and the type of AI technology that you implement.

AI in Australia

AA: What do you think is unique about Australia? What do you think really sets us apart from our global counterparts, and how do we retain our top talent so they don't leave?

JW: First of all, we are already right up there in the world rankings when it comes to AI – there's absolutely no question about that. Go and look at any global ranking for AI, and Australia consistently appears in the top 10, sometimes much higher than that. I think I saw one stat at one point not that long ago that said that, regarding the number of deep learning papers published per capita, we rank number three in the world.

There are some areas such as computer vision where you've got the **Australian National University (ANU)**, the **Australian Institute of Machine Learning (AIML)** there. I think AIML can genuinely claim to be number two in the world for computer vision. There's no question that we are genuinely up there with the best of the best globally.

In terms of our **unique selling proposition (USP)** or where we should focus, there is the AI Roadmap that was published and that we led on back in 2019. It identified three application areas for AI that Australia could take a lead on – health, smart cities, and environment and natural resources. Those are probably still accurate.

I think there is probably now a fourth one, which is responsible AI. We could really take a leading role in that. We were one of the early movers in terms of getting AI ethical principles up and running. We at Data61 particularly think that in the space of environment and natural resources, we could be leading the way. We sometimes call it AI for the great outdoors, given the nature of our country, and we're doing some really interesting work in that space.

We've got a partnership with Google, for example, where we are using AI in the Great Barrier Reef to track crown-of-thorns starfish, which is a starfish that poisons the reef. It's an interesting case because even today, the method for tracking these starfish is to send small boats out into the reef and have divers hang off the back of those boats to make notes. They have to stop every 2 minutes, so it obviously takes quite a long time. We're now working with Google to send in underwater gliders that have got computer vision technology on them. We've already proved that it's much more effective at tracking those starfish than any human diver could be. I'm often asked, "*Well, aren't you putting divers out of work?*" No, because the human diver's expertise is a scarce resource and it's still needed to dispose of those crown-of-thorns starfish, or for other work. We're actually just making the whole system a lot more efficient.

Discussing leadership

AA: Something you mentioned a moment ago, about needing greater understanding of AI in society, I think specifically also applies to the senior executives who often sign off and fund projects within organizations. There seems to be a greater appetite and urgency for them to upskill and build up their data literacy capabilities. Do they need to understand AI to make sure that they can make informed decisions and use it effectively, and be realistic about what they have their staff deliver?

JW: I think it is critical. I think the need for boards to be aware of AI is going to follow a similar trajectory to cybersecurity or, more recently, **environmental, social, and governance (ESG)** – non-financial values.

I saw some stats recently that basically asked boards, "*What keeps you up at night?*" During the pandemic, it was health, but it was cybersecurity before that, and it's now cybersecurity again. So, there's definitely been a market shift in terms of the governance of risks associated with cybersecurity, and all boards are on top of that now. I think they're going to need to be on top of AI as well over the next 5 to 10 years. I think we'll see AI come up as an answer to the question of "*What keeps you up at night?*" for a board member.

The first question I would probably ask them is the following:

> *"Do you know where AI is being applied in your organization?"*

The answer to that will be invariably *"No,"* and I would start there. The natural inclination would be to start with the risks and risk frameworks, which you need, but if you don't even understand where AI is in your organization, then you won't know what to apply those risk frameworks to.

Predicting the future of AI

AA: Overall, what are your predictions for the future of AI? I'd love to know what you are positive about and what you are a bit nervous about.

JW: I'm very excited right now about collaborative intelligence – how can we take the best of AI and the best of human intelligence and combine them to get the best out of both worlds? We are about to launch a $12 million initiative called the **Collaborative Intelligence Future Science Platform** at the CSIRO, which is all about that.

Just to give you a really simple example of this, there's a lot of excitement right now about this **Dall-E** system, which is an AI-generated art system. You can type in a query, such as, "*I want to see a frog riding a horse in the Outback*," and it'll produce an image for you. It's fantastic. It's a wonderful system. There are other similar systems, such as **Craiyon**, and there was recently a story where *Cosmopolitan* magazine used this Dall-E system to produce the world's first AI-generated magazine cover, only it's not the world's first AI-generated magazine cover once you actually look into it, because it's not really AI-generated. It was generated by collaborative intelligence, because the human that was operating that system actually spent about 100 hours trying different queries to get a result that was interesting and that captured what was in their mind.

I've used a similar system – this Craiyon system – and had exactly the same experience. You type in a query, and you somehow have in your mind what you want, but because I'm not an artist, I'm not very good at just drawing it. So, I type in what I want and I get something back. What I get back is very interesting, but it's not really what I had in mind.

So, you play around with the query and you try and phrase it in a different way, or you refine it to try and tell the computer what you actually want, and it becomes this quite painful iterative process. Each time you do a query, it takes up to 3 minutes to get the result back, so you could easily lose a whole day trying to do something and never get what you want.

I think the exciting thing in the future is having better ways of collaborating with a system like that, so that you have better ways of telling it what you want. I'm not sure exactly how to do it, but that's one of the areas where there's really interesting research and development to be done.

There's another area that I think is sorely needed. We talked a little bit about rules-based or symbolic AI versus deep learning – I think they've both got their limitations. Particularly with deep learning, people are beginning to realize the limitations now. Explainability is one issue, but it's also very hard to predict what answer you're going to get. It's not a very stable answer. It can change wildly, even if the input is only modified a little bit. Enter neurosymbolic AI, where you combine the best of these two approaches – there is lots of early research going on there now. I think we'll see that field advance very rapidly in the next couple of decades, and I think that that's probably where the action's going to be, for me.

Entering the industry today

AA: Last question. Consider the aspiring data scientists, and the next generation of future AI experts, leaders, and people transitioning to the field. What advice can you share with them about building a successful career in this space, in particular in terms of studies? A lot of people get confused: "*Do I need formal university studies? How much of it is hands-on learning? Can I use massive, open, online courses?*"

What is your take, having come from academia and now working in the R&D space? What advice can you give them on what's realistic when developing a career in this field, both maybe in general as a practitioner but also as a researcher?

JW: It's a very good question. Rather than answering that question directly, what I would say is that the first thing to do is to broaden your notion of what it means to have a career in data science or AI.

For most people, when talking about careers in data science and AI, they probably think, "*Go and take a computer science degree. Become a programmer. If you go and work for Google, Microsoft, or some start-up, you become a technologist.*" But coming back to the earlier discussions about interdisciplinarity, I think that broadens the definition of what it means to be a data scientist.

It may be that you are working in health but you're applying data science in a health context. It may be that you're an astronomer but you're applying data science. It may be that you're a lawyer but you're applying data science. I think there are going to be lots of really interesting opportunities out there that perhaps don't require you to have done an undergraduate and a master's degree in computer science, and that's where things such as micro-credentials and online courses come in. You can learn a lot right now if you've got the motivation and the dedication to do it. You can take courses from **fast.ai**, for example, for free. You don't necessarily need to be the world's foremost expert in data science to have a career in data science. It might be that you're an expert in a particular domain, but you know enough about the data science that you can actually apply it, and you can get massive productivity increases in that industry and build a great career for yourself – just not in the way that we might traditionally, and narrowly, think about it.

Summary

It was a great opportunity to speak with Jon in his capacity as Data61's director.

I enjoyed discussing the importance of entrepreneurship to help translate research into real-world impact, and the need to also foster basic AI skills and understanding in all disciplines to help enable university graduates to be AI-ready.

I agree with Jon's views that universities need to take the lead on teaching the ethics of technology across the curriculum, rather than it being a standalone course, and to ensure that the technologists developing AI have a solid foundation in it.

He also provides some pragmatic advice to operationalize AI ethics principles, including developing design patterns, guidelines, and assessment frameworks to support technologists that are building systems.

We discussed the career opportunities for inter-disciplinary data scientists; people whose primary qualifications are in another discipline (such as law or astronomy), who then add data science micro-credentials, and apply data science to their field. Jon's point is that there are many exciting opportunities in data science beyond the traditional, narrow view of becoming a data scientist.

One of the key takeaways of our conversation was Jon's advice on becoming a successful leader in the field. He advocates for the importance of defining goals having a clear vision, and communicating that vision broadly and repeatedly. These goals and vision become the North Star for the organization, inspiring and guiding people.

Jon's prediction for the future of AI is that neuro-symbolic AI is going to be where the action is in the next couple of decades. I agree that to develop AI that is capable of learning, reasoning, and abstraction, the current deep learning approach (which is focused on pattern recognition) on its own is not enough. A neuro-symbolic approach will help us develop systems that are capable of knowledge and reasoning – leading to **Artificial General Intelligence** (**AGI**).

12

Building the Dream Team
with Althea Davis

I was excited to have an opportunity to speak with **Althea Davis**, who has been an award-winning consulting **Chief Data Officer (CDO)**.

Her experience spans multiple industries and domains – currently she is the CEO and founder of ALSTRADA, a data monetization company at which she is a leading ambassador for women in data, and I was interested in learning how we can increase diversity and inclusion in the field.

Getting into data

Alex Antic (AA): Were there any particular people or events that inspired you or changed the course of your career trajectory early on?

Althea Davis (AD): This is not the answer that you're probably going to expect, but I will tell you because it's the truth.

I worked on really large portfolios of projects at banks, and I started to become aware of the power of data when I was working in retail, but I really got into it when I started working in the financial industry.

At that time, I worked on large portfolios. These were portfolios with tons of implementation projects. It was IT. But back in the day, IT meant doing IT *and* data. How I fell into it was that it was within my purview that I had to work on it, but a lot of the time when we tried to divide up the work between the teams, I noticed that a lot of the guys who were so-called "superstars" in doing the programs and projects steered away from the nasty parts of the project, which happened to be data. Because I'm a very curious person, I thought, "*Hmm. Let's see: it looks like fun. I can take this on.*"

So, I fell into the dirty work because other people didn't want to do it. Whether they said it explicitly or not, they didn't want to do it. They ran away from it because we know that data assets are totally different from technology assets. That's my point: it's not sexy. I literally fell into it. I fell into it, but I had my eyes open, knowing that it was hard, but I found it fascinating to solve because that's just my character.

That's how I really got deep into data.

Increasing diversity and inclusion

AA: As leaders in the field, we know it is important to increase diversity and inclusion, and in our field, it's obviously missing, as it is in most leadership roles. How have you seen this change in your time? Have you had any personal experiences where you've found this to be an issue? Is there anything that's been positive? What should we do as leaders to change things for the better moving forward?

AD: In all of my roles, not just as a data leader but as a management consultant and when I was working primarily in portfolio and senior program management, of course, in every profession, you'll experience a certain amount of discrimination, be it gender bias, color – whatever combination. Maybe you're a foreigner. I'm not talking about anything that most people haven't experienced. But, yes, I have experienced it.

One particular time, it was very in my face. First, I'll explain the specific situation, and then I'll talk about how things have changed.

This particular situation happened to be two middle-aged white gentlemen – European gentleman – and one of them – love him to death, a wonderful man – got pulled into something. He was hijacked into it. He was as shocked as I was, but he couldn't really say anything in the moment: I was called to a meeting to present my strategy. I presented the strategy, and they both said, "*This is great...*" Then, one of the gentlemen turned to the other one, as if I wasn't in the room, and said, "*OK. Let's get this strategy from her.*" He proceeded to talk about giving it to a particular guy that he used to work with and who he was bringing in to take over my role. He said it as if I wasn't even in the room.

What was happening here was an obvious old-boys network, and I wasn't a boy, and I wasn't old – well, at least I wasn't as old as I am now! Was it specifically racism? Personally, I think it was more that I wasn't part of that power network, in his eyes. But when working with this gentleman, I did feel that there was a certain amount of envy or discomfort, as though he was thinking. "*You shouldn't have such a powerful position.*" Was it because I was a lady? I'm not really sure about that. It could have been. There's a little bit of sexism (maybe more than in North America) when you're working in corporate in Europe, but I think it was less about me being a female. I think it was more that I was a *black, foreign* female. So, with that combination, he thought I had too much power and he was going to knock me down a level.

Now, I knew what was going on, and at the time, I realized that the universe was telling me I needed to be in a better place. I got that. But in the moment, it really was unfair. As I said, the other gentleman was as shocked as I was, but because he was on the board (these were two board members), he couldn't really take him down a notch in front of a subordinate, which I understand. But it was wrong.

I still have contact with that particular good gentleman. He also gave me a reference for that role, and he's a wonderful man. But I'm just trying to say to you that it does happen. It is real. I am sure there are even more stories than that. I've had other incidents, but I just wanted to give you that one in particular.

The other point that I wanted to talk about is how have things changed. Society has absolutely changed. One of the things that I have noticed in getting older is that realization: "*Oh, my gosh. I'm 10 to 15 years older than the people on the floor.*" Going as a contractor from job to job is when you feel ageism, even though you're very good at what you do. Sometimes I've noticed a bit of ageism, but I was lucky because I had such rare competencies and abilities that it didn't affect me as much.

How have things changed? I'm now here in the Middle East. I don't have that much of a different color from everybody else. Most of the people here, as foreigners, are from Central Asia or Asia. The locals are mixed as well, so there are different types of power dynamics. I wouldn't say it's as much racism as color. Sexism is as vibrant as it is in Europe. People might say that it would be more racist and sexist here because you're in the Middle East, but that's not necessarily the truth. Throughout my career, I had the most amount of sexism against me in Europe. I've felt it in every single role, almost. So, things have changed, and if I look at my colleagues in Europe, yes: the landscape has changed there, but not that much.

AA: What more should we do to change things? We can't let things continue as they are. How much control do we have as leaders in the field to make positive change?

AD: Take your seat at the table. Apply for that job.

> Use your voice, be seen, be heard, and be present.

Participate in the larger data community. There are a lot of opportunities to do that. Speak, write, and participate in books. Do campaigns as part of your data strategy. In your leadership program, you can do something. But even outside of your immediate role, there are many things you can do. Let your voice be heard as an individual. Be present and partner up with other people just like a company partners up and collaborates with other companies. I really feel like what you and I are doing right now is partnering up. Do we look and feel the same? No. But that level of diversity and co-creation is a good thing. It's a good message to send out to the world.

In my case, I could become a part of, for example, **She Loves Data**. It's a non-profit organization. It's targeted at both women professionals like myself who are more in a mentor/coaching role and women who are up and coming, looking up to the rest of us, saying, "*Hey, what are my possibilities?*" Get engaged and get involved. You might get involved in black leadership. I am promoting the **Middle East and North Africa** (**MENA**) and the Africa channel for *CDO Magazine*. Why? Because I'm trying to get the voices of this region and Africa heard on a larger scale. There are so many ways to do it, and I think we're fortunate now because even five years ago, there were not as many outlets.

Working in consulting

AA: Speaking of leadership, you're one of the very few consulting CDOs I know. How did you become one? What drew you to the consulting aspect of the role? What are some of the real strengths of someone progressing into this rather than a traditional CDO-type role?

AD: First and foremost, you need to do the job, or you have nothing to share. Just like I said before, with leadership, you need to look at generic consulting. Read the classic pieces on consulting, such as *The Trusted Advisor* by David Maisters, Charles H. Green, and Robert M. Galford, which is an amazing classic – a real pillar for anyone who wants to get into consulting. I also learned about being a consultant at the then-Anderson Consulting, which honestly was one of the best educational experiences I ever had.

Get your data leadership experience. You've got to be credible at what you're doing, and then you also need to pair that with the fundamental ways to be a good, trusted advisor. There's other material out there on consulting, but if you pair those two together, you have your credibility, and you understand the consulting role versus doing the job yourself, it will be much easier to step into the space.

You also need to know how to prepare to go to market. That's the hard part. The need is through the roof, but your customers don't know they need you yet. They think you should do them a favor and tell them stuff for free. It's not like writing a blog where you just say, *"Download this checklist."* We, as CDO consultants, need to do a lot of tasting. Don't just show them but go and let them experience you. After doing that for a while, which might be a lot of free so-called work, they're going to understand your value, and then that's when you're going to be able to put a price on it. I'm not saying that you should never ask for money upfront. I'm saying that the market is so green and unformed that it's not even recognized as a service offering.

AA: When working for an organization, what do you like about the consulting aspect, about coming in as an advisor and working with different groups over the course of a year or two?

AD: I love being challenged and having new things. I've never been a person that is content with being in one organization, knowing exactly what's going to happen, and just getting on with keeping the lights on. That's not something that makes me excited. Although, I love being in an organization where you're starting afresh, or you have to turn it around and grow it like crazy.

I love growth situations. I love growing things, and I like taking things to completion. So, with my last event, I'm not happy that I could not grow and complete something. That's the biggest thing for me: that's my personal interest. I'm also saddened that I couldn't see the product of my growth in the people around me because that's who I'm there for. I want to be able to create something.

What gives me energy is that I'm a very curious person, so I love seeing and being in all sorts of different situations. So, it not only keeps my brain going but it also keeps my excitement for the field going, and that's why I prefer consulting to being in one position.

AA: Speaking of organizations, when you normally consult with an organization, especially those that are starting their journey in becoming data-driven or artificial intelligence (AI)-driven, what do you typically suggest in terms of where they should begin? For example, do you first help them focus on strategy, culture, technology, or people? Of course, it varies from organization to organization, but how would you typically engage with a client in that regard and help them commence their journey?

AD: Thinking about it now, this is the first time I've articulated it this way: if some organization is starting, they're just starting. Everybody knows that there's already a certain amount of data-related things going on. Rightly or wrongly, something is going on. So, if they're just starting to say, "*OK. We want to be more conscious or cognizant of it,*" then I would say, "*Take the asset that has the longest life and the longest impact, and start with that one.*" That is not a technology asset. The technology asset is the thing that has the shortest life. It's constantly changing, but that's what people grab onto. They don't even acknowledge that there's a whole range of soft elements when it comes to data. They grab onto things that they've been told are data: "*We need technical enablement.*" You're absolutely right; you need technical enablement – but who is going to decide how you want that technical enablement to support you? Who will decide on the approach once you have the vision? Who will decide how the whole organization will work together to do their data work and be supported by their technical enablement?

All of those things have nothing to do with the technology asset. They are to do with the human asset. They are to do with the data. They are to do with methods and principles: capabilities. So, I'm not saying, "*Bake the whole business model before you start.*" I'm saying, "*Have a very good look at the business of data: the people, the process, the organization, and the methods.*" Then, you plug in the enablement. The enablement is actually the outcome of what you need to facilitate that business data model so the data can thrive.

You need to do it the other way around.

> I see is a lot of organizations throwing money at a lot of technology, looking for some individual players that they think are going to come down from heaven and save them.

You're a data scientist, so you know exactly what I mean. They put a lot of weight on the data scientist and the CDO, but they forget that it takes a village. They forget that data is a discipline, just like finance, marketing, or IT.

So, that's my point. Look at the thing that will have the most impact on your data journey, which is the people, the process, and the whole data model – and plug-in amazing enablement where it makes sense. When I say plug-in, I don't mean go out and buy a load of technology. Look at your existing technology footprint and see where there are gaps to enable you to manage that business model. That's my two cents on where you should you start.

AA: That is very wise advice, and I completely agree.

What would your suggestions be for organizations who've been at this for a while but are still struggling to gain traction after a number of years? Redesigning themselves? Pivoting? What would you tell them to focus on in that regard?

AD: We know that there are a lot of companies that have that profile. They've tried this and that, they've even got whole data teams, and it's still not working. It could also be that they haven't got momentum with the business: it's still a compliance thing. Nobody cares about it. Nobody understands the value of it.

Well, the answer is in what I just said: people don't understand it. People don't own it. So I think what you need to do is assess. When I say assess, I don't mean assess purely the capabilities. Take a look at *Figure 12.1*:

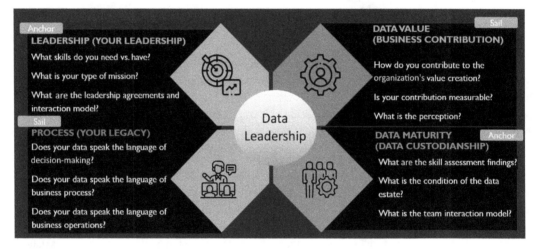

Figure 12.1 – Althea's four dimensions of data leadership

When I say you have to assess, most people think, "*Oh, I'm going to assess. We'll do a data strategy assessment.*" (See the bottom-right corner.) They look at foundational data management practices, capabilities, and maybe some advanced analytics and **business intelligence** (**BI**). They do an assessment only of capability. That's just one element. But when I say, "*Do an assessment as a leader,*" I mean, "*Do an assessment on four different quadrants.*" That means that you need to assess not just the data capability where you are and data-value creation (see the top-right quadrant); you need to ask yourself some fundamental questions. How are you contributing to the organization's value creation? How are you measuring it, and how are you perceived? Where do we think we are right now? You're going to see gaps there.

The next question is, "*How is value perceived in our organization, and how does the organization measure things typically? What value is considered relevant to the organization?*" That should take you forward.

You should assess yourself as a leader: your skills, ability, and other competencies. Earlier, I talked about competence as a data leader versus general leadership skills. You should also consider the following question: how are you, as a leader, anchored into the organization, both from a mission perspective? What mission are you on? Maybe you think you should be on a compliance mission, but the organization needs a totally different mission. Maybe it needs to be a mission where you're much more focused on the customer, looking at business outputs or outcomes.

When I say, "*Do an assessment*," I'm talking about these four areas. Another part of leadership is seeing your leadership agreements and interactions at the leadership level, both above your head and across to your peers. Why? Because if you notice you're not gaining any traction, it's probably because you've got some major deficits in your leadership agreements and interactions. It might not be that your team is not doing the right thing – it might be that you are not doing the right thing, or your management is not allowing you to have that interaction. You would then need to go back to your management and ask to make sure that you can get access. It's a gap analysis.

We've talked about capability assessment. We've talked about value contribution. These are very high-level questions. You can go very deep into each of these areas. We've talked about leadership: what your leadership is like and how it's perceived.

The last area is in the business of data. What is your business model for data? How are you anchored into everything: governance of the organization, the financial budgetary process, business processes, value streams, decision-making, and operations? These are the types of key questions you should be asking yourself. How are you anchored in there?

Say there's an organization that says, "*We're not making traction. Nobody knows we're here. We're ignored.*" You're probably not in the budgetary process as leadership at the table.

For example, there's a change management process, and nobody thinks about the data. Go and anchor yourself into that enterprise change management governance structure. Go and establish yourself there as a data representative. When I say "*you*," it could be you at a board level, and people from your team are in the overall change management process for software life cycle development.

I've mentioned budgetary processes and other types of key decision-making processes in an organization. If you are not at the decision-making table and you don't have the data, and the products from your team are not allowing decision-makers to get relevant information, then that's a telltale sign that you are totally irrelevant to the business. Try to identify where you could be relevant and provide the business with the data, products, and services so that they can collectively – including you at the leadership table – make use of data so that you're relevant.

I wanted to highlight these types of things. When I say you need to take stock, do not stop at "*OK. We've got the **Data Management Association** (**DAMA**) and some data standards, and I'm just going to use the checkboxes for every one of these standards.*" Maybe you have standards for BI, analytics, and knowledge management. That is great, but it doesn't stop there. You need to look at the performance part, value creation, and how you're anchored. It's about having a holistic check and not just doing it yourself. You do portions of it with your data team, and you do portions with your peers because you're interviewing. You know what you think, but they might have different perceptions.

I'm not saying you have to cover the whole organization. You just need to start somewhere. Start with your business strategy. If your business strategy or governmental organization strategy considers certain things very important, go to those stakeholders in that important area and try to see how you are relevant or not relevant to them. Do not go to areas of the business that are not strategically important. There might be 10 things strategically important: take the top 3. Just start from there. That's the way I would navigate that.

Many things can be done, but I think the advice I gave before about your data leadership capability versus your generic leadership capability and how you are anchored is very important.

Establishing a data service and culture

AA: How important is a data service catalog for most organizations, and how do you actually establish and manage one successfully?

AD: First of all, there are different data services in a catalog: IT, finance, data, and so on. The concept of data services is not new, but it's new in the sense that there are some things that people have only just thought to make services. For example, I recently saw somebody come up with a "business glossary as a service." That's a great idea. You can make many things a service. Obviously, out of that service, you will get concrete artifacts. They might be digital, but they're still artifacts, so they're, in themselves, data products.

That's the background in terms of the content of these catalogs. It can go right across the range from traditional things, such as data catalogs as a service. Even newer things such as DataOps could be a service, depending on how big the organization is. A knowledge management workbench could be a service. Off of that workbench, you would have multiple artifacts. So, there would be many services – some traditional, some non-traditional – in a data service catalog.

When you're servicing somebody, you need to know what is relevant to your customers. You need to know who your customers are internally and externally, and you need to think about appropriately comprising your service catalog. It needs to be relevant, not just to the day-to-day services that keep the lights on but also potentially to services that will make the organization grow faster. For example, you might want to do a data lake service, which an organization can use to quickly set up a data lake (I'm talking about larger organizations, obviously). Why would you do that? So that the organization can grow faster and better.

You might have something similar in the innovation area. For example, you might curate a unique dataset for research. I'm giving examples of services catering to different strategic business objectives. There's performance as usual, there's growth, and there's also innovation. Think big when you think about a service catalog. It does not have to be just the mundane day-to-day stuff.

Once you've decided on a set of relevant things, how do you actually manage that? Well, do not reinvent the wheel. Other disciplines have already thought about how to manage a service catalog. Go and look at that. This is not coming just from me. This is what I've learned from my mentors, and one of the best guys on this is Doug Laney. He wrote the book on infonomics (*Infonomics: How to Monetize, Manage, and Measure Information as an Asset for Competitive Advantage* by Doug Laney). He references purchase order management and library science and data people borrow from existing disciplines that have got this down pat. So, I won't explain how you set up a service catalog and manage one. Others can do that really well. Borrow from them: that's how you become successful.

> **You are a data artisan. You are a designer.**

You are not going to reinvent the wheel. The elements that go into your designs will be borrowed from multiple different places. Bring existing things together – think like a designer. If you think like a data designer, it's a much easier task. If you think, "*Oh, I've got to do everything*," you will never get anything done.

Part of your art, as a data artisan, is to go beyond keeping people engaged and making them a champion so that you get this community of people pushing the boulder up the hill instead of only the data leader and the data team. To create this effect, you need to be clear to always take a two-pronged approach. Get the strategic stuff (such as the assessments) done along with the tactical stuff. Give them crumbs of results and impact right now so they continuously live the experience of impactful data. Do this with day-in-the-life experiences, pilots, inclusive multidisciplinary meetings, and town hall discussions.

A data artisan who does this does not get exhausted because they have taken the organization along with them. You have evolved as a data leader because you know you shouldn't do it alone. You ensure that the organization turns on. They are coming up with ideas because you have them participate. They are taking on their data citizen to-do's day to day. Most importantly they make the connection between what they are doing *to* the data and what the business is getting *from* the data.

AA: What advice would you give to a senior data leader, someone who – say – reports to the CDO or the C-suite, who's having trouble gaining traction in gaining investment in data science/ analytics from people who don't seem to value innovation or understand what data is? How do you convince them that they really should be investing in this and learning about it to make sure that they're astute when it comes to all things data and analytics?

AD: That's a really interesting one. You need to take the bull by the horns. It can feel like you're in a glass box, your nose is up against the top of the glass, and you're trying to breathe. Your management gives you oxygen at the end of the day. If they don't create that space for you, you can't make it happen – you're dead in the water. A big part of your job is educating your management. How do you respectfully tell somebody, "*You are ignorant of your reality, and I'm going to teach you*"? Obviously, you can't say it that way, but I wish we could.

You need to tell it to them in a way that is relevant to them. Put yourself in your management's position. Your manager doesn't get it. There are many levels they don't get. First of all, what's in it for them, first and foremost? Why should they even care to change this organization or champion something? Well, they should champion it because, guess what?

> They're going to become a dinosaur executive if they do not embrace data.

Give them the empirical evidence. Do your research. Show it to them. It's their survival of the fittest, not yours. You're doing them a favor by telling them, "*If you're not aware that you're going to be put aside, you might want to think differently about your being an executive because this is a new requirement for* **chief executive officers** (CEOs) *and board members. It's no longer just the realm of the chief data officer.*"

They all have to have data in their DNA. They need a certain level of understanding. They need not only to understand but also to sponsor and champion. Those are their three caps. They need to understand enough strategically to allow you and everybody else to get on with their job. They need to appreciate that data is a serious asset. It's a serious discipline. They need to understand that they are going to be antiquated if they don't understand, champion, and sponsor. Also, they need to know what it means to support. What is support? Well, support is the obvious things: give me money, bodies, and support for my mandate. Then, there is championing. How do they champion? You need to explain to them what is needed for them to champion. Think about things such as data culture, awareness, and infiltrating peers and other leaders of corporate governance. Those are the kinds of things they need to support you on.

AA: Say they agree to that, and they invest in a new capability. How would you then go about building a data analytics dream team in terms of the structure and the different roles and skills that you'd be looking for in that team? This is for a generic organization that's new to this.

AD: I like heterogeneous teams. Here is a slide to illustrate my dream team (*Figure 12.2*). I made this a few years ago because I'm always designing teams.

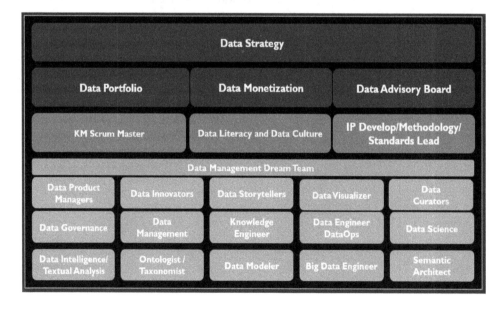

Figure 12.2 – Althea's data strategy dream team

Here, I thought about what exactly it is that forms the elements of a dream team beyond just the traditional titles. One of the things that I thought about was that data people are misunderstood, and business people need to be clear about what they need and also educated about data. We need people who are translators.

Then, we need people who actually understand the DNA of the data: people who can scope the metadata to the critical business elements and create selections for catalogs. There's also a need for catalyzation and democratization awareness, as well as there's storytelling and finding new ways of using data.

Then, there is the data scientist area. I know I just used the title, but I think of that area as unique because it's multi-faceted and wide enough to hold its own. For the engineering area, I didn't want to use engineering. I wanted it to be more about data operations – the actual pipeline. That is a unique area of structuring things and managing multiple pipelines.

Besides that, the traditional data modelers and architects bring it all together, but you also have a new breed of knowledge graph engineers. From taxonomies and models to ontologies and semantic layers, these people need to be able to tell stories. So, these are the general areas that I like to talk about.

As for who is in the dream team besides those generic areas, I think cross-pollination is great. That is, you should have people who are really from the bedrock. You must have people from traditional roles. You then need to complement them with another group, the newer, innovative, up-and-coming roles. My favorite up-and-coming roles are people who are data curators: taxonomists, ontologists, knowledge engineers, semantic architects, and information and data product managers. You can even have data service managers. Outside of the running organization, I think you need people involved in data research and data innovation. They're not in the day-to-day team. They should be cross-pollinating into the day-to-day team and trying things out.

Besides that, I think you should have a layer above because everybody is getting it wrong in the data space. You should have an advisory group telling you, "*Guys, think outside the box here. We see what you're doing in operations. You need to start moving in a different direction. Why not invest in this?*" Why am I saying that? Why would it even make sense to have something like that? I think it makes sense because we know how wrong people are getting it and how immature people are, but this is a hard sell because if you're a CDO and you advise having a data advisory board, your management might think that you're not capable. I would beg to differ. I would say, "*I am very capable, but it takes a village to raise all these data citizens. There are many more data citizens than my data team and myself, so we need to have a couple of layers in this organization.*"

That's not all of what I would do, but I do feel that from a dream-team perspective, you need to have a couple of different weights. For example, I just advised an organization with many CDOs that were CDOs in name only, not in substance. You need a group of very experienced CDOs and a manager for them.

There's a certain amount of things you can support in an organization via its design, methods, process design, performance management, and technology. You should bring all of those things together. The organization's design should not be typical. It's a hard sell, but there are ways from a monetarization perspective to show the value of having a different type of data design. We, as data leaders, should stop just doing what people think that we should do. As I said at the very beginning, I am sick and tired of hearing other people talk about us. We should take the lead in talking about ourselves. We are the ones that are having the experience, we're the ones that have the pain, we're the ones that have the beautiful gains, and we're the ones that have the global community echoing our own experiences. We should use our voices much more.

AA: As an expert in the field and someone who's worked across different sectors, industries, and countries, what are some of the mistakes, pitfalls, or challenges you've seen organizations face that you feel are quite common to many organizations that are not well versed in the data space and are still learning how to apply data successfully to their domain?

AD: There are two big violations. The first big violation is thinking that you're going to be able to throw technology at it to save the day. You cannot buy your way out of it. That's a big violation; it sets all the alarms off.

The other violation is simply not taking it seriously. If you underestimate what the value of data is, or you underestimate what the work effort required is, how are you ever going to be successful?

I won't name names, but I spoke to a colleague once. We agreed that they needed so many resources, and then they said, "*No. I'm only going to ask for a third of that.*" I asked, "*Why are you going to do that?*" They replied, citing something about that being how the culture was. I said, "*OK. I understand, but you're actually setting them up to believe that data's just not worth it.*"

It's similar to this: think of a couple. One partner says to the other, "*Let's go to a fancy restaurant,*" but the other says, "*No. I can't go there because it's too good for me.*" That would be crazy. It doesn't make sense.

AA: For some organizations, data governance can be an afterthought. How important do you think data governance is in getting it right from the get-go, and what are some of the key elements to successfully develop a data governance function in an organization?

AD: I'm taking this from one of my mentors, Robert S. Seiner. There are two key things.

Firstly, when you're kicking off, never think that nothing is being done because that way, you just insult and marginalize people. They're not going to listen to you because you've just ignored the efforts they've made – maybe silently – in the corner. You won't be able to tap into that knowledge base, and it's a shame for the company. So, don't come across as arrogant. You might not think you're being arrogant but don't come across as arrogant, as if nothing's being done. Something is being done; it's just not being pulled out.

The other thing from Rob Seiner is being non-invasive. Governance should not feel like a burden. It should flow in like water into lemonade. People shouldn't think, "*Oh, we're doing all of these extra backflips.*" Every organization has a different level of governance and way of working. Understand their way of working, and try to mimic that with the data governance work. If you are off-pitch and too heavy compared to the rest of their types of governance, you're not going to strike the right level.

There's also a third feature, which is my own: inclusion. When people talk about data governance, they traditionally talk about data owners, stewards, and custodians. They'll even talk about **subject-matter experts (SMEs)**. That's all well and good, but then there are the data citizens: all the masses out there as the consumer groups. How could you set up governance and not think about those you're governing? You're governing data for a reason, and the reason is your data consumers. Data consumers can be your regulators and internal and external consumers. So there should be a whole concept of having formal governance people thinly infiltrated into the organization in a non-impactful way. You should hear the voice of the customer – the data citizenry. Some of them might be superusers. Some of them might be temporary. It doesn't matter – have their voice be heard.

Alation, with its cataloging, talks about this a lot. There are other concepts, such as communities of practice. Have the data citizens be heard, and do not exclude them from the governance. Observe them non-invasively. Respond to them. Give them a voice in the arena so it's easy for them to speak. Don't make a bunch of rules. Don't make it difficult. You could even say to them, *"Don't type. Just record a video, and we'll listen to it."* That's what I mean about making life easy for these people. If they're consuming a report, you could put a star rating system in for them to rate it, for instance. Make it easy to observe and listen to the customer. That inclusion is super-important.

The other aspect of inclusion is making sure that the formal governance people map non-intrusively into the data team because the data team is doing things.

Make things easy. Make things not so top-heavy for the data stewards and data owners. Really look at your design because it needs to make sense for your organization.

But don't think that you have to run out the gate and get it perfect. Start, and start iterating. Make some decisions about not having it as a finite thing. Never assign a data owner or a data steward to a project. It needs to be something that's more permanent in the organization and makes sense. That's my advice about governance and getting it right.

AA: I think some of my clients would definitely appreciate that advice because they've really struggled with this.

AD: There are also some more obvious things, such as the enablement factor. If you want to be streamlined, you have to have some technology there, but it's just as important to have a method and a way of working.

Managing projects

AA: When it comes to project management, how different is managing data analytics projects from managing traditional IT projects?

AD: That's a good one.

Data is a totally different beast. It's got a very long life cycle, so you should never handle data like a technology asset. They simply don't have the same life cycle. Technology assets are really short-lived. Data life cycles are really long. They'll be here longer than you and I! That's the first thing.

The second thing is that data requirements should be considered from the start. Traditionally, they're not. People forget it, and then at the end, they say, "*Oh. We forgot to scope the data and get the requirements. The application/platform is sitting there, and there's no data in it.*" When you're anchoring data into your organization as a data leader, one of the areas you need to anchor into is the **software development life cycle (SDLC)**. If you're looking at value streams and business processes in the business model, that cycle is one of the big business processes for change. You need to anchor into that formal process of change and development as a company and figure out what it means. Change and development may be one thing or two separate things: it doesn't matter. Whatever it is, anchor into it.

For example, I'll take just change. If you're anchoring into change, what kinds of potential changes could you have for data? It could be anything. You could be changing the data model. You could change a business glossary: artifacts, products, and informational products. You could be changing reports. You could be changing a catalog. You could be changing the data lineage. You could be changing a data quality requirement. That's a traditional one, but that's one of the ones you should be checking.

Some of these have no big impact, but some have a huge impact. For example, say you're changing the reference data tables from an external provider, which happens to impact 60,000 tables in your data landscape and a whole bunch of informational reports. Certain artifacts or changes are very high-impact.

The list goes on. For example, you could change the role and responsibility of a data steward, but did you account for that with the rest of the interaction model? An impact analysis always needs to be done for those changes. If you want to anchor into project management and change management, you need to come up with a formal set of the types of typical changes for data and look at what you're doing right now formally in the organization. What's the process for kicking off a project, for example? You might be using agile and big-room planning at the portfolio level. You need to go in there as a data person to anchor data to the overall portfolio and the productized project.

Don't think that as a data person, you are a silo or sidekick. You have to incorporate yourself into everything that's already there. You have to think about what's unique to you and get into the organization's DNA.

AA: How important do you think a formal change management process is to all of this? Do you think it's necessary for most organizations?

AD: Yes, because formal change management does not necessarily mean that it's top-heavy. All it means is that the change is being managed. So how you manage that per organization is where the nuance is.

Imagine there's a report going to a key corporate governance meeting where people are going to make decisions, and there's a whole bunch of changes to the data that would actually make the insights incorrect. Wouldn't you want to manage that? I don't think people want to make decisions on incorrect data.

AA: In your opinion, how important is it to stand up data monetization as a discipline within an organization so that there is a repeatable process for value creation and business models that support it? Do you think that's integral to success in this field?

AD: I am such an advocate for data monetization. When I say data monetization, please don't think I mean that you have to generate revenue as a part of monetization. I do mean it in the natural sense of indirect and direct monetization: you can still improve your partnerships and customer engagement.

Value creation is tangible and measurable. I'm a formal proponent of that because I think that is exactly why data people have such problems gaining traction. Everybody else is measured by the corporate measure. If you're a core business, you know what that is. If you're a support organization, you know you're a support organization. But data, in the beginning, was seen like support, similar to IT – but is it really? No. It is not.

Remember when we talked about the dream team, and I talked about product and information managers? Those are the goods that we sell as data people, and we need discipline and practice to manage that, which is the monetization of our goods and services. It's not about the actual goods and services – you also need people doing performance management. It's not just performance overall but performance management and data-value calculations. It's really important.

Stand it up as a practice. Stand up a model of doing it to make it repeatable. Whenever people come to you and ask you for something, you can compartmentalize it and figure out, "*OK. This is an intangible, this is tangible. This is indirect, this is direct.*" You should understand all that and encourage it in your team. It shouldn't just be the CDO. Value is something that the whole team needs to appreciate that they are a part of creating, and they're critical in that creation. It's not just something for the business. If the team doesn't do something, that value will not be realized.

Why does AI fail?

AA: We hear a lot about organizations struggling to affect change with AI and data science. You hear about all these statistics of 85 percent of projects failing. Do you think this is true? And if it is, why is it happening when there's an abundance of data? Technology is quite cheap these days. Where are organizations going wrong in actually trying to generate value from it?

AD: You just said it. Technology is abundant, data is abundant, but is literacy abundant? No. Yes, AI projects are failing massively. There is also migration to the cloud: there have been huge problems with maintaining it and getting value out of it. AI and the cloud right now are the It Girls: they're failing massively. I'm not saying nobody's successful, but there are a lot of failures. Before them, it was advanced analytics that failed massively. Before that, data warehousing failed massively, or maybe data lakes.

Why does history keep repeating itself? Because you can't just throw technology budgets at something and think that it's going to resolve itself: it's not going to happen. In the case of advanced analytics, you had a bunch of overwhelmed data scientists spending the lion's share of their time doing what they should not have been doing. It should have been data engineers and hardcore data people who understood data, such as modelers, quality people, and metadata people. It should have been those people, right back to the business glossaries. All of that should have been happening there, not on the shoulders of the data scientists. At the end of the day, you need to remember it's not the "data" in "data scientist" but the "scientist" in "data scientist" that data scientists are amazing at. They should not be doing fundamental data preparation work. It's not good value for money, and they're not trained to do it.

That's why advanced analytics failed. The problem for data lakes was the same – you had people doing other people's work. Data warehousing and BI also failed because fundamental data work wasn't being done.

Fast forward to AI and the cloud: people are failing at an even higher rate because the stakes are even higher. You've got data scientists doing the statistical AI model. You've got those same data scientists trying to prepare the training data. It can't be done – not with a statistical model. Why? Because all they're getting is a bunch of dumb data, and that dumb data will not make it easier for the algorithm to do its job.

So, you get a vicious cycle where people get overwhelmed, and there's no outcome for the business because the AI is not working. Again, it's the same problem. You have people doing other people's work. You need fundamental data management people in there, from modelers to taxonomists. I would argue that we shouldn't be doing big data and AI. I'm not against having a lot of data. I'm for a lot of relevant, good-variety data, and I'm for lean data concepts, not big data. We need lean data that is rich and linked. I advocate for semantic AI, meaning all of that semantic stuff is done upfront with the right people from a data team, such as modelers, knowledge engineers, ontologists, taxonomists, and those who take care of data curation, glossaries, and catalogs.

Engineers are also way too overloaded. Some engineers need to be involved in terms of pumping things through the pipelines, but many other flavors of data people are under-tapped and overshadowed. They need to be there, and then you will be able to use that statistical model built brilliantly by the data scientist. Then, if you pump that rich linked data in, what comes out the other side? Something that you can use is a knowledge graph to check how accurate it is. Furthermore, you could use that knowledge graph to fine-tune the algorithm.

So, my advice is to stop doing big data AI. I'm not against the volume. You could have all the volume in the world, but is it rich and linked? Combine lean data concepts with rich and linked data: that combination is absolutely the formula for success. Of course, you could use some brilliant and innovative data people and a brilliant data scientist too. Inclusion!

Summary

Althea had so many important tips, insights, and advice to share. I agree with her on the importance of senior data leaders needing credibility and to be seen as trusted advisers. This helps win support from both colleagues and the broader organization. Leaders need to continually make sure that the data function is relevant and providing value to the business. As Althea points out, data *is* a team sport, requiring input and investment across an organization. To be truly successful, it can't be seen as an independent or siloed part of the business.

Regarding increasing diversity and inclusion, Althea encourages women to use their voices to be seen and heard and to be present. It is the responsibility of senior management to support and promote this.

Althea also raised some important points about leveraging technology to help deliver on your data strategy. Specifically, she advises you to look at your existing technology stack and capabilities and identify gaps before deciding whether you need to purchase new technology or whether you can simply reconfigure and reuse your existing capabilities. As she warns, you cannot simply buy your way to technology success.

Her advice on establishing a data governance capability is pertinent to establishing the broader data capability – focus on making a start, rather than waiting for perfection, and iterate and improve as you go.

Some key takeaways for me were as follows.

- Data is a serious asset and a serious discipline

- You should never handle data the way you do a technology asset, as data assets have a longer lifespan

- The data life cycle needs to be anchored into the SDLC

- An analytics function needs to have data scientists who are standing on the shoulders of data modelers, data quality experts, and data engineers – they need to be doing the fundamental data work that enables the "science" part of data science to happen. Data scientists should not be doing fundamental data preparation work.

13

Igor Halperin Watches the Markets

Having started my career as a **quantitative analyst** (**quant**) – a specialist who applies mathematical and statistical methods to financial and risk management problems – it was exciting to speak with Buy-Side Quant of the Year winner Igor Halperin.

Igor has experience working for leading financial organizations, from his current role at Fidelity Investments, as well as previous roles at JPMorgan Chase, and Bloomberg LP. He also has a strong academic background, including a PhD in theoretical physics. Igor is well versed in the technical underpinnings of **machine learning** (**ML**) in finance, including both the opportunities and challenges in this sector.

I was eager to get his views on the adoption of ML in the quant space and its broader applications to financial services.

Coming to AI from another field

Alex Antic (AA): I'd like to understand how a nuclear physicist ends up becoming Quant of the Year. What made you pivot into finance, and were there any particular moments or people throughout your career that inspired you?

Igor Halperin (IH): I was brought to finance by very common things, such as there being no jobs in theoretical physics. I had done two postdocs by 1999. Exactly a year before that, a huge research program was closed in the United States and a project for a bigger particle accelerator did not move forward. Many people who built that kind of thing found themselves without jobs.

I was a postdoc, so my chances of landing a faculty job anytime soon were close to zero, so I started to think about other things. I was lucky in the sense that around that time, physicists started to look into finance, and a couple of years earlier, a subfield called **econophysics** emerged from a few good physicists, including Jean-Philippe Bouchaud and Eugene Stanley. Eugene Stanley actually coined the term "econophysics."

The way they understood it was like applying the methods of statistical physics, mostly to understand the dynamics of markets as complex systems. In this sense, I entered finance through the back door, because I had just read articles by some physicists and mathematicians, and that drew my interest.

For example, there is a very wide class of methods called **maximum entropy methods**, which have deep roots in physics. I first read about these methods without actually understanding the kind of motivation for people to use them in the first place. I first learned alternative theories, and then I came to the initial financial theories. That's why I called it getting in through the back door.

Since then, it has been a combination of strategy and chance, in the sense that my strategy was to find something interesting and fun and useful to do in finance – not just making a buck but also trying to combine making something practical and useful with interesting research. That was almost immediately what I started to look into in finance. Initially, there were all sorts of non-parametric Bayesian models, which is already something pretty close to ML. And then, pretty quickly, I started to read books on ML because I had the sense that it was related to so many things that we did.

First, I went to Bloomberg, where I worked as a quant developer, until I moved to JPMorgan, where I stayed for nearly 14 years. The maximum entropy model that I mentioned earlier was a workhorse of so many things that I did. Every time, people said, *"Wow! It works. It calibrates automatically."* That was by design.

Then, I decided to pivot my career a little bit. I wanted to do research. I didn't see as many applications of ML at this time in the particular business unit that I was in with J.P. Morgan. So, I decided to do this thing in academia.

I took up an opportunity with New York University (NYU) to create a course on ML in finance. Nothing like that existed back then, so I had to start from scratch, which was very interesting and challenging. One of the great physicists, Richard Feynman, once said that the best way to understand something is to try to teach it. You asked me about the role of chance. Indeed, it had a huge role because, as with most things, my thinking about this was shaped by particular interactions and particular projects at work.

My idea was to find some simple models for reinforcement learning. I didn't want to take the route that pretty much everyone else took of jumping immediately into deep reinforcement learning, because deep reinforcement learning is a recipe for a mess. If you want to learn how to drive, you don't jump into a Ferrari. You start with something simple, and the same goes here. Physicists tend to play with certain methods to learn about the world.

AA: Why do you think physicists make great quants and data scientists these days? We see a lot of them pivoting into data science. I think it's more about the way they formulate problems and try to understand the world as an abstraction via a model.

IH: I think it connects to what I already mentioned briefly before. Physicists have their own way of understanding things. Physics is, first of all, the science of nature, which makes it very different from mathematics, for example. In mathematics, you don't deal with actual nature. You deal with abstract notions – such as abstract spaces – and therefore, in order to convince yourself that you are doing something reasonable, you have to prove all sorts of existence theorems and uniqueness theorems.

So, mathematicians think differently. They try to bring rigor, but sometimes, they try to bring rigor to the problems of finance, which are much closer to physics. You cannot deny the existence of money, financial contracts, market prices, and so on. You don't need to prove existence theorems. That's exactly why I think training in theoretical physics is highly useful, because of the way it looks at the world.

The best example would be thermodynamics and statistical physics, which are actually the theoretical foundation of many things in ML.

Think about it this way. In the 19th century, people did not have computers, but they had to work with systems that had astronomical numbers of particles – **Avogadro's number**, which is 10 to the power of 23.

If modern-day computer scientists were given such a problem, I wonder how they would perceive it. If he were alive today, Amedeo Avogadro would build huge supercomputers with tons of GPUs to research the property of gases, for example, under certain conditions – measuring velocities and positions, making empirical observations, and so on. But think of it from a historical perspective.

> The genius of people such as Boltzmann, who created statistical mechanics, was that they found the correct ways to think about highly complex systems by stripping out unessential details.

There are tons of examples of the most successful models in physics keeping the most relevant parts of the phenomena that they want to describe.

For example, the most classical model in statistical physics is called the **Ising model**. It's a model of binary spins (where spins can be in one of two states, +1 or -1). People use it to describe magnetics and magnetic materials. Obviously, it's a super-stylized description of complex systems. Nevertheless, it captures a substantial amount of all the relevant properties of such systems.

My view is that natural sciences and physics bring into data science or industrial engineering an important and unique mental discipline – an approach. I see computer scientists who tend to believe in brute compute power, and some of them believe that when brute compute power is not sufficient for any particular problem, more compute power will solve it. But as scientists, we know tons of examples where this approach is provably wrong. No matter how much resource power you put into it, you will not succeed.

Applying ML to problems in finance

AA: It seems to me that the application of ML to finance is still relatively young. There are pioneers, such as you, looking at areas such as reinforcement learning to really try to shift how people use ML in a structured way.

Where do you see the industry heading? What are some of the key areas that you think are making a difference, and what really excites you about the future when it comes to the application of ML in finance?

IH: I think ML is already at the stage, even in finance applications, where there is so much work going on with different applications that it's hard to keep track of all of it.

I'm mostly interested in reinforcement learning, but even in this space, there are various applications. So, you can separate it into different methods that you should use. There's using reinforcement learning for intra-day trading or high-frequency trading, and there's using reinforcement learning for systematic portfolio managers; some cases involve trading a few times a day, others a few times a month. That imposes lots of constraints on you, even before you look at the data.

If you want to apply reinforcement learning to trading commodities' futures or options for commodities and intra-day trading, then your dataset would be in megabytes, maybe in gigabytes. That's something that already justifies the use of deep learning methods. But if you deal with investment trading, then your dataset is very small, by the standards of deep learning. You have no options. What you are doing is bound to be something small, and because of that, you can also try to make it more interpretable.

That's what I was trying to do. My reinforcement learning models were simple, where the algorithm is interpretable – an actual white box, not a black box – and the formulation of the model is also interpretable.

AA: Now for the second part of the question, which is off the back of the quote by Marcos M. Lopez de Prado: "*The most insightful use of ML in finance is for discovering theories.***" (***Machine Learning for Asset Managers***, 2020) I'd love to hear your thoughts on that, especially from your perspective as a theoretical physicist .**

IH: When I worked on this course for NYU, I came up with the concept for reinforcement learning that I wanted to apply to model the behavior of individual portfolio managers, but I didn't have data. Then, I thought, "*Why don't I apply the same model to the whole market, aggregating the behavior of all individual portfolio managers into a single agent?*" The agent is analogous to the theory of the "invisible hand" – the unobservable market force that helps the supply and demand of goods achieve equilibrium in a free market.

You cannot map thousands or millions of portfolio managers or agents to one agent with a particular goal in mind. In reinforcement learning, you specify what agents do by specifying what is called the reward function, which is what they have to maximize. If you define this reward function, you give the agent a prescription of what they have to try to attain. If you aggregate many agents, you obviously cannot have one agent with a well-defined reward function; instead, it should be something such as bounded rational agents and stuff.

So, when I worked out the details of this model, I found that it produces something pretty different from conventional financial models. All these classical financial models are linear models – linear in terms of the deterministic drift that they predict. This drift function is a linear function, and everyone believes that it's perfectly fine. I used to believe that it was perfectly fine for almost 20 years until I worked out this reinforcement learning model and then compared it with existing models. I found that actually, yes, it does correspond to something that is known to physicists but not to financial mathematicians.

More specifically, it corresponds to something known as the non-linear Langevin equation, which was actually discovered almost at the same time as Einstein and Bachelier discovered Brownian motion. Bachelier did it in 1900. Einstein did very similar work for free Brownian motion 4 years later, and this part is well known to financial mathematicians.

Somehow, what they missed is that 3 years later, there was another piece of work done by Paul Langevin, a French physicist, who generalized this theory of free Brownian motion to the theory of Brownian motion under the influence of some external, non-linear potential. If you look at financial models from that perspective, you will see that they are missing non-linearity.

There is an amazing documentary about one of my favorite physicists, Richard Feynman, called *Take the World from Another Point of View*. He talked about how important it is to try to see things from another perspective. It's easier said than done. If I tell you, "*Find another perspective and tell me something interesting,*" it's not always workable. Using this ML approach and view helped me in this particular case because it gave me another point of view.

My research since then has branched into two, related directions. So, one is research into reinforcement learning for portfolio trading and things like that, and the other one is modeling the market using this new understanding and new ideas that were drawn from this previous experience.

The current question for me is which direction I like more. I like both of these directions for several different reasons. That's a real-world example of what I think Marcos was referring to.

There might be another case that I could add – there is a very exciting new development in ML called **scientific machine learning (SciML)**, which is the application of deep learning methods to solve scientific problems.

Say you have a partial differential equation. It used to be that scientists of the past undertook a two-step procedure. First, you build your theory. You arrive at some sort of differential equation – a partial differential equation, or something like that – and then you have to find ways to solve it. SciML lets you solve those equations using neural networks.

Making AI explainable and trustworthy

AA: Something you've discussed is this notion of explainability in the use of ML. Particularly in finance, how important do you think explainability is, and how do we actually achieve it? How do we also ensure there's enough explainability, especially for fund managers and others who want to have a better understanding? There are also requirements from risk, auditing, and regulatory aspects – how do we meet these needs?

IH: I have changed my view of explainability a few times in my career. Initially, my interest in ML methods started with various non-parametric Bayesian statistics, such as maximum entropy. Maximum entropy methods are essentially considered part of ML these days. They're something very flexible that can fit any data, essentially.

I believe that one of the challenges of financial models is appropriately fitting the market data. If you have non-parametric models, it's easy to fit the data. The same goes when people apply deep learning to financial predictions.

Feature engineering is not hard. Avoiding overfitting is a tricky thing. Sometimes, people do what Marcos calls "test data overfitting." When you use the test dataset more than once, then what you're doing is not kosher anymore. So, on that count especially, explainability- and interpretability-related terms are highly important to assist in identifying issues, such as overfitting.

From my experience of talking to actual portfolio managers and traders, explainability is super-important. They will not act if you come to them with some sort of black-box model and say, "*Do this because my model tells you to do it.*" No one will talk to you after that. You have to explain "*This happens because of this and that.*"

Now, explainable AI is a huge topic. I don't follow it much, mainly because my version of AI is already explainable. When you don't have models and you have tons of data, you have to resort to black-box architectures, such as deep learning, and then your explainability becomes a more technical thing. You rely on the sensitivities of your black box. You try to develop intuition, basically by playing with it, which is OK. It's short of a full mathematical understanding of what your model does, which is not always possible, but it's the next best thing.

> I think explainability is 100% important for financial applications but not so much for applications that are not critical.

Take the most classic example of using ML – the US Post Office uses convolutional networks to read zip codes. It's been in production for 20 years. No one asks to interpret it, because it's a secondary task. No one cares. In finance also, there are some tasks that do not necessarily require you to explain what it does, but for investment decisions and trading decisions, it's 100% important.

Planning for successful AI

AA: When it comes to organizations applying AI, we often hear in the media that failure rates of 85% are common. If we take that statistic as a fact, would you say that the same failure rates apply to finance?

IH: I think that the 80% failure rate is probably a more universal number that goes beyond data science and the financial industry. I would estimate that that's probably correct for any new project in science, and in business too. From my personal experience, I know that about 80% of ideas that I have had at different points in time went wrong.

Also, if you look at the success rate of technology start-ups, it's probably even less than 20%. Even more than 80% of them fail in the end, and the ways to fail can be many. There was once a show on TV called *1,000 Ways to Die*, which was a kind of horror show recreating all sorts of stories about how people accidentally died. It may not be quite that grave, but there are still many ways you can fail. Your initial idea may be wrong in the end. It might sound reasonable but be wrong on many different counts. It might be wrong in a business sense – you might incorrectly estimate market gaps, for example. It might fail at the level of the organization, where you may find that there is no one who is interested in collaborating. Something can even fail scientifically. Also, especially in finance, there can be so much noise in discussions with your colleagues.

Navigating hype

AA: One thing I've seen through my consulting business and through people I've spoken to in the industry is that there seems to be a lot of hype around AI, especially when senior leaders in organizations want to implement AI in some capacity. They buy into the hype. If you had a room full of CEOs from financial services, what advice would you give them to paint a more realistic picture of what AI is and how it applies to their domain?

IH: My advice is that technical executives are not supposed to be experts in AI. They're not supposed to discuss with data scientists the use of a random forest model versus a neural network model for a particular problem. It's not what they're supposed to do.

> However, executives with technical capabilities are supposed to have a good common-sense understanding of what the business needs are and what is feasible and not feasible.

My favorite analogy is that of driving a car. To drive a car, you don't need to know the exact details of how the engine works, but you have to have a general-level understanding of how to operate the vehicle. The same applies to AI.

Sometimes, people will make a comment, and you will discover that they actually don't understand what AI does because they're pretty serious when they say it. The key thing is education, of course.

Discussing the role of education

AA: One thing you spoke about earlier is the course you developed at NYU. I'd love to hear your take on what you think universities – and educators more broadly – should do to better prepare the next generation of AI data science professionals, no matter which field they go into.

IH: With soft skills, there's no doubt, yes. I think they're highly important, and yet universities do not pay enough attention to them. In an industrial environment, it's super-vital.

> People have to have the ability to express themselves in a clear way, both orally and in written form. It's important for efficiency because, sometimes, you can spend huge amounts of effort just trying to understand what a person is saying.

On the more formal side, I was very happy to reunite again with physics recently in my latest research. I'm happy that when I went into finance, I thought that only a tiny part of my knowledge in physics was relevant, but it turned out that it was actually much more than that. In general, it would be very good for many data science and ML programs to add to the curriculum some courses in physics, especially in statistical physics, statistical fields, and maybe quantum physics, for at least two reasons.

The first is that many many financial models can be mapped to models in quantum physics. That's one thing. So, if you have knowledge in such areas, that definitely can help. The other thing is obviously related to quantum computing. There are so many people who believe that quantum computing is the future of everything. Whether you believe in that or not, it's still a good idea at least to keep an eye on it and get some basic education.

Maybe there should be more emphasis on math as well. What I notice is that modern-day data scientists, especially people who do NLP, talk about all sorts of transformers and so on, and it turns out that they don't know basic linear algebra because they don't use it. Maybe they heard about it very early in their program and then totally forgot it, because they think it's irrelevant. That's bad practice.

Maybe things about quantum physics are more optional, but I think there's a consensus that linear algebra is the mathematical foundation of ML, and no matter what you do, you have to know that, as well as basic statistics and calculus.

AA: How important do you think formal studies are for someone becoming a data scientist as opposed to doing online training or short courses? And following that, how important do you think postgraduate studies are? Do you think it's only important if you're going down a research path, or is it advantageous in this competitive space to also consider a master's or PhD in a related quantitative field?

IH: There are many skills you get from academia that you can bring into the workplace.

> The value of graduate degrees, especially PhDs, is that on top of the formal knowledge, you also gain skills in doing independent research, and I think that's highly valuable in new areas especially. Formal studies and academic research are exciting in that most of the problems faced are open problems. It's not that you go to a book, go to a particular page, and implement whatever's there – it never works like that.

When I started in finance, we looked into ways to adapt some academic research papers to practical needs, and even that involved lots of open-ended questions because things never work as written on paper. That's the value of graduate study – giving you skills to explore problems in new ways.

> One good piece of advice given to young physicists (I found this in one of the books that I read in college) is that when you are given a new project, spend a few days trying to solve it without looking into literature and resources. Why? Because chances are that you will not be able to solve it, but at least you will be able to understand the relevant questions, and then your next steps will be more directed.

You will know what to look for.

Considering the future of AI

AA: I'd like to hear your thoughts on the future of AI. Do you have any advice for business leaders to help them understand where the future is going and how they should best prepare their organizations?

IH: Finance is a huge field. Obviously, there are different strands – there is trading, banking, and so on.

My favorite topic is reinforcement learning. One of the reasons I like it so much is that out of all areas of ML, it best aligns with what we need to do in most finance applications when it comes to trading. In finance, your ultimate goal is to act. We're not in finance to seek an abstract understanding of the world. If we use predictive models, it's because we want to get signals about how to act now – how to trade – and in reinforcement learning, that's the main goal. All the rest of your tasks are subordinate to this main task. Sometimes, your future may be very complex, but your actions – what you should do right now – are clear.

So, I hope and anticipate that more algorithms of this sort that are based on reinforcement learning (and also on something called **inverse reinforcement learning**, which is another of my favorite research topics) will become more widely used. I think they are already well received in some applications. There have been successful applications of reinforcement learning to many tasks. Some of them get into the public space, and some of them don't.

In the public space, I've heard about successful applications to trade execution. I've heard some distant rumors about the successes of some hedge fund that applied reinforcement learning to trading. So, at least in the short-term future, I think these methods will be more widely accepted.

Andrew Ng said something that I fully agree with. He said that "*if a typical person can do a mental task with less than 1 second of thought, we can probably automate it using AI either now or in the near future*" (https://hbr.org/2016/11/what-artificial-intelligence-can-and-cant-do-right-now). So, if your time horizon is short, then it makes it easier for the algorithm to do something. This means that AI systems may be more autonomous in some applications, such as algorithmic trading and high-frequency trading, which involve less human intervention.

For asset management, my expectation is that it will develop more as a symbiosis between human intelligence and AI.

> AI will not compete with human intelligence but will provide helper tools.

That will probably be best because of the complexity of what we do. What we do involves so many aspects – the geopolitical, the macroeconomic, and so on. I don't see that being formalized and put into some sort of a super box, such as DeepMind playing Atari video games. It's not going to happen anytime soon.

Summary

I enjoyed discussing with Igor the uptake and increasing adoption of ML in finance and, in particular, the application of reinforcement learning. I think it's worthwhile for all data scientists who work in the broader financial services field to investigate opportunities for the application of reinforcement learning. It was also great to discuss the importance of explainable AI in finance, and the challenges involved.

Igor talked about the value of seeing the world and specific problems from different perspectives. This principle operates on a number of levels – it can be achieved by building teams that include people with backgrounds in different disciplines, but it can also mean recreating a model using a different class of modelling – for instance applying a ML model to a problem generally represented by a linear model.

Regarding high failure rates of AI projects, Igor's view is that as most ideas don't end up being successful (as in all scientific endeavors). This is a common theme emerging from the answers to this question. Is the focus on the high failure rates in AI perpetuating the unrealistic expectation that project failure rates should be close to 0%?

Igor also reiterated the same point made by most interviewees about soft skills being integral to building a successful career in the field, especially communication skills. We also discussed the importance of having a strong technical foundation (including the fundamental mathematics and statistics skills that form the foundation of ML), and how two of the key skills that all researchers and data scientists need to have are the ability to solve complex problems and identify the key challenges and questions posed.

One of the key takeaways for me was Igor's advice for senior leaders. He doesn't believe that they necessarily need to have a hands-on background in a technical field, such as analytics or computer science, but he stresses the importance of them having a solid common-sense understanding of what is feasible with technology in order to manage expectations. I can't stress enough the importance of all senior executives (and anyone who works with data) being data-literate. It's also important to be able to look at a problem from different angles, and multidisciplinary skills can help with this.

14

Christina Stathopoulos Exerts Her Influence

Christina Stathopoulos is a well-known data and analytics influencer who is passionate about sharing her expertise and knowledge. She is a data evangelist, advisor, and educator, as well as an adjunct professor at the IE Business School in Spain, a LinkedIn Learning instructor, and a former analytical lead at Google.

I was excited to speak with her, and I was especially interested to understand her recommendations for establishing and leading a successful data and analytics function, and the importance of data storytelling.

Becoming a data science leader

AA: What inspired you to become a leader in data science and want to share your expertise with others? What prompted you to move down this path, and what keeps you in the field?

Christina Stathopoulos (CS): I would say that my path to where I am today has not been a straight line. I had a unique journey to get to where I am today, but I always knew that I wanted to work with data. I grew up with a love of numbers. Even as a kid, my favorite class was math and statistics, and I later went on to study for an interdisciplinary bachelor's degree with a focus on statistics, and later still, a master's in business analytics and big data. For me, it was a very natural path to pursue a career in analytics. I don't think that's the case for many people today because the idea of "data science" has only recently become a hot topic; 10 to 15 years ago, its existence was limited to variations of it at the intersection of statistics, computer science, and engineering. During that time, data science programs did not exist, and yet now you see these types of programs all over the place. So in my case, I entered the field by studying statistics, the original "data science."

In establishing my career, the most pivotal moment for me was the moment I decided to pursue my master's in the field and also to pursue it while living abroad. I completed my master's in business analytics and big data at IE Business School in Madrid, Spain, and this really helped me pivot and launch my career in the right direction. But I do recognize that there are other ways of going about it.

I took the formal education route, but it is definitely not the only way. In my case, though, it worked wonders. I took advantage of all of the opportunities that the university offered me, not just academically but also for networking purposes, which is crucial to newly establishing yourself within a field.

AA: How important do you think postgraduate studies are in the field?

CS: I think they are definitely important and can help you stand out from the crowd.

There are two very different paths that you can take, and one of them is the aforementioned formal education path: pursuing a master's or a Ph.D. and becoming formally trained. On the other hand, you also see many people today taking the less formal path, which is through self-paced learning: what I would call YouTube University. There's nothing wrong with the latter; it is a matter of personal preference and the resources you have at your disposal. Formal education typically costs more and is more structured, whereas self-paced learning requires you to have markedly more self-motivation to reach your goals. You have to have certain skills if you are going to go the latter route to become a successful data scientist or data engineer, for example, but it is absolutely possible to go that way if you have the drive.

I think one thing that is very different when you choose self-paced learning, though, is that it's important to build a portfolio of work along the way. When you go through YouTube videos and online courses, they have minimal formal recognition, so you need to be able to demonstrate what you've learned, whereas, with a master's or a Ph.D., when you go for the job search, they are going to take that at face value. They trust the institution that gave the accredited degree, whereas if you completed an online course, they don't really know whether you have the knowledge this online course promises to deliver. A portfolio can help prove that.

AA: Students I've taught or people I've mentored who want to transition to the big tech companies often ask about how much more credibility formal studies have and whether it increases your chances of getting a job at those places, given how competitive it is. Do you think it actually gives you an edge?

CS: In a way, I am biased because I took that route and I saw the benefits of it for me. It definitely worked in my case, and I think it helped me stand out from the crowd.

Indeed, landing a job in a big tech company is extremely competitive. If you are among the top candidates and you have an advanced degree while another candidate does not have those formal studies, then, in that case, you are probably going to be given the advantage due to your formal qualifications.

> But I recognize that formal education is not for everyone, maybe due to a lack of resources or time. With the right drive, you can become just as educated in data science topics through self-learning programs.

Although this is where the importance of having a portfolio comes in to demonstrate your knowledge.

So even if you don't have an advanced degree, I think you can impress hiring managers without having that formal education piece, but I would say that that's done in the interview process. You might be able to go and blow them away during interviews, but you need to get noticed first. If your CV includes formal education, such as a master's degree, versus someone who has a CV with a couple of courses from a platform such as Coursera, accredited universities are still given significantly more weight than online programs.

> Formal education can work in your favor when recruiters are sitting there comparing CVs, but it's not all about education. You've got to get noticed in the first place.

There are a lot of other things that you can do outside of your education that can make you stand out in the job search, especially for FAANG.

> It's about finding a way to make yourself shine within the field. That could mean hosting a podcast, writing a blog, or working on another relevant side project. You can find other ways outside of your education to make your CV stand out; get creative to show you have a real passion for what you do.

AA: It can be a touchy subject for some people, but I think it's important to try and address it because it'll be one of the questions top of mind for some of the junior folk looking to enter or grow their career in the field.

CS: Yes, and I would highlight that there's no ideal educational path.

You can definitely go both ways when embarking on your learning journey, but you have to recognize the benefits and the weaknesses of both. We haven't yet discussed the cost factor. If you are going to go the formal education route, there might be an upfront investment – a large one – whereas, with self-learning, you can do it on a budget and space out your investments over time. If you don't have the resources to make a large upfront payment or you don't want to take out a loan, you can definitely learn in a self-paced manner and still find success; many people already have – especially in data science, I would say more so even than in other fields.

AA: Like you, I have a bias toward formal education; however, the accessibility and lower barriers to entry via self-learning have been a game-changer in the field of data science. Either path requires a significant dedication of time and effort.

Observing changes in the field

AA: Apart from the proliferation of online courses, how have you seen the field change throughout your career? What have you observed, and where do you see it going?

CS: Online courses, as you just mentioned, have changed the landscape a lot, making it more accessible for the general public to get started.

Outside of that, I have seen a continuous state of change in the data science and analytics field. Technology has, of course, changed. It is improving to be more affordable for all and, at the same time, more powerful. This is great news for everyone because it means that we are experiencing data democratization whereby advanced data applications are not limited to the digital tech giants as they were in the past; they are more open and accessible for small or even family-run businesses.

I'll give you an example with regard to technology. When I studied for my bachelor's degree, we used SAS software from SAS Institute. It was a wonderful analytics solution yet extremely expensive. Now, this solution is mainly limited to use within the banking, finance, and insurance sectors, whereas, with broader data science applications, you see more use of Python (the most used language in the market today), R, and other open source technology. There has been a complete shift toward open source and thus the technology that we are using is available for anyone, regardless of budget.

> We are seeing an acceleration in the adoption of low-code and no-code technologies, which is automating a lot of data or machine learning processes to the point that we can work with more drag-and-drop tools rather than manually programming everything.

It is making the field more accessible, even to the citizen data scientist, who can use tools such as DataRobot or Dataiku to create and put machine learning models into production, which, in the past, would have required a very dedicated specialist team.

The last thing that I think is interesting is that as I was finishing my bachelor's, data science was becoming a hot topic, and the data scientist was replacing the traditional statistician role. *Data scientist* sounded a lot more "sexy" at the time, so it was considered the new role for the 21st century. I am starting to notice the same type of hype driving the data engineering role today.

> I think that companies are realizing that data engineers are the backbone of any analytics and data science team. They are realizing the importance of the role and giving data engineers the recognition they deserve.

This all goes along with the fact that companies are realizing the necessary touchpoints for making a data strategy a successful reality. It is not just the data scientist that performs miracles – there is an entire team behind them. And indeed, you are only as strong as your weakest link; disregard the importance of data engineering and the rest may be for nothing.

AA: I couldn't agree more. There have been huge changes in Australia too with the popularity of data engineer roles and with there being a shortage of them.

CS: There is a global shortage, which is reflective of what happened with data science at the beginning. Now, you are seeing the data science shortage being filled, but, in parallel, we're seeing a new shortage for roles such as ML engineer and data engineer. As these engineering roles become more popular, there will, in turn, be more interest and more such talent churned out by the universities. It seems to all follow a cycle.

Increasing diversity and inclusion in the field

AA: Speaking of changes that you've witnessed throughout the industry during your career, have you also noticed much change in regard to diversity and inclusion among people actually working in the field? Things used to be very different when I started off as a quant many years ago in investment banking: it was unfortunately very male-dominated. I'm seeing a positive shift in data science, but I think we still have a long way to go. I'd love to hear your thoughts, and what you think we could and should be doing to change things up.

CS: It is no surprise that when you joined the field, it was very male-focused: it continues to happen in technical fields.

Regarding diversity and inclusion, there is still an enormous amount of work to be done in data science, engineering, and related fields. I think it's incredibly, *incredibly* important.

> I cannot emphasize enough the importance that diversity and inclusion need to be given within this field, more so than in practically any other field.

This is because data science and analytics teams are oftentimes working on analyses and developing algorithms that ultimately make decisions that affect our lives. It could be banks deciding whether they are going to give us a mortgage or car insurance, or algorithms deciding who is going to be passed to the next stage in a job interview.

> We need to ensure that the teams that are working on these data applications and algorithms are diverse because a lack of diversity in the team can end up instilling unconscious biases directly into the algorithms themselves.

You explained that when you joined the field, it was very much male-dominated. This still happens more often than it should in data science, analytics, data engineering, and so on. You have more male-majority teams and a lack of diversity, not just in relation to gender but in every way imaginable. For instance, I was the only female on my data teams during most of my 5+ year tenure working in FAANG.

There are certain cohorts of the population that work on these data-driven technologies. When a person is surrounded by peers of a similar gender, race, culture, and so on, they tend to share and propagate the same unconscious biases. If you have a single team that shares all the same types of unconscious biases, there is a high chance that these biases are also going to be reflected in the algorithms they develop.

> A lack of diversity on a team has been shown to lead to a higher probability of a less diverse dataset too. If a team lacks diversity, then they are less likely to notice those weaknesses in themselves and in the data that they are working with.

This can cause significant problems later down the line when bringing a data product to market, especially if done so with a product that is meant to be used by all spectrums of the population.

How do we fight for more diversity and inclusion in the field in general? Of course, we need to do better when it comes to hiring diverse teams and encouraging people from diverse backgrounds to get involved from an early age. In terms of gender, we should encourage girls when they are young and show them how exciting this field can be. Don't push them into more traditionally female-held fields. Show them that they can do whatever they want. Expose them to female business and tech leaders. We keep going back to the gender equality argument, which is the one given the most visibility, but it also comes down to culture, skin color, language, and so on. The more diversity you have across every sort of spectrum, the better.

As for incorporating diversity and inclusion into the work that we're doing – I recommend following the work of Dr. Joy Buolamwini. She is an MIT researcher and founder of the Algorithmic Justice League, fighting to expose data and AI harms and biases. I would recommend anyone interested in the intersection of data and ethics to follow her work closely. It would help if we were all more aware of risks and ways that we can fight against bias in this field, outside of improving hiring practices or increasing diversity in the field.

Advising new organizations

AA: Given your expertise, if you were to join a new organization and they were quite new to the whole field of data science and analytics, what would be your general advice for them to get it up and running?

CS: When you are just getting started, I would say be careful; do not take on too many costs from the start by immediately hiring a full-on data science and engineering team. It can be quite pricey. The right talent is going to cost you, and analytics also usually sees more long-term benefits. You may not show enough short-term benefits to justify its maintenance if it is very expensive right from the get-go. Build up slowly with a small but solid team that can pursue different pilot projects for production – projects that show promise and steady growth opportunities for the business. Then, you can start growing that team as is but don't try to take on too much at once. Narrow down what can bring value to the company through a data value identification process and focus on gradually scaling your pilot projects alongside your growing data team.

Also, I would recommend that as you are starting out with your small team, you need to ensure that your business is prepared for what is to come. Data should not be siloed, and it should not be limited to the use of one team. It needs to be integrated throughout the entire business. Consider how you can make it readily available to those who need it when they need it. Everyone within the organization needs to have a minimum level of data literacy, and they also need to learn to recognize the value of data-informed decision-making for all, whether it be in marketing, human resources, operations, or more. Proper data use can benefit everyone. Airbnb is a fantastic example of this with the establishment of their internal Data University to upskill all staff on data literacy.

Everyone has to understand the need to move away from the old-fashioned way of thinking, which assumes data and analytics is here to automate tasks and cut jobs. We need to move toward having fully data-literate organizations where data is used to augment opportunities and empower workers.

AA: How important do you think a formal data strategy is for an organization, and what do you think are the key elements to include as part of it? I think that data literacy is one of the standouts, but I'm sure you've got many other elements that you'd include.

CS: Any time we talk about data strategy, it boils down to data literacy.

> If you don't have data literacy at a large scale, then you are setting yourself up for failure.

I have actually read a lot of books lately on the topic of data strategy. I recently finished *Minding the Machines* by Jeremy Adamson, which is about building analytics teams and how to appropriately place them within the organization: that is critical. Will you establish a separate, divisional data team, possibly led by a chief data officer? Will data needs fall under technological requirements, reporting to the chief technology officer? Or will data teams be split functionally, with siloed teams reporting to different parts of the organization? And how will you spread data-driven thinking – from top to bottom, or bottom to top?

In the last few months, I also read *Data Means Business* by Barry Green and Jason Foster, which touches on data strategy too. I have been trying to read about different perspectives on the subject because it is top of mind for companies today, and different frameworks are being tested out to see what works best.

But I will say that time and time again, no matter what book you read and no matter whose perspective you're getting, the same thing comes up, and that's the people and the culture.

> It doesn't matter what strategy you put into place, what amazing technology you have adopted, or what processes you have put together: if you don't have the right people and the right culture, then none of it is going to bring value to the company.

You have to have the right team and the right people to put it into practice, and that goes into attracting and hiring the right talent for your company, as well as driving an open data culture.

Circling back to the placement of data teams within an organization, it helps to not be limited to siloed data teams but somehow integrate them within the entire organization. You should also make sure that there are touchpoints going all the way up to the top of the organization – all the way to the C-suite. Data teams shouldn't be left behind, so their placement and the value of data for the business should be a constant area for consideration by management and leadership. This is incredibly important for data strategy, but in order to do this, the business as a whole needs to realize the value that data brings, because if they don't, data is not going to make its way up to the top of the conversation in the C-suite.

AA: In general, a lot of these initiatives won't get off the ground without senior leadership support. You need buy-in. You need leadership to support and invest in the people and the culture. They have to embody the data-driven culture; otherwise, people will look up and think, "*They're not taking me seriously.*"

CS: Absolutely. You need leadership or stakeholder buy-in, and that connection all the way up to the top of the company helps to make sure you are getting the right leadership buy-in for the analytics or data science projects that you're trying to implement.

Understanding why projects fail

AA: We often hear statistics being thrown around about projects failing in organizations. We could delve deeper and discuss what they mean by "failure," but if we just take that as a given, do you think this is a reality in many organizations, and if so, what's the reason for this?

CS: I do hear this statistic a lot. It is usually stated that 80 or 90 percent of data science projects fail, and I believe it is a pretty good representation of reality.

I think it comes down to many different factors. Again, it depends on how we define failure.

> I don't like the word "failure" because, usually, you learn from your mistakes, so it is not a complete failure.

When you fail, you should learn from what you did wrong and implement changes to be successful the next time around. If you do it that way, then it is not really a failure – it is a learning journey and a benefit for you and the company.

Usually, projects fail due to a myriad of different factors. A lot of these factors are reflective of the fact that we are in the early stages of the field. We still have a lot to learn. We have not matured in our data science or AI journey – we are just getting started. We are hopefully going to get better at it and learn from our mistakes – but why are these projects failing to begin with?

Maybe it is about not having the right data or not having clean-enough data. We are seeing an explosion of data monitoring solutions popping up on the market to combat these data quality issues. Data observability solutions such as Monte Carlo and Soda.io are finding improved ways to track the health of our data throughout its entire life cycle. You always have to be careful with the data you are using. As the data science GIGO phrase goes, "*Garbage in, garbage out.*" The quality of the input determines the quality of the output. If you don't have good data, then everything after that point is not going to matter.

Another factor could be just not understanding the problem that you are solving from the start, or not planning accordingly for the integration of your solution within the organization. You might not be putting the right maintenance plans in place to keep the model or the application working properly. You need to consider everything that could happen once it goes to market; putting data applications or models into production at scale is wildly different from when they are tested in a controlled environment. Thankfully, the rise of the MLOps engineer is helping place importance on proper deployment at scale.

Another thing, going back to data culture and the strategy that we talked about, is not having the right stakeholder buy-in to keep the project alive and well. There needs to be people assigned to ensure data projects and data applications are performing as expected. Stakeholder buy-in is always needed to prioritize the maintenance of these projects – even after launch, a data product needs to be continuously monitored and updated.

Also, another contributing factor to the failure of data projects could simply be that everyone is drowning in data. In the past, we suffered from a lack of data, but today, the exact opposite is true. We have too much data and too much information, and we don't know what to do with it.

> An effective value identification process is needed to connect the relevant pieces of data to feasible business use cases.

But to go back to how I began this answer, I don't like the word "failure" because I think that if your project "fails," you should learn from what went wrong, go back to the root of the problems, fix it, reiterate what you have done, and strive for a successful run the second or third time around. I don't think that we should look at it as a failure if we are learning from it. Even multiple iterations are fine, as long as we are using the lessons learned to improve. A lot of these data projects are unique. You might be doing something that is being done for the first time in your organization. Perfection should not be expected on the first attempt.

AA: How important do you think a good data culture is in an organization? It can be seen from two perspectives: one is to enable projects for success, and the other one is to not only attract quality staff but also retain them, which can be a challenge in some organizations. More generally, what are the key aspects that you think help make a data culture successful?

CS: Interesting you mention that. Let's circle back to retaining staff in a moment because I think that is incredibly important for a data team.

Going back to things that I've spoken about before, data culture is key, and you are probably going to hear that from many different people in the field implementing their own data cultures. I think a data culture requires the right mindset and characteristics scaled across the entire organization. These characteristics include agility and flexibility, welcoming change and being able to respond to uncertainty, data literacy, and digital fluency, among others. These are all abilities that you can train staff to acquire – in some cases, requiring a shift in mindset.

Also related to a data culture is the avoidance of a siloed data team. There needs to be the realization within the organization about how you can use data in any department and any sector. It is not just limited to the typical technology or marketing teams – it can be used anywhere. Having people realize that and realize the benefit that it can bring to them on the job is important. Data and AI are not there to replace them. We have to get this thought process out of our heads: this is not purely automation or the "rise of the robots."

> Data is here to amplify your abilities and help you do your job better.

This reminds me of one of my favorite quotes; Pedro Domingos said "*It's not man versus machine; it's man with machine versus man without. Data and intuition are like horse and rider, and you don't try to outrun a horse; you ride it.*" It is up to you to find out how you can use data to empower you on the job. Building an effective data culture within an organization is key to bringing out everyone's full potential.

Earlier, you mentioned retaining staff. The way that I have talked about data culture up until now has been in relation to the entire organization and not just the data staff. But when we consider the data team itself, it is important for them to be accepted and respected for what they can offer. Having an established data culture can help.

> The company needs to realize the value that the data team brings and needs to give the data talent the right resources to grow. They are not supposed to be treated as a back-office IT team. They can be incredibly valuable to the company, but they need to be given the right environment to flourish. That might mean opportunities for growth.

Historically, your opportunities for growth were that you would grow into people leaders and follow a managerial path to move up the organization. But nowadays, what is happening often within tech and data science is that a separate track has been opened for what we call **individual contributors** or **ICs**. These are people who don't necessarily want to go down the people-managerial path. Instead, they want to continue contributing through hands-on work. Maybe they are exceptional at what they do and they enjoy it. You still have to give them a way to grow because, if not, they are going to grow out of what they are doing and you will likely lose them.

Give your data staff space to grow, even ICs, and give them a voice at the table – allow them even to sit in on management meetings. It doesn't have to be just people managers. If they are senior data engineers, for example, who can add value to particular meetings, let them bring their voices to the table and share what they are working on.

> Make sure that the data team's voice is amplified as much as it needs up through the organization. I think recognition and the feeling that you can continue to grow and contribute at a large scale can definitely help with retaining these types of employees.

Using data storytelling

AA: Speaking of giving data scientists a voice, I think one key way to do that is through storytelling. We hear a lot about data storytelling: the ability to talk about what you've done to a broader audience in a way that isn't filled with jargon. How important is storytelling in most organizations, and how would you recommend someone develops the relevant skills?

CS: Data storytelling, along with data visualization, is incredibly important. I call it "the last mile of analytics," because it is what you might use to pitch your idea, project, or your findings and ultimately influence your audience to take an action. You have to sell your idea to stakeholders, who, many times, might sit closer to the business side. They are not going to know all of the technical jargon. You need to be able to describe to them – in relatable, easy-to-understand language – what you are doing and how it can bring value to the business.

Something that is very important in data storytelling is connecting data to the business. It is great that you performed your analysis, but how does the final output affect the business? Always include a clear business link within your data story. I think it is incredibly important because, without it, your data, your ideas, and your projects become siloed. They are not going to get anywhere. It is a vital skill to have, to be able to tell stories with data while connecting it all back to the business.

I also firmly believe that exceptional data storytelling skills can set you out from the crowd.

> For those working in data science, if you get good at this, it can help you stand out and grow in your career because there is a shortage of data storytellers.

I see a common problem of people getting stuck in their work with data and not being able to translate the value of what they are doing to the business side. Refine your communication and data storytelling skills; draw that connection to the business.

You asked for advice on how to improve this skill. Data storytelling comes down to your communication skills, which is something that you can work on in many different ways. In my case, I have worked on my communication skills from early on. That has meant learning to get comfortable in front of people, learning how to explain complex technical topics in more understandable words, and learning how to be clearer with what I am saying (especially in cases where I am working with not only non-technical people but also non-native English speakers).

> One of the methodologies that I always follow when I communicate, whether it is data-related or not, is Grice's conversational maxims, which I translate into be brief, be clear, be relevant, and be true.

This is a framework from linguistics that you can study in more detail. The maxims help you achieve more effective communication – especially important in a data storytelling situation.

Also: practice, practice, practice. Volunteer to be in front of people to get more opportunities to put these skills to the test. I did so by teaching. Outside of my corporate work, I teach in higher education and this has been incredibly valuable to my career because it forces me to take these complex technical topics and prepare them for a wide range of audience members. In order to do so, I first need to make sure that I really understand what I am speaking about, and then I break it down into simpler terms. I usually craft a story around each message to draw in the audience.

There are fantastic studying resources out there. *Effective Data Storytelling* by Brent Dykes is a fabulous book that can give you some clues on how to work on your own data storytelling skills.

AA: As an introvert, it took me a long time to step up my game in storytelling, but since doing that many years ago now, it's been a game changer. It's something I'm constantly working on.

CS: I can relate. As an introvert too, it is like acting. You have to put yourself in an acting or theatrical position, so it does take practice. You really have to push yourself out of your comfort zone.

Understanding the fundamental skills of data science

AA: Apart from communication and data storytelling skills, what do you think are the other fundamental skills that a modern data scientist needs? In particular, how important is a good-to-deep understanding of the mathematics and statistics that underpin ML to becoming technically proficient and standing out? I'm seeing a lack of that in many people, especially those who mainly focus on short-term online courses. I don't think they're getting the depth they need. They have breadth, but not depth. Would you agree?

CS: I think there are two main questions here. I'll start with the skills that you need to be a successful data scientist, and then I'll connect that to the math and statistics part.

As a data scientist, there are hard skills that you need to master.

> I always say the number-one hard skill that you need is SQL.

Any data scientist, data analyst, or data engineer should know SQL. I think it is pretty easy anyway. It does not have a sharp learning curve like many of the other hard skills in the data domain. A data scientist should also be proficient in something such as R or Python. I am seeing more demand for Python nowadays. There are other languages as well, but with Python, you are safe for sure. There are additional hard skills too depending on your specialty – for example, cloud certifications.

You mentioned statistics and mathematics fundamentals. Do I think that is necessary? You are asking somebody who comes from a statistics background, so there is some bias there for me! But absolutely, I think it is incredibly valuable for a data scientist. You might be able to get away without it for some time. But eventually, you may make simple mistakes in your data work that could blow up into larger problems because you didn't understand the fundamental math or statistics behind it. As a novice in the field, you can get away without knowing much more than the fundamentals, but eventually, it will catch up to you.

> I would recommend studying mathematics and statistics fundamentals for sure. If you don't, it is going to hold you back sooner or later.

Those are some of the main hard skills, and we have talked already about a lot of the soft skills. I do think data scientists should work on their communication skills, particularly if they want to grow in the field, as it can open a lot of doors.

Another thing that would be considered a soft skill is getting into the habit of continuous learning early on. Just because you completed your master's in data science does not mean that your learning journey is over: far from it. You are going to have to continuously update your knowledge, and it would help if you got into that habit early on. If not, you can get left behind pretty quickly in the ever-changing world of tech.

An additional soft skill is networking. Successful data scientists are also those who excel at networking. There are a lot of opportunities out there. You might be comfortable where you are, but you don't know what you could be missing out on. Networking and getting your voice heard within the field will help you if and when you are ready to make a career move.

AA: On that continuous learning aspect, one group I've seen struggle with that is people who come from software engineering who are looking to, say, transition into ML engineering or data engineering. I've seen them become very comfortable in the languages and tools that they're familiar with but struggle to move beyond their comfort zone. I think this could be pertinent advice to them: start learning new languages and new techniques, and start working with data and ML, even though they might be a bit scared by the maths and stats that underpin it.

CS: Good point. I think it might also help if they can find their transferable skills. Once you have a programming language, you should be able to find similarities with any new language that you are trying to learn. Finding those connections can help you in your ramp-up.

Getting hired in data science

AA: When it comes to data scientists interviewing for roles, what advice can you offer? You've touched on some of these key elements already, but it's so competitive: how can people stand out? Have a profile of work they've done? Have strong communication skills? How can they communicate their strengths in a nice, succinct, easy-to-understand way to the recruiter?

CS: I would split this into two pieces. First, there is the element of getting noticed. You are applying and applying; you want that callback. Then, you have the second piece, which is after you have gotten a callback: landing the job in your interviews.

We'll start with the first piece: getting noticed. As you have mentioned, there are abundant opportunities out there, but there are also a lot of people competing for good roles, so you need to get noticed. Of course, every time you apply for a job, you need to tailor your CV. This should be obvious, but I feel like it is not taken seriously by many job applicants.

> You should tailor your CV for every single position that you apply to. Make little tweaks that connect your experience or your studies to what you are applying to. Never lie, but find the little tweaks that can make those connections clear.

On your CV, find things that can make you stand out. This means that, early on, you need to start thinking about the future of how you are going to stand out from the pool of candidates. You might be able to complete a master's. That's great, but what else? Think of other things that you could do that might help you stand out from the crowd. The way that I have seen a lot of people do this is through side projects, such as starting a podcast or building a portfolio on GitHub or Tableau Public. Starting these side projects might show a side passion and a little bit of extra motivation outside of the typical studies and work that you do. It also helps show another side of you: what you are doing in your free time. These little things can help you stand out from the pack.

There is also networking, which I mentioned earlier. You can generate a surprising amount of opportunities just from networking: getting involved with local communities; attending local events; or using LinkedIn, YouTube, and so on. Don't overlook the power of networking.

Now, say you do get noticed, you get a callback, and you get into the interview process. For a data science role, they are probably going to test you on your technical skills. Before you go into an interview, always ask about what you can expect. They should be able to give you details, and then from there, you can prepare. I also highly recommend studying the company and the role prior to the interview to understand the business in depth. What is their business model? How can you help bring value to the business? Where does your role sit within the organization?

When going to the interviews, go with a smile. Go with confidence. Remember that they are not there to see you fail. On the contrary: they probably want to see you be successful, so don't be nervous. Be yourself, and don't be afraid to ask questions and ask for more clarification before you dive into answering. Never make up an answer. If you don't know the answer, it is OK to say you don't know. I wouldn't always expect someone that I am interviewing to know the answer to everything. Instead, you can offer how you might move forward with solving this and gathering the necessary information if you were presented with the same problem in a real work situation.

AA: How interviewees think about the problem and the process they go through in solving it is what interviewers often care about most.

Progressing into leadership

AA: For someone who's looking to grow in their area, either to become more technical or move up into a leadership/managerial role, what general advice can you offer to help them stand out as they progress, not just in getting their job but also as they progress through the ranks? Competition among their peers can be quite high. How do they stand out to their manager or senior management?

CS: First of all, take a step back and understand what your end goal is. Then, make a plan for how you are going to grow and achieve that. You need to realize what you want in the short, mid, and long term, and then take it step by step and make a plan for what can you do today to get on the right track and stay on it.

Your question was about those already in the field wanting to grow, wanting to be noticed, or wanting to grow in management. Definitely pursue things such as mentorship or coaching within your organization. It could be with your manager or a parallel manager if you don't want to do it directly with your own manager, but look for feedback and advice from someone who is more senior than you and can give you tips to keep you moving in the right direction. A form of mentorship or coaching is always beneficial. Ask them for actionable feedback: what changes can you make to get closer to your goals?

Also, show initiative so that you are not there just being a reactive employee but rather a proactive employee. Demonstrate true interest in things that you are doing, but also in what others are doing. If possible, find ways to get involved with opportunities outside of your job scope. You may be a data scientist but you want to get involved with another project in progress on the engineering team: ask whether you can get involved – to help them while expanding your own knowledge. This is what is called a 20 percent project at Google, where you dedicate a small amount of your time to something that is not directly related to your job scope but can help you and/or the company grow. And it doesn't necessarily have to occupy 20 percent of your time – negotiate for what management will allow, even if that is just 5 or 10 percent. Finding a way to grow in other skill areas can show initiative from your side, and you can in that way also have a larger impact within the organization while getting noticed.

Summary

I greatly enjoyed my conversation with Christina. She offered some pragmatic advice for organizations that are beginning their data-driven journey, such as managing your costs from an early stage as "the right talent is going to cost you" – this aligns with Charles's views (in *Chapter 5*) that data science is not cheap.

I couldn't agree more with her idea of starting small with the right people in the right roles and remaining focused on delivering value to the business before growing your data and analytics capability – this is a common theme discussed in this book. Christina suggests doing a "data value identification process" and working on ensuring a minimum level of data literacy in the organization – including establishing a common understanding of the value of data-driven decision-making. A key part of this is democratizing your data – in other words, making the organization's data available to all staff.

She stressed that stakeholder buy-in is imperative to developing a successful data and analytics capability. In addition a plan is needed for productionizing and maintaining models – including looking beyond the development stage, and thinking about the entire data and analytics life cycle. Poor data quality as well as a lack of prioritization is a roadblock often faced by organizations who are struggling to achieve success with their data capabilities.

It was refreshing to hear that many tech companies are now offering career paths for technical experts, often termed "individual contributors," for those not wanting to pursue the traditional management path. This is an important initiative to not only help address the staff retention issues that many are facing but also to support staff to develop into roles where they can have maximum impact.

For aspiring data scientists, she believes that even though postgraduate studies will help you "stand out from the crowd" in such a competitive field, both the formal education and self-learning paths are valid and can help you develop a successful career. If going down the self-learning path, it's important to create a portfolio of work to help prove your skills and knowledge.

Christina also stresses the importance of being able to tell a story with your data, and her three tips are as follows:

- Have a deep understanding of what you're talking about
- Break it down into clear and simple terms
- Craft a story around each key message to draw in the audience

Beyond developing strong technical skills in SQL and Python coding, and an understanding of the mathematical and statistical fundamentals, her advice for all aspiring data scientists is to focus on the following essential soft skills:

- Communication
- Continuous learning
- Networking
- Mentoring

The key takeaway, though, for those entering this competitive field and who may be looking for their first or next role, is to get noticed – so put effort into your CV, LinkedIn profile, portfolio of projects, and networking.

15

Angshuman Ghosh
Leads the Way

Angshuman Ghosh is currently the VP of Data and Analytics at Sayurbox. He's an experienced data science leader and has previously held leadership roles at Disney, Sony, Target, Grab, and Wipro.

I was keen to learn his views on what it takes to ensure success with data science projects, and how to build and manage a productive team.

I was also very interested to understand how he's seen the industry change over his 15-year-plus career. Finally, I was looking for him to share his advice on developing a successful career.

Getting into AI

Alex Antic: What inspired you to become a leader in the field of data science and AI? Were there any particular people that were your inspiration? Was it a passion for maths, stats, and problem-solving that you had when you were young? What keeps you excited and fascinated to stay in the field?

Angshuman Ghosh: This is one question I get a lot, and it's a very good question.

Honestly, for me, it was not planned. When I learned some of the basics of the field, data science did not exist. There were maybe some of the fundamental disciplines. I think you also did your PhD a while back, in 2002. At that time, you maybe learned math, stats, programming, and more, but nobody called it data science. It was not as sexy as it is today, but the foundational disciplines were all there.

In these foundational disciplines, there are three core pillars that are essential to data science. One of them you already mentioned – mathematics and statistics. The second pillar is computer programming.

Then, there is a third pillar, which is business or domain knowledge, because data science is an applied discipline.

Now, coming back to the question of how it happened for me – I got a bit lucky because I did computer engineering a while ago. In 2005, I graduated in computer science. I worked in software engineering for 3 years, supporting some Fortune 100 clients and building software programs for them. I had a background in computer programming for a very long time, for more than 20 years. But I did not know much about business. I did not know statistics well enough. Since I did not understand business well, I thought, "*Hey, IBM may be a good option to learn about business, and it will be something cool as well.*"

I did my MBA coursework at one of the top institutes in India, and I followed it up with a PhD. The MBA gave me a theoretical foundation of business knowledge – some ideas about marketing, HR, finance, and so on – but I would say that I was still limited because, in the business and practical world, theoretical knowledge is not enough. I understood more when I joined the industry.

So, I had some understanding of business, and during the PhD, I learned a lot more. The PhD was all about mathematics and statistics and getting deep into research methods, sampling, probability, and so on. That gave me a good solid background in terms of mathematics and statistics and some business knowledge. I completed a PhD in 2014, and for the last 8 years, I've been working in corporate. I worked for Disney. The last company was Sony. Then, in between, I worked for Target and Grab as well.

I've worked in different industries and companies, and some of them are very well-known, large companies, so I've learned a lot about business, and that's why I'm saying that theoretical knowledge in business or domain knowledge is not enough. Practical knowledge is essential.

As I mentioned, there are three pillars of data science. I gained the science background, and then I did a PhD. As for computer programming, I've been doing it for more than 20 years now. For business and domain knowledge, I learned theory during my management coursework, and then I practiced it during my work experience. My path was not very well planned – it evolved.

I'll finish this answer with a quote from Steve Jobs: "*Sometimes you may not be able to connect the dots looking forward.*" When I did computer science, I did not know it would be useful for data science. When I learned statistics for my PhD, I did not know it would be useful for data science. But I thought that it was essential to learn. Now, when I look back, I can connect the dots and see how my journey evolved.

Watching the field evolve

AA: How have you seen the field evolve over your 20-year career? How have you seen it change in terms of attitudes in organizations toward leveraging analytics to change decision-making? I think it's still quite new for many organizations. Would you agree?

AG: I completely agree with you. I think attitudes toward analytics have changed completely over the last 20 years. I'll talk about the fundamental reason first, then move to the technical, and then answer the specific question on organizations and the business front.

I think there is a fundamental reason why it has changed, evolved, and grown so much. The reason is not to do with the foundation of statistics and mathematics because, as you are well aware, some of the models or theories were already there. Neural networks are not something new.

Over the last 20 years, we saw the growth of computers, with the number of chips multiplying along with computing power and storage. We have so much data now. Everything is connected to the internet and digitized. Data is generated by whatever you are doing – even this conversation. We have a lot of data; that's why we call it big data.

Then, once we have the data, we can store it, because storage is also cheap and efficient. Once we can store it, we also have huge processing power. Nowadays, my mobile can probably run a neural network model, whereas I remember when I started using my first computer, it used to struggle to play very simple games or do very small tasks.

So, because of the ecosystem of processing, storage, and computing power, and the proliferation of devices, we have a lot of data. We can store it and we can process it. On top of that, there are a lot of free tools and packages that have become popular, such as R and Python. Python is not new, but nowadays, we have so many packages. For anything you want to do, you can just go and search for a package for it. There will be somebody who has created the package you need, and people like us in the industry can just download or import it and use it.

I think those are the fundamental reasons why the field has grown over the last few decades, and it will continue because these fundamental trends are going to continue.

Now, coming to the technical aspect and how it has changed – back in 2000–2005, the focus was on software engineering. Google used to say that software was eating the world. Now, in software engineering, the mindset and approach were quite different, because you would think, "*OK. I as a human will create a solution to a problem. I'll create an algorithm or a flowchart to solve the problem. Then, I'll codify my thinking in the form of some code and model.*" A human created the logic. They implemented it in the code. Basically, you are translating a human language to a computer language and then optimizing it.

But in the last 10 years, the big shift has been from a software-based model or statistical model to a machine learning-based model.

In machine learning- and deep learning-based applications, you don't program in the traditional sense. Instead, you put the data in and the machine tries to learn. In the case of deep neural networks, they go for iteration after iteration. When you're starting with the model, you don't know what the solution or equation will look like, but the computer is learning from data.

Sometimes, when a model is created, we don't even know why we get certain output, but that's what's wonderful, especially in the case of computer vision or NLP. That's the fundamental shift from the human-based algorithm to a statistical model to a machine learning approach.

Now, from the industry perspective, there have also been changes. There used to be a very human-focused approach. Data was very limited, and free tools such as Python, R, and so on were not available. Computing and storage were costly, though big companies were still using them. When I started my career, I supported some Fortune 100 companies. We used many of the new mainframe applications. Only companies in finance and retail had sufficient resources to use mainframe computers.

On my first job, I was programming in COBOL, using databases such as IBM Db2, which I don't know whether many people use today. Now, technology has become democratized. Anybody, even a freelancer or a small start-up, can use Google Cloud, AWS, Python, R, and other packages. Things have become democratized from a company's point of view.

The challenge is that there is too much hype. In industry, a lot of people assume that AI is magic. It's not magic, and I hope you'll agree with me. There's a lot of science to AI, there's a lot of input to it, and sometimes, there's a bit of an art to it as well. People have to have some basic understanding of AI.

Becoming data-driven

AA: Apart from the challenges you just mentioned, based on your expertise in the large organizations you've worked with, what do you think are some other common challenges when it comes to organizations wanting to become data-driven and develop data science capabilities? Do you think part of it is a lack of data literacy at the senior levels, or having senior leaders who don't have enough of a technical background or technical appreciation, leading to unrealistic expectations?

AG: Previously, the interest was not there. Ten or fifteen years back, people didn't think that these things could be useful; they thought it was just academic. That changed a lot with that article in the *Harvard Business Review* by DJ Patil, saying that "*data scientist*" was a sexy career. Then, there was all the consulting work by McKinsey, PWC, and others. That consulting promoted data science at the top management levels.

The AI field involves more mathematics and statistics. There are a lot of concepts that senior management need to have at least a high-level understanding of.

I saw a nice infographic a while back. It said that a lot of senior company stakeholders think, "*We have data. We'll add AI to it. We'll get huge value – it's as simple as that.*" Or they think, "*Let's hire one data scientist. They'll be like a superhero. They'll do some magic. All of our business problems will be solved.*" But it doesn't work like that. You need to have a high-level understanding that data quality, for instance, is very important. If you don't have the right data, you can't use a model to solve a problem.

You need to have an understanding of what is possible and what is not possible, and I don't think machine learning is a solution everywhere. I have worked in companies where, in some use cases, even a rule-based solution would have been much faster and more cost-effective and would have given a good business result, whereas if you were to invest in an AI model, you might see an improvement but wouldn't be able to justify the return on investment (ROI).

> A decision has to be made about where to use AI and where not to use AI.

That requires some technical understanding. At some point, it should not just be senior management making that decision.

The way I look at it, we are living in the era of computers. Computer programming and AI are essential. They're already part of your day-to-day life. You have Google Home, which is listening to you and responding to you. You have online search engines that understand what you type. You can use a transcription service to transcribe things directly and even translate them into a different language.

I think data literacy is definitely one of the biggest challenges, especially for big companies. They have an interest. There is no lack of monetary investment, but we may see a challenge in a few years because organizations are investing a lot and hoping for magic. People who apply AI and data well will see a lot of ROI. People who don't apply it well may not see a result. I think people who see results will invest more and see more value, but people who don't see results may develop skepticism about it.

AA: I've seen some companies in the past get very excited. They realize the benefit and importance of data literacy. They bring in a team of data scientists and engineers. They have some understanding of how to get things moving, but then they hit a roadblock when it comes to taking what they've developed as a proof of concept into production. What do you think are some of the challenges that companies face in that transition, and how do you think they can overcome them?

AG: This is a very common problem. In fact, I've read some research that said 80 or 90 percent of data science models don't make it to production.

If we want to understand the reason why, we cannot start with data science – we have to go one step before that to understand what is happening. Data scientists need data. The science part they can bring in, but they need data to solve problems. In many companies a few years back, there was not much focus on data engineering. They thought, *"Hey, maybe the data is there, maybe not – doesn't matter. I'll bring in a data scientist and the problem will be solved."* That didn't happen because we did not have the foundation of quality data. We did not even have a lot of the features that are required for a model.

Now, most data scientists come from the statistics field. If they are coming from statistics, they may not be so good with engineering, computer programming, or putting together a system, but these are not so essential when you are solving a small statistical problem with 100 data points.

But if you want to scale it up, it's not going to happen if your code is not optimized or you're not using the right packages and services. You may need cloud computing to solve it. I think data engineering is one of the foundations that has to be there for data science to build on.

Another big challenge is that for data scientists to create models, they need to know what features are important. Maybe there are a thousand features related to one particular dependent variable, but then some exploratory analysis is needed to find out which have higher relations, how they are linked, and so on. That is when a business intelligence or data analyst can help a lot. They also bring the business perspective, because sometimes you may have a thousand variables. You might do a correlation. The correlation may or may not be there, but different things have different business implications.

So, just by doing statistical analysis, you may not understand everything. A lot of data scientists don't do good exploratory analysis. To give an example from my own experience, we once tried to create a model at one of the largest Fortune 100 companies where I worked. There was a brilliant data scientist who came from a statistics background. He was brilliant in terms of statistics, he could understand Bayesian, and he could understand a lot of complex statistical theories. He put together a very nice model, but he did not even consider what data features were most relevant. He just took whatever variables he had and applied the best statistical model.

In terms of results, we got almost close to zero-percent accuracy, but he was very happy because he thought, "*Oh, I ran my model perfectly. It gave this answer, so it is the correct answer.*" That's mathematical thinking. You have to start with the input.

To come back to your question, there can be problems with how the data science team is set up.

> Data science is a team sport.

Sometimes, organizations want one person to do everything. You cannot have one person doing brilliant stats and brilliant programming and then implementing it. There may be a few such people, but they are the unicorns of the data world – everybody is after them.

Business leaders should have data literacy and realistic expectations. Put together a team of engineers, analysts, and data scientists. Structure the project in a proper way. Look at the objective of the project, and then think, "*How do I create a data foundation? How do I do an exploratory analysis? How do I do data modeling? How do I productionize it and implement it?*" For a business, the final value comes once you implement something in production – when it increases revenue, reduces costs, and so on.

Organizing a data team

AA: Speaking of teams, some organizations ask themselves, "*Should we be centralizing the data science function? Should it be distributed and based in different business units?*" **Do you have a preference that you've used in the past? Of course, it depends on the organization and the analytics maturity, but do you tend to favor one over the other with how you structure the teams that you manage?**

AG: When you're first starting out, if you're a small company, you can't have a lot of different teams. Generally, you start with a centralized approach. Because the team is centralized, all the data is together; you have one source of truth. And generally, those people are good with the technical areas. They can solve technical problems quickly, and they can talk to each other. Knowledge sharing within the data team is very good. It definitely has a lot of benefits.

The challenge is that domain knowledge is also very important in data science. If the team wants only data scientists and engineers, many times, they will miss out on the domain knowledge; even the analysts may miss out on the domain knowledge. To do your analyses or modeling, domain knowledge is important. If your data team is working with your marketing team, they'll need to know, "*What are the problems in marketing? What are the things we should be doing? What are the features?*"

Another model is to have completely distributed teams, where each data science team works for a different function or business unit. Now, that model has problems because, for each team, you'll need a data engineer, a data scientist, and a data analyst. You may need more people, and the knowledge sharing between teams may not be so good.

Plus, there is no one source of truth. Everybody will be trying to look at the data in their own way, and sometimes, they may have their own biases or agendas. The marketing people might say, "*Hey, my campaign is always successful.*" Finance might say, "*Hey, I always do perfect financial models.*" And if things are not going well, people will point fingers at each other. So, I think that model also has certain problems.

There is a hybrid approach, which I am trying and which I think is a better solution, where there are certain things that everyone has in common. A lot of the data engineering work is common. MLOps, or at least those parts of it that you don't need every day for every functionality, can be common. There might be a certain dashboard or certain things that are required across a company that you should definitely centralize.

Then, for each of the departments, you can use just a couple of analysts, mainly focusing on the day-to-day operations. It's better to use analysts there because the day-to-day operations will involve more analysis rather than data science work, such as the modeling of solutions. Modeling solutions may happen only once in a while per function. One day, there might be the creation of a churn model, which is relevant for marketing or **Customer Life Time Value** (**CLTV**), but once that modeling is done, you don't have to do it every day.

I think it makes sense for data science to be centralized and centrally funded. Then, use data analysts to support each business function in their day-to-day operations. That's how I go about structuring a team, but there was a learning process. Probably, there is no standard solution, but if you or others know a better solution, do let me know. I would also love to learn more about this one.

AA: Yes, I agree, I think hybrid approaches tend to work well in most places – pulling out the common elements as best you can for reuse, but also maintaining some coherence between the team for knowledge sharing and standards setting. It also helps with training, attracting, and retaining staff, which can be a big challenge, especially in Australia. We're facing a lot of churn-out.

AG: That's also a big point. Data scientists like talking to each other and discussing how they will conduct research.

AA: Also, some of them would rather report to a leader who has a technical background like you and I, rather than a business manager.

AG: That's absolutely correct. I have also faced that same thing, and I see it on a day-to-day basis. If a data scientist's leader is technical and understands technical things, they feel more comfortable and happier with them. Nowadays, a lot of people don't work just to earn money, like we used to do before. They want to be interested in the work, they want to be part of the team, and they want to be understood and appreciated. That's important to them – not just the salary.

Every data scientist should at least once in a while have interaction with the central data team and their leader. They should also have exposure to conferences and so on.

Building a good data culture within an organization

AA: Something we just touched on, which I think is important, is having the right data culture within an organization to attract and retain the right people, understand how these people work, and align their work with the goals of the organization. Do you have specific views on what a successful data culture looks like in an organization? What do you do as a senior leader to try and enable the right culture throughout the organization?

AG: I think culture is extremely important. Peter Drucker said that culture eats strategy for breakfast.

A culture is not just for the data level. It should be at the company level. You can then translate it to the data team. Fortunately, many of the companies I have worked for had a very good culture and took it very seriously. Those companies had cultural values. Everything you did would be based on those cultural values.

To me, culture is what I do every day.

We get to a point, once we get up the hierarchy, where we may not do the technical things on a day-to-day basis. So, how do you set the vision, mission, and values of a culture? How do you make sure that people are motivated? How do you make sure that you get the best talent but also retain them and motivate them on a regular basis? I think these are the core things that a leader has to take care of at the company level. That might be the CXOs or the head of the data science team. Their most important job is to make sure a team has the right culture, motivation, and goal.

So, with my team, I have a weekly meeting, and I try to make sure every week that I spend some time on culture. It's not about what we say but what we do. Initially, of course, we have to explain things to people, but after that, we can just call out the things that are going well. That's the best example. Say somebody in the team goes the extra mile to help the team, or helped another team – that's what collaboration is to me.

In regard to taking ownership, maybe something was not strictly your work. However, it may have been related to your work, so rather than saying, "*Hey, it is not my work. I will not do it,*" you went the extra mile and did it. That's taking ownership and solving end to end. When that happens, we should try to celebrate the action then and there.

Leaders have to take culture seriously. You should not just do a 6-month round-up to say, "*Hey, culture is important.*" It should be a day-to-day thing; it should be a habit for everybody on the team.

AA: I completely agree, especially with your points about how it has to come from senior leadership.

On that note, what do you think makes a great data leader, based on your own values and what you've seen in others as you've progressed up to your position today?

AG: I don't think there is one type of great data leader. That applies to most business leaders as well.

When I look at data science, I don't look just at the data. We use data and tools, statistics, the cloud, Python, and whatever else, but these are just tools. We use these tools either to solve a problem or get a business impact. A good data leader should be able to use tools efficiently to solve problems.

At the same time, you need to maintain a good culture within a team. Try to get the best talent, but also make sure that you retain and motivate them properly. If you can do all of that, you are a good data leader. If you do only one or some of those things, it's not enough. I say that there is no one-size-fits-all approach here because different companies will have a different goals, and the data leader has to be aligned with them. In fact, the selection process itself should be based on that. I worked for Sony, where the role had an R&D focus. There was more focus on the research publication patent than on implementing a model. If that is your role as a data leader, you have to be very good with research and publication, and you should be able to do that through your team as well. But if your role is more focused on business, you have to have a business impact. If you are leading just the data science team, getting deeper into that is OK, but if you are leading a data team where engineers and analysts are also important, then you need to take a broader view and not just focus on data science.

If it is a very small company, you may also have to be multi-hatted. You may have to do everything – a bit of engineering, a bit of programming, and so on. Given the role, what is your objective? What is the final impact that you are looking for? What is the problem you are trying to solve? If you can do that well while maintaining team harmony, motivation, and culture, then you will be a great data leader.

Understanding the value of data storytelling

AA: Do you think the ability to be a good storyteller is important for data leaders and others working in the data space broadly? How important do you think it is, and what are some of the ways that you think people can develop the necessary skills?

AG: The three core skills for data science are maths and statistics, computer programming, and domain knowledge, but if I had to add a fourth one, it would be communication and storytelling.

Storytelling is important because it's not just communication. It's not just speaking good English. At the end of the day, you want to convince somebody. Sometimes, data models don't go to production because you've created the POC, but you were not able to convince the stakeholder that it was the right model to implement in production. If the business leader doesn't understand what you are saying or you don't understand what the business leader is saying, that can be disastrous.

Right now, both of us are speaking English. We understand each other. But suppose you were talking Hebrew and I was speaking some other language – we might be trying our best, but we would not understand each other. The business leader has to also have some technical understanding because if they don't understand what accuracy and precision are, what should be used when, and what a neural network is, it makes things very difficult. Data scientists and data people in general have to have a business understanding. Especially in a data team, data and business intelligence analysts have to have more understanding because they'll be dealing with stakeholders on a day-to-day basis.

The importance of storytelling increases as you go up the ladder. People who interface with the business have to be very good at communication and very good at storytelling, and if you're leading the data team, then it's critical. At some point, I would argue (and this may be controversial) that communication, storytelling, management, and software skills become more important than your technical skills because, at the end of the day, your stakeholder may be the marketing head or CEO of the company.

> Understanding their world, understanding their problem, and then finally, communicating and telling the story and convincing them are actually the most important parts of your job.

Those who want to progress their career in data science by leading a team or a project need to understand the business and work on their communication and storytelling.

Hiring new team members

AA: Are there specific skills and attributes you look for when you hire data scientists and engineers for your teams? Do you put a lot of weight on formal studies as opposed to online courses? Do you look for particular attitudes and aptitudes? How do you try and vet different people to make sure they're going to fit into your team?

AG: I'll look at three things here. Firstly, skills; secondly, software (including languages and tools); and thirdly, qualifications.

In terms of skills, it's clear that maths and statistics skills have to be good, especially for a core data scientist. For a business analyst, not so much, but basic descriptive stats should be there. For a data engineer, not so much either.

Now, coming to the second skill – computer programming. Computer-programming-specific skills will vary based on the role. If it is a data engineering role, the cloud and SQL databases become very important. If it is a data science role, since most of the companies I have worked for use Python, I think Python is very important. If you are using R, then go with R – whichever language the company is using. For the data analyst or business analyst role, then a bit of Python is OK – not so much skill is needed on the cloud side – but you will need an understanding of visualization and dashboards – maybe Tableau or Power BI.

Those are probably the most popular technical skills. If the role is very senior in the data science domain, a knowledge of statistics, machine learning, and deep learning is more important. If you are hiring for a research role, deep learning, NLP, and an understanding of papers and conferences become very important. When looking to fill a research scientist role, seeing a candidate who is engaged in conferences, publications, and research becomes really important.

Domain knowledge becomes difficult sometimes if you're hiring for a junior role. When I joined the industry after my PhD, I thought I knew everything because I had a lot of degrees and education. But I realized it was all very theoretical; my practical knowledge was not great. Domain knowledge is something you can compromise on a bit, but if you're hiring for a particular role, such as a marketing analyst, some knowledge in marketing may be helpful. If a trade-off is necessary, I would probably be more focused on math, statistics, machine learning, and Python compared to domain knowledge.

Now, qualifications – this again will be a debatable answer because a lot of people nowadays say that skills are more important than qualifications. I argue that both are important. Without skills, you can't do the job. If you wanted to get into the field 5 or 10 years back, it was rather easy. For you and me, if we had a PhD, even in a not-so-related field, people would ask us to join them. But nowadays, there are a lot of PhDs, and there are so many people who want to join the data science field, so competition has increased so much. It's the basic economic principle. Previously, the demand was there but the supply was not there, so anybody who wanted to join the field could easily join. Even if they were not data scientists (maybe they were financial analysts, for instance), they joined the field.

Now, everybody wants to join. The demand is still there but there is a lot of supply, so if somebody wants to join the field now, qualifications are important. If you have gone to a good college and have your bachelor's or master's, or even a PhD in a relevant discipline, it gives you an edge in the selection process. I'm not saying that will be good enough for your final practical work, because if you just have the degree and not the skills, it's frustrating for you, your manager, and your employer. I don't think it should be an either/or question.

Another thing I want to add is that even continual learning is important, because this field changes very quickly. I solved problems very differently only 3 years ago from how I would solve them today. For forecasting, we used to use time series. Then, I went to Target and used Prophet from Facebook. Nowadays, there is a service called Amazon Forecast that has Prophet, DeepAR, and so many models that are already coded. The cloud was not so popular previously. Now, it has become popular. Today, I'm talking about Python. Tomorrow, maybe Julia will become more popular.

If you want to grow in the field, you need a growth mindset. You have to continuously upgrade yourself. An important thing in today's business context that some people lack is flexibility and agility. Some people say, "*I'll only work on computer vision.*" But the business might require you to solve some other problem, such as churn modeling, that you don't find interesting. You have to be flexible enough to solve the problem that is given to you. Situations change, teams change, and technology changes. I think flexibility and agility are critical, and when I'm hiring, I try to understand whether a person will be adaptable because, on my team, projects may change and the role may change over time.

Those are the things in terms of skills, qualifications, and attitude that I give very high importance to. Attitude is equally important to skills and qualifications, because if somebody has the right attitude and wants to learn a lot, even if they don't know everything right now, they will figure it out.

One more thing I forgot to mention – I really like a go-getter attitude. Given a problem, you should go figure out what the solution is and solve the problem, rather than breaking your head or coming and complaining.

Summary

I enjoyed learning about Angshuman's impressive achievements and how he's seen the field evolve, including the shift from traditional/statistical modeling to machine learning.

It was refreshing to hear him speak about the importance of understanding what is realistically possible for an organisation, and that ML isn't always the best option, as sometimes traditional approaches, or rules-based models, are faster, cheaper, or simply the only option.One aspect that Angshuman raised that resonates with me is the importance of data literacy, not only at the enterprise level but also, more importantly, at the senior leadership level. Lack of data literacy among senior executives and mid-management remains a challenge for many organizations.

Another important point that he raised is the importance of domain knowledge – without it, it can be very difficult to create value for a business. He also stated the importance of culture in an organization and emphasized that a positive culture depends on the everyday actions of leaders.

The question of where an analytics capability should sit in an organization is one that involves balancing the various trade-offs of different structures – there is no perfect solution. Angshuman leans toward a hybrid model, where the function is centralized and centrally funded. Then, data analysts are assigned to support each business function in their day-to-day operations.

For aspiring data scientists, he strongly suggests a solid grounding in mathematics and statistics, programming and software engineering skills, and business/domain knowledge. He also believes that formal qualifications – and particularly postgraduate qualifications – give you an edge in the competitive employment selection process. No matter which path you take – traditional, formal studies, or self-paced learning – it's important to never stop learning, as the field and technology are continually evolving.

16

Maria Milosavljevic
Assesses the Risks

It was an absolute pleasure to have an opportunity to discuss so many important and challenging topics with **Maria Milosavljevic**, one of the most highly respected and experienced CDOs in the **Australian Public Service (APS)**. She is currently the inaugural Chief Data Integration Officer at the Australian Department of Defence, with previous experience in the private sector.

I was very interested to hear her views on the key differences and similarities between the public and private sectors, and what she thinks the key elements of a successful data science capability are.

Getting into analytics

Alex Antic: To begin with, I would like to get a better sense of what inspired you to pursue a career in analytics and how you ended up at the point you are at these days. Were there people who inspired you? Was it a passion for problem-solving?

Maria Milosavljevic: I think it was more accidental. When I went to university – a long time ago now – I wanted to be a biochemist. I had flip-flopped around, saying, *"What am I going to do? What am I going to do?"* as a confused 17-year-old often does. I only decided I wanted to be a biochemist the week before university started. Up until that point, I was enrolled to do pharmacy, and my dad kept saying to me, *"Why would you want to count tablets?"*

I went up to the office to change my enrollment to a bachelor of science, and I thought, *"Well, I'll do biology and chemistry – obviously – and I will do math because I'm really good at math."* I would have done something completely different, but my dad made me do computer science, which he also persuaded me to study in year 12. I was the only girl in the class. I absolutely hated it. Not because of the subject matter, though the subject matter was dry, but because I really wasn't that fussed about that. I hated being the only girl. So, when my dad said, *"Just do it in your first year,"* I said, *"OK, fine, but that's it. I'm not doing it after this year."*

I went to the office and said, "*I want to enrol in computer science, biology, chemistry, and math, please.*" Their response was that for biology, I had a choice of zoology and botany. I didn't actually know what the difference was. I also had never actually realized that all throughout doing biology at school, which I loved, I'd never held a scalpel when there was an animal. I had always done it in pairs, and I had just not noticed that I'd always make my partner do the animals! So, I said, "*Oh, zoology sounds like a cool word. Let's do that,*" which was a huge mistake because after seven weeks of freaking out doing all the labs (and I'm not a vegetarian or anything), I just said, "*I can't handle this,*" and pulled out of zoology. It was too late to take botany, so I took on more math because I was good at math.

I was down to chemistry, math, and computer science, and then I did thermodynamics. I really didn't like it, and also thought, "*I don't know that I like chemistry that much anymore.*" And so, by the end of the year, I said, "*Oh my gosh. I'm down to math and computer science.*" I really was quite disillusioned because I thought, "*Well, I don't really want to specialize in either of these areas.*" But in my second year, I did AI and I fell in love. I haven't looked back since.

So, that's actually how it happened. It was accidental more than anything else. My dad, I guess, is the reason why I'm doing it.

Discussing diversity and inclusion

AA: One thing you touched on that I'd like to tease out a bit, if you're comfortable with it. You talked about your discomfort at being the only female student in the class. It must have occurred many, many times throughout your career that you're in a meeting and you're just surrounded by all these males. How has that been, and what do you think we should be doing to change that? This whole notion of diversity and inclusion is a big topic. I would love to hear your thoughts on that.

MM: It is a really hard one, and it happens all the time. I'm far more comfortable with it now; otherwise, I guess I wouldn't be here because it is a common thing, and also because I work a lot with people in IT, and IT is the same as data.

It's been up and down. I would say that I've had some interesting experiences along the way. I have a daughter and a son, and I wanted them both to go into IT, and I tried so hard with my daughter. At the end of year 10, her top two subjects were computer science and math, and yet she still decided to not continue those into year 11. She said, "*I just don't enjoy it.*" I remember feeling like that myself, but when I found AI, it was everything – all the lights went on in my head.

So, I think it's really about finding the part of STEM that presses someone's buttons, and it's about communicating to girls that it's not just all the really deep tech stuff. There are actually a lot of different careers that you can have, and even in some of the very technical areas, some of them far more interesting to you than others will be. For me, that was AI, because it was a combination of technology and math and psychology and linguistics, and so it was a mixture of art, science, and math, and I love that about it. It was also that I'm a natural problem-solver, so it worked for me. So, one of the things that would help is really communicating what the career paths are, and what the options are.

People often say, "*You can't be what you can't see.*" Women, unfortunately, we do often go, "*Oh, yeah. We don't really like the limelight very much.*" We are very happy to step back and not be in the limelight; I've had to push myself to do that.

A long time ago, I was invited to a conference, and they had a completely male lineup. I said, "*Look, why would you invite me? You've got 100% men. You don't even have the token female.*" They replied, "*You know what? We asked a bunch of women and they all said no.*" That was a huge wake-up moment for me because I – and I still do – say no to most of the invitations that I get because I find it exhausting. I think really encouraging women to be out there and to be seen is another really important thing, as well as mentoring them along the way.

We've come a long way in this regard. There are a lot of programs. I have a son who's 19, and when he was going to high school, I struggled to find outside-of-school activities that I could send him to. There were loads for girls, but they actually didn't have enough for him. So, I also saw it from that perspective, and thought to myself, "*Well, have we got this right?*" He's in IT. He actually loves IT. Recently, he was going on an out-of-work team thing; it was a farewell for one of the staff members. I said to him, "*Oh, are any girls going to be there from your team?*" He said, "*I don't know.*" I said, "*Are there any girls in your team?*" He replied, "*Yeah, there's one.*" I said, "*Ah. Has she been invited?*" He said, "*I don't know.*" That was an opportunity for me to actually talk to him about what it was like for me to be the only girl in the team. So, it's also about mothers with their children, and other women, explaining to girls and boys about the good, the bad, and the ugly.

AA: That is fantastic. Thank you, Maria. I loved that last point around what you can do as a parent to instill the importance of inclusion in your children. I think men can often be oblivious to subtle forms of exclusionary behavior from themselves and others in the workplace – both professionally and socially – so it's important to raise awareness.

AI and analytics

There is something I'm asked a lot about, and there are two parts to this question. Firstly, how do you achieve success with data analytics and AI broadly in the APS? What do you think are some of the real secrets or pivotal elements of success? And secondly, how do you think it's different from what you do in the private sector?

MM: I've been in government since 2010, and before that, I was a CTO in effectively an incubator for start-ups in analytics, so I was building teams to create a capability that would then become a company that could be spun up if it was successful. It was great. I loved it. It was really interesting. But I left the private sector for a very particular reason – I'll come to it in a minute as I talk about the similarities and differences.

What's similar between government and the private sector is attracting and retaining talent, prioritizing limited resources, getting access to the right data, understanding the risks inherent in that data, thinking about how to solve a problem, and balancing the risks of using the data versus not solving the problem. In the private sector, that's not so much of an issue because it's more about making a profit, which I'll come to in a minute, whereas in government, it's genuinely about, *"Here's a problem we need to solve. What's the risk of us not solving that versus the risk of using this data in this way for this purpose?"* Designing a solution that's actually going to work and managing the data well for over its life cycle and munging it is the "cleaning the toilets" part of the role that nobody wants to do, but it has to be done. Those things are all really similar.

There are a lot of transportable skills between the private sector and government. What's different, though, is that in government, we are driven by public value rather than profit, and we have an authorizing environment, which is the government of the day. On the other hand, in the private sector, it can be either management or the board. In government, the authorizing environment is driven by the public, whereas in the private sector, the board is driven by the shareholders. By making this comparison, it helps you to understand that they're quite similar. There are just different drivers, different types of members in their authorizing environment, and so on.

But our priority is the problems that matter rather than the problems that will make a profit. I'm not saying that in industry, people don't solve problems that matter, but it's just not the primary driver; otherwise, they won't make a profit. Not-for-profit factors can still appear in the private sector and are often driven by the problems that matter. They're not profit-driven, but they still need to actually be sustainable. But in government, it's really about the motivation of, *"Does this matter to the nation, and is there something that the government should fund?"*

The reason I made the jump from the private sector to the public sector is that the problems that we solve in the public sector – in my humble opinion – are far more compelling, and the data that we have access to, likewise, is far more interesting than that in the private sector. It's something that hits me in the heart; I feel passionate about it.

Another thing that's really quite different in the public sector is that when I talk about managing data as an asset, I say we're managing it as a national strategic asset, because very frequently, that's exactly what it is.

> You'll often hear me saying that the data we collect today will outlast us, because a lot of the data we collect – particularly personal data – will have to be retained and managed well for a so-called lifetime.

That genuinely matters, and the integrity of data through its entire life cycle really matters. The decisions that we make based on that data can be national decisions. They can go to the heart of who we are as a nation and how we treat our citizens as a nation. The core value that data brings to public service is just an incredible opportunity, and that's why I love it. I find it so rewarding – I find it far more rewarding being in government.

I've also had staff who have left – who have come in, been trained up, and then the private sector says, *"We'll pay you two or three times what you're getting from the government."* They go, but what they find is that they miss that compelling drive for the purpose of why we're here, and frequently, they'll actually cycle back for that very reason. I think those are the key differences between the public sector and private sector, and they're also the things that are all part of establishing that successful capability, because you've got to be able to do all of those things – the things that are the same and the things that are different. They're all part of what you have to do.

AA: That's one of the best comparisons I've ever heard, so thank you for that. It's a really clear and unambiguous way to state the difference in the mission and incentives between the public and private sectors.

Becoming data-driven

AA: Where do you see many agencies and departments getting stuck in terms of the challenges they're facing in becoming data-driven or driving capability uplift? How would you approach, say, an agency or department and say, *"This is what you should do to become successful with data analytics?"*

MM: Great question.

> The first thing is, understand why. What are you trying to achieve? What's the problem you're trying to solve?

Every organization should be data-driven in the way that it operates, especially these days. Data is absolutely an incredible asset. When we make decisions in the public sector, because we operate under administrative law, we need to be accountable and transparent and be able to show how we've arrived at particular decisions. That is part of the "why." That's common to every agency. But then, there will be specific things that every agency wants to achieve beyond that. So, start off with the "why," and then have a strategy that has clear outcomes in it and shows you how you're going to actually achieve them and target them, and then continually report against that. Measure your progress and become data-driven within the data function itself. My nature means I often orient around thoughts such as, *"What's the North Star? Are we there yet?"* I'm constantly measuring how close we're getting. That would be the first thing.

> The second thing would be to appoint the right CDO.

I've spoken often with people about what I call the five hats of a CDO, and I think this is actually quite important. The hats that I will often talk about are as follows – the first is being a visionary. Not everybody understands what the art of the possible is and how an organization could actually use data to become so much better (more effective, more efficient, or some other outcome). The first hat is about being a visionary and challenging the status quo, as well as having a very long-term view of managing data as a national strategic asset.

Figure 16.1 – The five hats worn by CDOs in Maria Milosavljevic's metaphor

The second hat is often what I'll call the bridge. Companies – including organizations and even governments – often work in silos, but data flows, and *should* flow, across those silos. It's about making sure that data gets to where it needs to be so that its latent potential can be realized. Make sure that it's democratized and that everyone has access to the data that they need. You need to work through the reasons why data isn't flowing. Maybe there's a legal reason, maybe there's a cultural reason, or maybe there's a technical reason. There are lots of different reasons why data doesn't flow well.

That's all related to the third hat, which is the regulator. That is about making sure that data is managed well throughout its whole life cycle – managing it as a strategic asset. The regulator is basically the governance function. Quite often, CDOs go straight to governance, and that's all they do. Then, what happens is that they're just seen as a blocker rather than an enabler. People jump to the regulator hat as the default hat, and that's where they spend most of their time. This will definitely happen if a CDO doesn't have the technical experience to understand the potential of data but has only an auditing- or risk-related governance background.

The regulator makes sure that data is designed properly right from the get-go, it's managed well throughout its life cycle, and its value is derived or the latent potential of the data is activated. You need to be specific about how you want to be able to do that, putting in data standards and quality checks around the data and making sure that projects have KPIs around data. Too often, projects have KPIs that just say, "*on time, under budget.*" Today, it is actually all about the data in systems, so how do we make sure that from the get-go it is designed and managed well, which includes privacy and security issues?

Hat number four is the scientist, which is about the art of the possible, experimentation, and pushing beyond the boundaries. This hat is often at odds with the regulator hat, because the regulator is all about safety, whereas the scientist is about exploration. You need to be able to balance trying all those things and demonstrate that you can experiment safely or swim between the flags.

The final hat is the engineer. The key to this is scale and industrialization. A lot of organizations, when they first push into this space, will have a bunch of very technical people and data scientists, but what they don't then do is understand that analytics is not normally a solution all by itself. Very often, it needs to be integrated with a much broader solution to an important problem. To be able to do that, you have to be able to scale, industrialize, and sustain. Do we have the right data flowing through pipes, rather than people carrying buckets from one system to the next? When we do that, are we checking the integrity of the data to make sure that it's remaining consistent? Is the system going to scale to the number of users that we need? Is it going to be performant or not? Is our analytics accurate? Is it still achieving its purpose? How do we monitor that and make sure?

Those five hats are really important.

The third part of this answer is about designing an operating model that can bring this all to life. There's a term called **MECE**, which means **mutually exclusive, collectively exhaustive** – no gaps, no duplication. You need to look at how a function works with all of the others – both suppliers and consumers. A supplier example would be the CIO, which would entail getting that relationship right. What does the CIO do to support the data that's required to support the business? The CDO often sits between those two. If a business is trying to be more successful, they need data to be able to do that, and the data needs technology that it can flow through. That flow from left to right is really important to design well, and an operating model is how you do that – you tease out all the elements so that you can understand who's playing what role.

Then, you need to make sure that within the data function, the CIO area, the technology area, and your consumer area, the right people are in the right places, because it's not just about setting up a means to its own end. It's about how to enable the business more. Attracting the right people on all three counts is really important. You need to understand the customers, understand what the business needs, prioritize everything that you do in that function, design the governance for all of it, and then be honest about it. Within all of that, there is a whole load of pitfalls that I've talked about, but there are some others I would add.

For instance, we are not a means to our own end. We are here to support the business to be better, so we've got to do everything we can to better enable the business. If people replace an organization's interests with self-interest – for example, by thinking that the CDO owns all the data – that's a huge mistake. You've got to be more like a CFO, because a CFO says, *"I'm giving you your money. You spend it within these rules."* It's the same with any of the problems that we solve. Business areas should actually solve their problems. They should manage their data – as data custodians – and they should be able to solve their problems. A central function in a hub-and-spoke model should be about enabling them to do that, just like a CFO does, rather than saying, *"I'm taking everything into a center, and now we're going to do everything."* If that happens, you get choked up. The CIO needs to play an important role here. It is not just about the CDO becoming a more technical role where everything is maintained and sustained; building out a sustainment model is very important to get right with the CIO.

> Another big pitfall is the relationship between the CDO and the CIO. If you don't get that right, you may start off well and show some value from data, but ultimately, it will grind to a halt.

AA: That's a great point. I've had many discussions with people, both for the book and out of general interest, about how you establish a strong, successful working relationship between the two. Sometimes, an organization will bring in a CDO under a CIO – that's a recipe for disaster. Alternatively, they won't give enough authority to the CDO, and the CIO will be treated with greater responsibility. How should that relationship work?

MM: When I entered government, I entered as effectively a CDO, and I was told, "*Look. Just create your own technology stack and don't worry about IT. They just do all the desktops, the laptops – all of that stuff.*" I replied, "*Well, actually, no, I can't do that. I have a pretty small team, and we're going to get choked up with this. I need IT. I won't succeed if they don't help me.*" I worked with them on this. Initially, they were unsure, but then they did start to see that they played a really important role. This was a long time ago now, and I think we've come a long way since then.

Then what happened was I was so successful that the acting CEO at the time promoted me to become the CIO. So, I then pulled the data function under me and I was able to successfully navigate both. I've done that in multiple roles where I have been a CIO and I have had the data function, but I guess because I understand data really well, I could do that. I think it falls down when the CIO doesn't understand what a CDO does and why it is so critically important.

About seven years ago, when I was in my first role, I spoke at a Gartner conference, and I was a bit brave. I basically said to all the CIOs in the room that the reason why CDO roles existed was that they had forgotten that the "I" in their title stands for information. I'm sure you can guess some of the reactions that I got to that! But in some senses, it's true. The CIO role was created a long time ago, and things have changed dramatically since that role was created. The importance of data and information has never been higher than it is now. When the role was first created, it was all about paper and then going digital in terms of documents. Now, we talk about far more data and information and integrating it all together, and so a lot has changed. I don't think it's impossible for that to work, but it does mean that people need to understand that.

The way to do this is to understand what an organization's objectives are and the roles that both functions play in achieving them. Take the personal out of it. Focus on what needs to be achieved for the organization. Get everybody working on that objective, and then work out how to collectively achieve it. It is possible. The area where it often gets very sticky is architecture. We didn't really think about architecture outside of a CIO domain until the last five years. Now, we understand quite well that the mission and the vision of the organization should drive its business architecture, which then should drive the data and information architecture required to support it, and then that should drive the technology that the data and information flow through.

So, we're at a point – or precipice – right now where that "aha!" moment has happened for many organizations, and they are now trying to move to a more data-centric approach. It's being driven by data-centric security – zero trust and so on – but it can have multiple benefits in terms of understanding the data part of architecture. Gartner will say that a CDO should be accountable for data and information architecture, and I agree with that, but it doesn't mean that that's all they should think about. It's about three areas – business, data, and information (sometimes even process and technical architecture as well), all collectively creating enterprise architecture. It's a whole that's made up of multiple parts, so it's about stitching things together and getting people to work on it side by side. That's the key to getting it right, but it does rely on people putting an organization's interests first.

Ethical AI

AA: One thing that I've been increasingly asked to speak and write about recently has been the whole topic of ethical and responsible AI – it's becoming big. A lot has been written – there are a lot of policies and frameworks around it. How do you actually make it practical and put it into practice, especially within the APS? What do you think are some of the steps that need to be taken to realize it?

MM: Step number 1 is understanding the "why" of the problem. In the private sector, it might be profit, but in government, the question is, "*Why does this problem actually matter? What are the risks to government, to the community, to the nation as a whole, and globally, if we do not solve this problem?*" Then, you need to understand what data is required to solve the problem and what risks are associated with the data. People jump on ethics straight away, but it's not the only risk. I prefer to talk about *trust* and maintaining trust because there's just as much trust lost in not solving the important problems as there is trust potentially lost in solving them badly. You need to have conversations: "*Why are we doing this and why does this really matter? If we do this, what are the potential risks and things that can go wrong?*"

I think more about trust as an umbrella with a range of different pillars, and one of those pillars is ethics. Another pillar would be legal – for instance, privacy. Everybody has to comply with privacy laws. That's a legal requirement. In the APS and public service, we also have to comply with administrative law, and so things about fairness in decision-making, the rule of law, the right of reply, and being able to actually see why a decision has been made and question it are legal requirements. Then, there are regulatory requirements. They can range from soft regulation, such as self-regulation – your policy as an organization – through to really hard regulation. In the APS, we've got the **Australian Cyber Security Centre (ACSC) Essential Eight** and an information and security manual that we use to guide all of those decisions. Then, you've got things such as human rights.

There are a whole bunch of other things as well, but let's focus on security. People will often say, "*Oh, we can't do that because of security – #becausesecurity!*" Well, a lot of the regulation around security is about best practices, but it doesn't mean that you can't accept risk. People will too frequently say, "*No, that's the way it has to be,*" rather than, "*Well, hang on a second. What is the nature of the risk that we're talking about here? What is the nature of the risk of us not solving this problem, and is it that high? If we fail to do this as a nation, is that more of a risk than the risk of the security of this data? How do we manage that? How do we actually understand it rather than just making a decision and pressing ahead anyway?*" Well, you said there was a risk there – are you monitoring it? Are you ready to act if things go wrong?

It's ultimately a risk initiative. It's all about understanding and managing risk well, and then communicating it very well to those involved. If you talk to people about what they expect of government and you explain to them, "*OK. Here are the options and here are the pros and cons of all of the different options,*" people will say, "*Oh, OK. We understand now what the pros and cons are, and we want this at the expense of that.*" People do this all the time with apps such as Facebook; they put their data out there. If they understand that someone else may be able to access that data, they can make the decision about whether they put it there or not. Sometimes, people are not informed, which is not great, but if people are well informed and good decisions are made and good practices are put in place to manage those risks as they go forward, I think we can navigate it quite well. I'm not afraid of it. I think it's about understanding, taking the required time, and doing the right consultation to get it right.

AA: Why do you think there's so much focus on it at the moment?

MM: Because things have gone wrong. It's the same with cybersecurity. Things have gone wrong because organizations may have documented a risk but then not mitigated or managed that risk, or haven't rehearsed how they're going to respond if that risk is realized. We've seen risks being realized, and then we've seen organizations behaving in a less-than-optimal way. That's why there's a focus on it.

AA: How important do you think a formal data strategy is for organizations broadly, especially in the APS? I've had some people say to me, *"We've been told to produce one and we would like your help, but we don't think we really need one."* **What are your thoughts on that? How important do you think an actual strategy is as a document for senior executives and a whole organization to be able to refer to, and how do you actually develop a sensible one that means something?**

MM: There are a couple of questions in there. First, how important is it? I think it depends. We could go back to what I said before about why you are investing in this capability in the first place. If that's actually already very clear, then maybe you don't need a strategy. If there's not a whole-of-organization uplift that you're trying to achieve, then perhaps you don't, but you do need to ask yourself the question about whether you should be going for a whole-of-organization uplift.

Today, organizations rely more and more on data, and we want to be far more evidence-based in the way that we make decisions, drive things forward, and prioritize. If an organization does not want to do that, it needs to ask itself why.

> There are many things around whole-of-organization data literacy that should be a focus right now.

I would hope that in 20 years' time, everyone's so data literate that we don't need to worry about that anymore, just like we don't teach people how to email anymore. That was a thing at one point in time. Maybe at some point in time, the term "data literacy" will cease to exist because everyone's so data literate already. If you're not focused on that, then you have to ask yourself why you're not; maybe there is a reasonable reason why. If you don't have whole-of-organization objectives, then you may just have a plan within an area where you're trying to do something.

An example would be when I joined the government 12 years ago. When I joined the then Crime Commission, there wasn't a data strategy for the entire organization, but there was a funding proposal that had been approved for creating something called the National Criminal Intelligence Fusion Capability. There was a plan for that, and I was responsible for fleshing it out and delivering it. That was something, but it was more contained. It just depends on what you're trying to achieve.

As for the second question, about how you do it successfully, you've got to have buy-in – and not just buy-in, but genuine leadership from the top. If the head of an organization doesn't see the point, then don't bother, because you're never going to get the resourcing you need. You'll just have a piece of paper with no ability to execute it, which actually will cause more trouble than it's worth. If senior leaders didn't want to do something, the question I would have is, "*Why?*" Start with "*Why?*" Focus on the vision of the organization. Consider its purpose, and work out what's blocking it from achieving that – does that purpose need better data or better analytics to achieve it?

If you genuinely want to take a whole-of-organization approach, another thing would be actually looking at everybody's role in the organization and making sure that everyone can see themselves in the strategy. What are the success factors? What does a good outcome look like? When we get "there" (wherever "there" is), what will the organization look like? What will be different? What will be better? How will it be better? And as I said before, we need to continually measure ourselves as we get "there."

I was asked at an audit committee what I thought a success factor might be for a data strategy, and one of the things that I said was, if they, as the audit committee, as well as other chairs of every other committee, started saying to people coming to the committee, "*Thank you for your paper. Thank you for your recommendations. Where's the evidence base? Give me the evidence behind this conclusion.*" That's one of the success factors. If you drive down from the top and also have a ground swell from the bottom up, you can meet in the middle and achieve quite a lot, but ideally, you will have a strategy that brings all that together.

Establishing a good data culture

AA: Would you agree that in addition to having a strong strategy that you deliver on, having a strong data culture is imperative within an organization? How do you establish a good culture? I assume you've been in situations where you've had to face one that wasn't so positive, and you've had to try and change it. How do you deal with those cases?

MM: There are three pieces of advice I would give to a data leader who's having trouble gaining traction. These are critical to get right.

The first one is building curiosity. Every day as leaders, we make decisions based on information that is presented to us.

> Our decisions will only be as good as the information that we have access to.

If the decisions don't improve or evolve, if the environment we operate in doesn't improve, and if we don't explore things and the art of the possible, what can we do? If we don't enquire about the value of our data – the quality, the security, and how we're managing it – and don't explore the art of the possible with our data, we're never going to get the most out of it. Curiosity's the first thing.

The second thing is management. You need to make sure that your data is accurate, timely, and managed well; otherwise, your decisions will be bad – garbage in, garbage out. That can include some other topics, such as automation, integration, robotics, process improvement, and machine learning. There are a whole lot of things that we can do to get the most value out of our data.

The third thing – as you've just mentioned – is culture. If people are curious, they have a data-literate mindset, they're supported by excellent capabilities, and the governance is good, that's fantastic, but without culture, nothing will stick and nothing will evolve. We've all heard the saying, "*Culture eats strategy for breakfast.*" It absolutely does, so making sure that things stick is important. One thing that you can do, of course, is put in place a good policy, but a policy's only as good as its implementation.

> Culture is not just about a piece of paper, strategy, or policy; it's actually about implementation. It's about building data culture into the DNA of an organization, making sure that people can see its value.

It's very much related to the art of the possible. You can't have good culture without good curiosity. The two go hand in hand. When you demonstrate the art of the possible, the lights go on and people say, "*Oh, my gosh. Wow! This is so much better. Now, I can do these things.*" You make them genuinely passionate about what you're doing.

As I said before, the central function in the hub-and-spoke model needs to be an enabler for everybody else, because one of the ways that I've also seen a culture fail is where there's an idea of competition – where areas across the organization think that the central function is going to take over or compete with them. That's not what it's about. In a large organization, such as Defense, there are hubs and spokes. I'll say to some of the hubs, "*Just keep doing what you're doing. Tell me what you need when you need it. I've got a whole lot of other things I can focus on right now. I'm putting the flags up. Swim between the flags – in terms of governance – but you're a strong swimmer already. I don't need to teach you how to swim. Go for it. We'll keep the sharks away and rescue you if you need it, but just keep swimming.*" To get all of this working, culture and curiosity have got to be part of it.

Why do data science projects fail?

AA: We all too often hear of data science/AI projects having high failure rates. Do you think this applies to the APS, and if so, why are the failure rates high?

MM: Do we see a high failure rate? Going back to what I said before about experimentation in data science, start with "science." The word "science" is in there for a reason. We have to be able to experiment. I guess it depends on why they're failing. Maybe something failed because the problem wasn't well designed; it may have been really important but just not solvable right now, and that does not necessarily mean that people are failures. It means you have to say, "*OK. Put it on the shelf for a bit, let's see what else we can do in the meantime, and we'll come back to it.*"

But if something failed because it wasn't the right problem, then there's a prioritization issue. I will not proceed with a problem if I don't have an owner of that problem who is passionate about it. I will not do it; it is a waste of time, particularly for the APS. In the private sector, it'll be a problem such as, *"What's the likelihood of making a profit?"* If I don't have someone who deeply cares about it, it's just not worth the effort. Failure is something that is normal in science and experimentation, and it's not necessarily a bad thing. It depends on why it failed.

Discussing data leadership

AA: What makes a good data leader, and how do you gain traction and buy-in from fellow executives?

MM: There are three things that I've learned are critical to get right:

- **Curiosity**: Every day, especially as leaders, we make decisions based on the information that is presented to us. And the decisions we make are only as good as the information we have at our fingertips. Our decisions won't ever improve or evolve with the environment we operate in if don't explore what data we have and what insights or possibilities we can get from the data we have or do not have. We need to enquire about the value, security, and quality of the data we do have! Exploring the art of the possible with data is critical to ensure that decision-making is optimized and an organization evolves with the changing environment around it.

- **Management**: Accurate, timely, and well-managed data improves decision-making and business operations and opens up so many opportunities that poorly managed data cannot provide – opportunities such as automation, integration, robotics processing, and machine learning. The list goes on. Good data management also means ensuring appropriate and commensurate governance, security, and oversight of your valuable data, ensuring it doesn't get into the wrong hands, either intentionally or unintentionally.

- **Culture**: Leaders with a curious and data-literate mindset, supported by excellent data capabilities and governance, are good, but you need to ensure that these things stick and evolve with your environment. An organization as a whole needs to champion good data management and exploration, not just data champions such as myself. Therefore, it's important to build that data culture into the DNA of an organization to ensure that it can harness the value and opportunities of well-managed data and ensure that the organization looks after it as a critical asset and capability.

Looking to the future

AA: What excites you the most about the future of AI in the APS?

MM: Like I said at the start, the ability to make a genuine difference in our nation is phenomenal, and we are just at the start of seeing the potential for it. I've been in government now for 12-and-a-half years, and in that time we've gone from small, bespoke projects (at a few different organizations in narrow areas) to everybody understanding the value of data and AI and trying to do it. I'm so excited about the potential, but I'm also nervous about people charging in without actually understanding how to do it well and without understanding what might be solvable. I have definitely seen AI done badly. I've seen AI being used for what I would consider to just be a plain math problem. 2 + 2 is always 4. It's not somewhere between 3.8 and 4.2. It is always 4. If it's math, please use math. Just use a normal algorithm.

I've also seen issues where someone says, "*Can we apply AI? Let's have a look and see. Oh, let's apply it to that problem over there.*" But hang on! Wait a minute! What are you trying to achieve? You need to make sure it is actually achievable and sustainable. Take that whole relationship with the CIO and the process of making sure that something is actually an enduring solution. People think, "*Oh, it's just a quick experiment and you're there.*" Well, no. There's so much more that goes into succeeding.

I think the future for the APS is going to be about making an incredible impact on our nation. I love the engagement-with-risk part of my role. I love thinking, "*It is so important that we solve this thing, but my gosh, it's hard to do it, do it well, and avoid the risks that could undermine our efforts.*" I love that part of it because it's challenging but so rewarding, all in one.

Summary

It was truly an honor to speak with Maria about so many aspects that are integral to enabling success in data and analytics.

Her five hats of a CDO offer a clear description of the ingredients for success in one of the most important, challenging, and (relatively speaking) newest roles in the field, including highlighting the importance of having strong data literacy skills.

Maria prefers to talk about trust, which she sees as an umbrella, under which there are a number of pillars, including ethics. She points out that there can be just as much trust lost in not solving problems, as in solving them badly.

It was also enlightening to hear Maria's views on key differences between the CDO and CIO roles, and how to ensure that both roles can support one another.

Maria also provided some salient advice on how to establish the right culture to support data-driven decision-making while navigating challenges that are often faced, such as self-interest versus organizational interest.

Her insights on the difference in incentives between the public and private sectors also helped clarify some of the key differences between the two. Her excitement about the potential to make a positive impact on the nation in the APS is infectious!

17

Stephane Doyen Follows the Science

Stephane Doyen is an innovator and entrepreneur. He is the co-founder and Chief Data Scientist of Omniscient Neurotechnology, a successful medical start-up. Stephane has used AI to create groundbreaking brain-mapping algorithms. Previously, he led Oliver Wyman's machine learning and advanced analytics division in the Asia-Pacific region.

I was particularly interested in hearing his thoughts on what makes a successful start-up founder, and also about some of the unique challenges of applying AI to the medical field.

Given his experience across consulting, academia, and start-ups, I was keen to hear Stephane's advice for establishing the right data culture, and how to avoid common pitfalls and challenges when implementing and scaling data science.

Getting into data science

Alex Antic: I want to start by getting a good idea of your career trajectory. Were there any pivotal points along your career path? Points where you thought data science and analytics were things you needed to pursue? Were there people that you were inspired by? I would love to hear your story.

Stephane Doyen: Interestingly, I never thought of myself (at least when I left high school) becoming what we call a data scientist. At the time, it wasn't even a term – I think it's something that came about more recently.

As a young student, I was pursuing my own interests, and one of my key interests back then was trying to understand how the human brain works, how thoughts come to be, what processing is involved, and so on. So, I started a psychological science degree. I became a licensed psychologist and then went into neuropsychology, because I was interested in how neuropsychology views the brain: it's about the mechanics and cognitive information processing systems.

At that point in time, I met one of my mentors, who then became one of my thesis supervisors. His name is Axel Cleermans; he's a professor at the University of Brussels. He wasn't very well known to students. He would spend most of his time in the lab, writing, or talking around the world. He gave one class, which was a very short class called Cognitive Science. It's a bit of an obscure term; you could put anything in there.

I remember very clearly that one day, he came in and taught us about Newell and Simon, who developed some of the earliest AI programs in the 1950s, and there was a famous quote of theirs that was mentioned. As students in 1956, they came back to their peers and said, "*Well, over the Christmas break, we invented a thinking machine.*" This was in the 1950s, so there were IBM mainframes everywhere. The thing that really struck them back then was that essentially, they found how a numerical program was able to think non-numerically: that's what they had with their knowledge structure system. Their computer program, Logic Theorist, had the ability to model the way humans proved mathematical theorems.

So, they found a way to use a non-biological organ to generate thoughts that are usually attributed to biological organs, and that resonated so well with me. It speaks to who you are, and it spoke to my quest about understanding how the human mind works.

After that, I engaged in my Ph.D. A lot of it was about trying to understand how we can simulate processes, what's going on in the brain, and how we can take all of that apart. I found that a lot of it is actually a data problem. If structured properly, a lot can be unpacked using experiments, analysis, coding, and the extraction of measurements and the organization of information into knowledge structures.

So, I think if I have to put my finger on one point, that's truly it: the time when I found out that what we started looking at in the 1950s, we're still working on today. The Logic Theorist was the start of general AI as we know it today.

Becoming a leader

AA: I can see some connections here, but I'm still going to ask anyway: how do you feel your background in academia, and then later in the consulting space, helped you prepare for various leadership roles, such as now being a chief data scientist and one of the leaders and co-founders of Omniscient? Specifically, how do you think that's helped you prepare to launch a successful start-up?

SD: Academia and consultancy are very complementary.

> Academia gives you technical depth, and it tells you how to approach a problem. Consultancy gives you breadth, and it tells you how to craft an answer to a problem.

I think there was a quote from Neil deGrasse Tyson, who said, "*It's not enough to be right.*" You have to be able to convince the person on the other side that you're right, and only then can you get progress. I think the latter part might just be my interpretation of his quote there, but the point is that with consultancy, you learn how to convey what you've gained from your previous analysis.

The two operate in very different paradigms. In academia, if you map out a complete project to explore a problem, then divide the project into 80/20 slices, most of the impact is delivered by the time 80 percent of the project is completed. That last 20 percent is often exponentially more costly. In academia, you have the funds and the time to actually explore that remaining 20 percent. That's something that I've found immensely useful because you understand what is going on down the rabbit hole.

Now, with consultancy, it's the opposite: it's about how you can deliver as much impact as possible with the limited resources and time you have. That means that you learn how to summarize, take valuable shortcuts, and organize your work a lot better. When you combine the two, you can handle the two ways of working, so that's very useful.

I would also say that the added advantage of consultancy as opposed to academia is that you get to apply your technical depth across many industries. That's related to the breadth point I was talking about earlier.

> Your technical skills become a toolbox that you can carry from one set of problems to another, which makes you appreciate subject-matter expertise and how you need people from the field, and you can augment their skills with your toolbox.

It's quite useful as well, if you want to shop for a new tool, to realize what works, what doesn't, what works only in certain industries, and how you can apply certain patterns from one industry to another.

AA: I really like the depth-versus-breadth approach. I also like how consulting forces you to learn how to communicate technical concepts to a broader audience, including executives who just don't have the data literacy or time to understand technical concepts in depth.

In your opinion, what do you think makes a great data leader? In particular, what are some of the challenges you've faced in your journey from technical expert to leader at Oliver Wyman and now Omniscient?

SD: I would say that a data leader should, using technical means, extract information from data that you wouldn't see otherwise. There are a couple of traits I can think of.

Everybody's got their own organizational context. Technical means today are not the same as they were years ago, but one timeless theme is having a vision beyond the hype.

I'll give you an example. There is a great site called Kaggle (it was bought by Google a while ago). Kaggle was the Mecca of ML practitioners. You got a ludicrous prize if you were able to solve a problem and become a Grandmaster. It was six figures sometimes, plus a job in one of the tech companies of your choice.

At the time, there was an ML algorithm that came out. It was called XGBoost, and we all know that algorithm; it steamrolled everything. People back then joked, "*Data science is about cleaning data and then applying a bunch of algorithms to see which one wins, but XGBoost always wins in the end.*" That was the hype.

The algorithm is irrelevant; what is more relevant is actually what it is used for. XGBoost was just one tool in the toolbox, and what matters is the appropriate use of your toolbox for the appropriate problem – whether it's XGBoost or the scikit-learn toolkit or you're using R is not important. It's about seeing beyond the hype.

ChatGPT-4 was recently released. The hype is definitely there, and the hype cycle is an inherent part of ML developments in general. What you need to realize is that after summer comes winter, which is when you see limitations. Then, you overcome them and summer comes again. So, seeing beyond that summer to understand what the winter will be will help you to live better through the winter because you're prepared in your practice and your own business.

Another challenge is selling your ideas, and that requires simplifying them and starting from another perspective. That simplification is instrumental to getting other leaders' buy-in – who cares about your AUC being a few decimal increments higher when what you're thinking about is tens of thousands of customers and the costs related to that? You have to translate the technical details into business impact: the potential impact on cost or customer behavior.

Sometimes, you'll have people in your team with such a wide array of skill sets; if they don't understand your vision and why they should be doing things, you're probably going to lose them, and that's a missed opportunity as a leader.

I'm very hands-on. Leaders work in different ways. You can be a thought leader and a public speaker, which I am sometimes, but that's more about getting tech to the real world, which you do a lot in a start-up. Most of my time's actually spent alongside the team coding.

> **As a leader in this space, you have to understand your technical landscape. You have to understand how you should organize your teams and the skills they have. You need to know what type of process you should use with your team, depending on the stage of maturity the business is in and the type of question you're trying to solve.**

Some require a lot of rigidity about production and boundary conditions, whereas others are very much freeform and in early-stage ideation.

> **Understanding how to work with other people is such a key component of what makes a great leader, at least in the space where I'm currently leading.**

AA: How important do you think that is: data literacy among senior data leaders across all industries? Would you agree that it's something they need to commit to if they don't come from a technical background?

SD: I've thought about data literacy, and I think before that I would put another type of literacy, which is technological literacy. I think that's a precursor to data literacy, because nowadays, I don't think we can dissociate the data we're getting from the systems that are generating them.

There was a time when data was less ubiquitous – when it came from credible sources, official sources, or a handful of experiments – but now, data is everywhere. Your robot vacuum cleaner at home generates data, and that data ends up somewhere. It gets analyzed and it becomes useful. I think tech literacy is such an important component, living in a technological world as we do today, as a precursor to data literacy.

From there, if you want to get deeper into your data literacy or develop more understanding of your tech landscape, coding is actually pretty good. I'd recommend it to anyone. At least try it once or twice, because it tells you how the world is structured from a tech standpoint, and it gives you the tools to analyze your problems with data. But more than this, it also helps you to understand what data sources you should actually trust.

AA: Also, it forces you to think logically in a step-by-step way to organize your thoughts.

What advice would you give to a data scientist who's looking to progress their career to a leadership or management level? What should they be doing to prepare themselves for leadership roles? How important is mentorship, for instance?

SD: When data scientists ask me, *"What are the skills I should have for the next step in my career?"* I tend to dwell on three key skills.

Skill number one is understanding how to put together a plan to solve a business problem. Say you're faced with a database that shows traffic between servers, and you need to understand how you can weed out some of that traffic to streamline your load on the server and decrease costs. That's an interesting challenge, and so you have to structure your problem in a certain way so that you can apply a set of steps to get to your answer. That's something you learn by doing the job. In becoming a leader, you then understand what it takes to do that job and solve that problem, which will be incredibly helpful when you need to put plans together.

The second skill is about selling your ideas. It is not only about coming up with a plan to solve something – it's about getting it resourced as much as you think is necessary: money, people, or whatever is needed. If you can't sell your wonderful plan, it's never going to happen. That's a skill you should develop. Work on how to communicate very complex ideas in very few sentences, how to speak to certain audiences, and so on. One way of developing this is just by doing it often. Just talk about your work. Be proud of it.

The third skill I tend to recommend developing is essentially to foster collaboration and collaborate with others on your ideas. How do you steer a ship? You know how to work in the engine room. You can tell the captain where you want to take the ship. You've got to have all your teammates working together. That's probably one of the toughest skills to develop. It takes a lot of trial and error to learn how to foster collaboration, fail in certain areas, learn from it, and then do it again.

If you have these three skills, which obviously are underpinned by strong technical ability, from a solution delivery standpoint, you're bulletproof.

Becoming data-driven

AA: You mentioned earlier that a data leader should extract information from data that wouldn't otherwise be seen. When it comes to organizations becoming data-driven, what do you think some of the common challenges and pitfalls they face are?

SD: It's quite interesting, actually. It speaks to the history of the role of the consultant.

There was a time when Google was not there and the internet was not there, and if you wanted to get access to information, you had to go to a library, or you had to find the experts, and that's what consultants were for. They were individuals that went from one company to another, with all due care and ethics, sharing knowledge across the retail sector or banking sector, and so on.

Then came the information revolution, and that role became absolutely redundant because organizations could just access the information from the internet. The task of the consultant then shifted from, "*Tell us what we should know,*" to, "*Tell us how we should read what we're seeing. How do you interpret this?*" You have to elevate yourself a level above, and so that changed the role from being about declarative memory to being about the actual semantic joining of dots and extracting information. The phrase "**Actionable insights**" is something you often hear in the role of consultant. You're basically wired to gather all the knowledge you can and get information out of it.

As a consultant, it is a really interesting position because you come in as an external agent. You have some prior knowledge. You bring specific toolsets. Say you're pretty good at statistics or coding – that's your skill set. You have some prior knowledge of having worked in various sectors. Maybe you worked in the aerospace industry across a few organizations, but you're coming fresh to a new organization. You've got to have advice that is useful and that's based not only on the data that's available internally but also on data you've gathered internally *and* externally, and that you connect to your external point of view and therefore bring potentially useful insights to the client.

So, that's one way I think about the role of the consultant in the data-driven world. It's more about joining the dots and filtering the music from the noise, rather than taking a database and counting things as fast as possible.

AA: While you were a consultant, did you see any particular issues that were systemic across many organizations when they tried to invest in data science and analytics but they struggled? Would they often overcommit with technology, say, or did they struggle to get the right people? Did they have a misalignment between the problem and the solution?

SD: It depends on what client you're working for.

> I would say the largest problem I found was organizations suffering from the curse of their own data.

Around 15 or 20 years ago, they started gathering data in larger amounts, and then they found that this data now sits in several different systems. There have been changes to the team, strategy, regulatory landscape, and so on. The data is scattered everywhere and they can't reconcile it.

Also, they will have heard about the AI revolution and thought, "*Oh, well, we've got this amazing war chest with all of our data in there. There surely are a few good million dollars in there.*" But then, the processes by which they can actually unlock that value fall flat.

These are problems I faced several years ago, and I would assume it's still the case in some organizations, although the cloud revolution perhaps has made things different. You might find that you've got a much easier structure to deal with beyond the data lake and warehouse, where you can start putting everything together in a structured manner and a system where you can run serverless experiments.

But often, the curse of data was a reoccurring theme. I'm sure the latest AI revolution will excite some business leaders and make them think about how they cannot be late to the game, because it is truly a game-changer to be able to go through some machine's unstructured information and extract meaningful value. It's happening at such a pace that as a leader, if you haven't got there, you're probably going to be left behind.

That said, I think we shouldn't discount the actual value of lower-tech analytics. Often, there's an emphasis on AI in the form of general AI, so suddenly you find all these businesses investing in chatbots. That was a good few years ago; it was the thing. There was a lot of value to leverage from that data with lower-tech analysis. Can you predict when a customer is going to churn? That's a pretty common theme when you're working with various businesses. So, getting to that point using data is absolutely lower tech than ChatGPT-4, but it has a massive impact on your value if you can save your customer or extend their lifetime value.

So, another thing to consider is the role that lower tech plays in that space.

AA: I completely agree. I think it can be driven by hype: looking at complex solutions when simpler ones are much better suited. Most organizations won't find the greatest business value in cutting-edge AI solutions, such as ChatGPT – there will be less technically advanced "low-hanging fruit."

Developing AI solutions for the medical field

In your opinion, why hasn't AI achieved its promise in medicine and the broader medical field?

SD: I can speak specifically about the medical field. I've been very active in this field for the past 3 years.

> I've found that the medical field is one of the toughest places to apply analytics, ML, and AI.

Why? Well, there are multiple reasons.

The first is that in most cases, someone's life is on the other side of your code. You have someone's life in your hands, which makes it very tough.

An additional element is that it's extremely specialized. If you want to be proficient in the medical field, it often takes a medical degree, which is hard to combine with a statistical or ML type of degree. So, you find that you have to heavily rely on domain experts.

As a result of the first component I mentioned – someone's life is in your hands – there are a lot of regulations in place, and so that means how you can actually do AI in the field is bound by all sorts of guardrails. That is the right approach, I believe, but those guardrails also have to allow innovation.

Since you rely on domain experts, it's often complex for non-medical experts, such as statistical and ML experts, to come in and propose a change because a change that you might perceive as positive might have a negative impact somewhere, and so you don't have the whole picture.

That's an outline of why it is hard to see anything new in the medical field, but I can see we're experiencing a positive trend. In the next couple of years, we're going to see even more of these start-ups really delivering game-changing therapies to patients.

For instance, there's a series of challenges that relate specifically to how ML specialists approach the medical field that are actually very solvable. One of the first issues is really about solving the right problem.

ML and AI products can be found all over the internet. You might find that somebody very recently posted an absolutely wonderful application on Reddit that uses a phone camera to look at brain scans and tell the location of a tumor from a brain scan. That's a good piece of tech; it's impressive and works off your phone. If you're not a doctor, you will think that's pretty cool.

But if you think about it from a neurosurgeon's perspective, finding where the tumor is is not the hard problem. In most cases, it's the easiest one. If you look at a brain scan, you see a mass, and that mass is a different color from the rest of the brain, and therefore you can safely say, "Well, that's likely to be it." There are other things that are far more difficult.

One of them is identifying what type of tumor it is, which often requires a biopsy. There are some ML projects coming out now that can help do that. Then, there's thinking about how you're going to approach that tumor generally. What else should you consider when looking at a scan and thinking about treating a patient?

That is about solving the right problem. You have to target your skill set and your craft at what experts need it for: so, not identifying there is a tumor, but telling them how they should potentially think about it or approach it.

Another reoccurring problem in the field is about bringing practical solutions. You find wonderful projects online that use dockerized images, which you can throw a scan at and it comes out with all sorts of metrics that are useful for a neuroradiologist. But that dockerized image requires a GPU to run and all sorts of extra-beefy RAM, and then you get all those useful metrics. Well, if you go to your doctor's office and see the type of computer they're working on, they've got laptops from 5 years ago that are connected to servers even older than that, and they're passing laptops between whoever is in the room at the time. So, there's no way your solution is going to run on their system. You've got to think about how practical your solution is. Often, some of the greatest things are just not quite practical, and they don't make it.

The next problem is actually very well documented, which is the lab versus field problem. In the past 2 or 3 years, I've identified some very promising projects that burst into flames as soon as they left the lab. One of these was a project where you could use your phone to take a photo of your skin and it would tell you whether a mole on your skin was cause to go and see a doctor or whether you had nothing to worry about.

That app was actually pulled back just a couple of weeks after release because they'd not accounted for the fact that not only do people not always take a perfect photo but often, they also have very different skin tones from the ones they had tested on in their lab. That particular system did not work properly on darker skin tones, and it was leading to false detection and other issues.

So, that's another major issue when you're working in your ivory tower as a medical AI practitioner: you often tend to work with the data you've got at hand, which should be data from the field, but most often you find yourself with data from the lab, which is the perfect brain of a 20-year-old medical student or neuroscience student, or the perfect picture of the perfect skin of younger folks.

Another problem is **quality assurance (QA)**. We published a paper about this very recently, which was quite a finding for us. We used a certain Docker image to compute streamlines in the brain (streamlines are pathways in white matter within the brain). Docker essentially simulates shipping your hardware and software together so that you can run your code in the exact same manner, whatever system you're using. It turns out that our developers used one system and the cloud used another system, but they thought they were safe using Docker. As a result, when they processed images on the server, the results were completely twisted in a very problematic manner: one that would have seriously impacted the output being delivered to a doctor.

The challenge here was that the underlying math at the CPU level was slightly different when using the developer's machine versus the cloud machine, and therefore there was a rounding error. That rounding error was scaled however many times for however many streamlines, and that ultimately led to this situation.

Now, I'm using this example to illustrate the importance of QA. Especially in the medical field, you have to manually check the results, and you have to manually check them with the eyes of an expert, which are not available to everyone. Indeed, if you think about a developer in ML and AI, you might picture a geek in their bedroom or the new hot start-up on the block. Without the right people in the business with the domain expertise and the time to look at these things, there can be a lot of issues.

That's the importance of knowing you have someone's life on the other side of your code. Sometimes ML models behave strangely, and sometimes they make mistakes in a very weird fashion. The language models that have emerged in the past few days or months are actually very good illustrations of that because their mistakes are very obvious. They say something that is just plain wrong. Numbers can be expressed in unknown units and unknown ranges; checking them without expertise is actually very complex. That reduces the extent to which you can apply AI to medicine. If there were going to be such an error, you'd need to make it obvious and catch it, or else your AI would fail.

Lastly, on the medical side (but this actually speaks to a lot of AI solutions across the board), implementation is key. If you build a system with the value proposition that you're going to replace your user, it's very unlikely that your user will embrace your system. However, if you release a system with the value proposition that you're going to augment your user and then you actually do, it is quite likely that your user will endorse your solution and rehire you for the next one.

That was a series of elements that you find specifically in the medical field but they do, however, apply to other fields in some instances and represent big breaks for the progress of AI in patient care.

AA: I think your insights will be of great interest to our readers in learning about some of the key challenges in the field of medicine (and AI more broadly).

There's the assumption that there's a lot of data out there and it shouldn't be that complicated. But you've articulated a lot of those challenges extremely well, so thank you very much for that.

The next question touches on an issue you've already highlighted. When developing AI solutions in general, I've noticed that many organizations can struggle with moving from R&D to developing a solution and productionizing it. What do you think some of the key steps for ensuring that this process works in a repeatable fashion are?

What often happens is that things get passed from a data science team to an engineering team. I've noticed issues can occur there. There's a lack of communication and different skill sets, and things just fall over.

SD: I'm going to share some perspectives here, and I appreciate that my perspectives are not everyone's. Depending on the problem, you might take different approaches. But my approach to these things is really about iteration.

One common issue I've seen in organizations regarding churning something into production is that you create a big milestone, and you call it productionized assets or a system in production. You tie everything to that ultimate deadline, which you should. But by doing that, you're creating a little bit of a challenge along the way for yourself, and I think the "along the way" is what matters most.

The first principle still applies, which is to iterate and work piecemeal so that all of those steps actually get you on the journey to where you want to be. As a leader, keep that destination in sight, but get your teams to walk along that line without being frozen by the end goal. Obviously, depending on where you sit in an organization, development and production can be different. For data scientists, when building concepts, they tend to scale well and use the right data upstream, whereas engineering is completely different. There, scaling from your laptop to a cluster will not only scale the problems you have but also create a new realm of problems that you have to account for. Suddenly, you've got a distributed system because you've got too many inbound connections to your model, and so on and so forth. There is just the same model running in different places.

One corollary of this, for instance, is that the specialist skill sets required from dev to prod create silos in the process, and sometimes you're way down the path toward production, and some of the choices that were made by specialists upstream are causing you issues, which will then see you go all the way back to the start. That reinforces my point about a step-by-step journey, because think about being at the bottom of a long waterfall and you've got to go all the way back, as opposed to just needing to step back one step. You're in an environment that's a lot easier to manage.

So, I would say use your experts wisely to make sure that they make the contributions that are necessary for them to make but do not create silos and black boxes. Do not create schisms in the way they work. There should be appropriate interfaces between the team members, and that requires knowledge to be shared across the board.

For instance, if you think, *"Well, my engineers only engineer my models so that they scale, and my data scientists are building models with no consideration for engineering,"* you can see immediately how that is going to create an additional cost between the teams. You'd have what we could call a translation cost from one to another. But if you allow them to work together from day one, then you remove that cost.

Lastly (and this is more of a challenge in large organizations, but some of the bigger tech companies have found ways around this), you need to consider ownership. Often, you find that things are better done with smaller teams, who are accountable and own a solution from end to end, than when you have a chain of specialists who only do their small piece with no full ownership.

I think these are a couple of guiding principles to avoid the freeze of going to production or crashing along the way and set up your teams to get there successfully and navigate that treacherous water.

AA: I was actually having a discussion earlier with a colleague about what you call the translation cost – I think that is a great way to summarize it – between the data science and engineering teams. We were discussing the need to have some data science understanding and capability on the engineering side, and vice versa. You can't consider them as two distinct and disparate steps in the development of a solution, resulting in a seamless path to production.

Putting the "science" in "data science"

Let's discuss the science component in data science – the process of iteratively developing and testing hypotheses. This is something you're well placed to discuss, given that you're a scientist by training. The word "science" is an operative word in data science. It's there for a reason.

How do you ensure that you keep the science element active and front of mind in your team and staff?

SD: That's a complex question. There's no perfect answer, but nevertheless, here's mine.

The term "data scientist" is often misused or misunderstood. There are two parts to it. If you take away all the engineering components, then one part is about counting things and the other is about running experiments. Counting things might appear to be the role of the data scientist, which it should be.

> Really, the true value of the work is in uncovering answers you didn't have before to questions that are useful for the business. If you do that with a problem that is oriented toward value creation for your business and you allow the right process to happen, you will unlock a step change.

The process may vary depending on where you are in the life cycle of your solution. Early work will require research, whiteboards, and a team that is hyperfocused on trying to find something new, whereas optimizing a prototype will probably require completely different approaches. So, that's one way of handling it.

The other thing is that "science" is not a vague word. It is a word that has been very well described by many scientists, philosophers, and others, and it is a set of established ways of working by which you make your idea obvious to your peers so that they can comment on, criticize, and test your idea with various data and other experiments so that it can be proven or disproven. As a leader, it starts with questioning your own ideas very openly and allowing anyone to criticize them. Should you find that your idea has some issues or won't float further down the line, leave it behind.

Beyond that, put your team in a position of testing: testing ideas and testing hypotheses as quickly as possible so that the cost of an answer becomes a lot cheaper and you no longer rely on trying to be the smartest person in the room, or the most eloquent, or the most convincing. Make sure you can say to everybody, "*Well, that's a great idea. We'll come back in an hour with a better understanding of it by testing it and looking at the data.*" That speed of iteration is one of the single most important success factors for a data science team: how much time it takes for people to get to an answer so they can start building a better picture of the truth at a lower cost and therefore create knowledge through scientific methods in the most effective way.

AA: Do you think that could be part of the reason why we often hear of AI/ML having high failure rates? Do you think that part of the problem is that, too often, data science isn't clearly aligned with an organization's strategic goals? Are there particular views that you have as to why there is a systemic failure to succeed across different industries?

SD: I came across that number, but I think that projects failing is part of the job. If you're a scientist, you probably see as much value – if not more, sometimes – in closing a door than opening one, because at least you know that was not something you should explore; that answer is plain wrong. You do consider that a lot of your projects/experiments will fail.

Now, where things are a little bit more complicated is when you look at production-grade projects. You think you've cleared a lot of the unknowns, and then you bring it to the field and you expect users to interact with it. That's where you don't want failure, because you think that you've answered most of the questions, and therefore there's less uncertainty and you think you should be fine.

The points I made from my observations in the medical field do apply quite a bit to that space: creating a non-product, solving the wrong problem, not thinking about details, not thinking about implementation, just doing data science because it's exciting and cool, and not applying the right quality checks.

Establishing a data culture at an organization

AA: On data culture, what do you think some of the key elements are, and how have you established strong, successful data cultures in your own organizations when you've led this capability?

SD: The first point about data culture is a well-established one. Often, you find an attitude of, "*Let's do Python. We use Python around here.*" Well, that's not a culture. That's just one of the tools you use. Or, "*We're running sprints here.*" Well, again, that's a process. A culture is beyond that. A culture is the rules that define things. They're often underlying and not always spoken, but they define the way people interact in a group.

The approach that has been successful for me in various settings is basically to define spaces and acceptable behaviors within those spaces. I touched upon this a little bit before. Say you have a system that needs to go from early-state ideation all the way to very stringent and hardened production. The stages of delivery require a very different mindset and very different ways of working. That means I can say, "*Well, there's an ideation phase and then there's a proof-of-concept phase, and then there's a proof-of-value phase, and then there's a demonstration-of-value phase, and then there's an actual production phase,*" and so on and so forth.

For every step of the way, the behaviors you have are clearly outlined and tagged against KPIs, and sometimes they're antinomic. So, when you get to the optimization phase, there are no new ideas. You're at a point where every single idea needs to be bulletproof and needs to work in all the known cases and – potentially – the unknown ones. But in the early-ideation phase, you don't bother yourself with implementation details. You just think about whatever you can bring and how it is a steep change from what you had before; implementation is no object.

That brings a culture and a way of interaction that allows for harmonious functioning between team members by defining the behaviors that are necessary to support what you need for your business.

I've outlined a pretty linear process for my business, but other businesses may require different processes. But the point of the matter remains:

> Describe what you expect people to do, measure their success against those expectations, and add value when they display that behavior.

There's a saying in aviation, which is one of my hobbies: "*Trust but verify*." It doesn't matter how many training hours you've had in a plane. We generally trust each other to be competent at the task at hand, should you be licensed for it. However, if necessary, you can verify, and there's nothing wrong with that. Given the dangers of flying aircraft, if somebody more senior than you has forgotten something critical, having somebody more junior than them pointing it out is actually doing a favor for everybody. It is likely to get you all safely to the ground.

In my business, we think the same way: don't let others get to a wrong answer because they're more senior, and be grateful if somebody is able to show you something you haven't thought of.

AA: That reminds me of the use of checklists in surgery for the same reason: trust but verify the details, given the risk involved.

SD: There are a lot of parallels between medicine and aviation – maybe it's no accident that I find myself in both!

Building the right team

AA: When you're hiring and building a team of data scientists and engineers, what do you normally look for in terms of their technical skills and their attitudes? What is it when you're interviewing someone that makes you think, "*Yes, I want them on my team?*"

SD: There's no perfect recipe. Everyone's different.

> One of the things I look for is intelligent people, and intelligence is about joining the dots.

You might be faced with somebody who is hypertechnical, which is useful, but then if that hypertechnicality is not linked to appropriate solutions or something that is practicable or will work in a production system, all of that potential is not going to flourish when put in a team.

Part of it is about how rich your life is in general. I find that obviously, businesses take up a lot of time and involve working long hours, but that doesn't shouldn't mean that you don't have side projects or interests, or you don't maintain your own GitHub or contribute to various other projects. That for me is a sign of not only intelligence but also resourcefulness.

Think about the process I was describing earlier of how you bring a very complex idea from a very technical and challenging landscape into an actual application that solves a problem for the user: you want people with knowledge across the board. You want people with progressive gearing in that process. Intelligent people will be the best at that because they won't say, "*Well, I don't want to know anything about DevOps or SecOps or MLOps. I'm only doing models here.*" They will actually be interested in finding out how to parallelize things, how to throttle things, and how you process batches and streams, and that will deliver far more value. The other thing I look for is how people answer questions. They might be boring – coming from the same place all the time. They might say, "*As an X, I would do this,*" and *X* is the only position they're camping on. That's when you realize that the required flexibility and richness of intellectual landscape aren't quite there, and often, it doesn't give a very good result when put in a team. But when it's there, it's extremely effective.

AA: How much weight do you place on formal university studies – say, postgraduate, master's, or Ph.D. – versus online courses or on-the-job training?

SD: In an ideal world, for me, interviewees and resumes wouldn't show any credentials for schools, names, or countries of origin. Having gone through some of the best universities myself, I find that what matters is when you are on the job. There was a time when people were saying, "*How do you ace a data scientist interview?*" It was all about knowing the differences between lasso and ridge regression and the L2 and L1 regularization parameters, knowing about cross-categorical entropy, and so on. What I found is that a lot of those things don't really matter because they're in code, so you can swap them and they become a commodity.

What matters more often is how people assess the quality of their answers. They say, "*I'm going to establish a strategy. We will require balance: accuracy on the one end but also those edge cases, and also practicality for the user. We'll create a good test environment, and then we can iterate very fast and get to that answer.*" It is not about declarative knowledge and whatever you can find online but how you assess the quality of your answer and make it fit into a wider application that is relevant.

AA: What advice would you give to, say, a practicing data scientist/engineer looking to found a start-up themselves? What makes a good start-up founder? What skills and attributes, such as resilience, do you need?

SD: Well, resilience is certainly one of them. You hit the nail on the head right there.

There are a couple of things I can think of. It depends on what start-up you create and how hard you want to play that game. Maybe you'll find yourself in front of a virgin landscape. Then, instead of following somebody's route to go to where you need to go, you actually have to create your own path. That, for a lot of people, can be quite challenging, because there is right and wrong, but nobody tells you which is which, and only hindsight can tell you that you were right or wrong.

Also, the cost of failure, since you're a start-up, is pretty expensive, as you have very few ways of pivoting back and getting back on your feet, which adds a fair amount of stress. I think it's been shown that start-up founders age quicker than others, and so you've got to balance that as well.

Recently, some very talented minds have decided to come and help us in our business, and it's helped us realize our own limits. You need to know what you're good at and be able to say, "*Let's just focus flexibly on these few things, and I'm going to strategically place some guys that are way better than me at these other things.*" Sometimes, you've got to accept that the folks that you're placing are actually better than you at things that are important to you or that have defined your career previously. That goes back to resilience and flexibility.

Looking to the future of AI

AA: That's fantastic advice. I love it.

What could generative AI, such as ChatGPT and GPT-4, mean for the future of AI? What are your predictions of the future and their broader use?

SD: Large language models aren't new but the recent release of ChatGPT-4 marked a milestone in the history of AI. It signals an age in which AI becomes increasingly cheaper.

> As a keen observer of this technological revolution, I can't help being very excited at the pace at which AI is reshaping our world, opening the doors to a future filled with possibilities. That said, society will need to adapt more quickly and radically than we might expect in order to ensure that these technologies empower mankind rather than impose limitations.

Drawing a parallel to the game-changing impact of the internet, that technology enabled us to share information at an unprecedented speed. The true scope of that revolution was attained when it was interfaced with expert systems unlocking radical human progress.

The same could be said about AI-driven language models – their potential will be unveiled when they are seamlessly integrated with real-life applications and experts. This fusion has the potential to lead to profound changes across various domains, such as healthcare. The prospect of doctors being able to access vast amounts of medical data, enabling them to deliver personalized treatments tailored to each patient's unique needs, is thrilling.

Placing experts and their domains and systems at the forefront can also drive progress in AI. Some estimates suggest that we may deplete all usable data for training large-scale models before the next decade arrives. Obtaining training data from specialized experts to create expert systems will not only be more likely to prevent such a scenario but they will also be highly relevant for solving real-world problems.

Talking about future applications, an area that I am actively exploring is the application of transformer models to decipher sequences of brain activity. This work will allow us to better understand not only how the organ functions but also how it dysfunctions, such as in the case of mental illnesses.

My background in cognitive neuroscience has led me to appreciate the convergence of this field with AI. We are increasingly borrowing language from cognitive neuroscience to describe the architecture and functions of AI models, which hints to me that we are paving the way for an exciting new era of AGI.

As we continue on this path, I eagerly anticipate the advancements we will make in blending AI with various fields of expertise, unlocking a wealth of potential for the betterment of society. The future looks bright, and as a passionate contributor to this ever-evolving landscape, I am very excited to be part of it.

In my dreams, I also envisage a life lived among AI entities living on distributed systems, but will this be a good life?

AA: What excites you the most about AI, in particular in medicine, beyond ChatGPT, as we head toward AGI?

SD: For me, there are two levels. One is the application level, and the other one is more of the future-thought level.

For the application part, if you think about the medical field specifically, we're now past certain teething issues that prevented patients from really getting the care they needed. It's not perfect – there's still a long way to go – but we've cleared some of the basics. Regulators, for instance, are thinking about AI the right way. I encourage you to read some of the direction the FDA has given on this. Their approach is actually very sensible, and it comes down to two principles. One is, *"We're never going to be able to know all of an algorithm, and we're never going to be able to check your code, but we trust that you do it correctly because you're doing it according to an ISO certification. Therefore, we only need to see results – not the black box, just results."* Principle number two is, *"Just keep monitoring, and as soon as you see something wrong, course-correct it immediately."*

> **FDA guidance on the use of ML**
>
> You can read the FDA guidelines mentioned by Stephane here: `https://www.fda.gov/medical-devices/software-medical-device-samd/good-machine-learning-practice-medical-device-development-guiding-principles`.

That gets around all the issues such as *which algorithm was used, have you optimized against this, what was the AUC here, how many folds in your cross-validation* and whatever else. They just say, *"Make sure it works safely and effectively."* Those principles will unlock a lot of start-ups. We're seeing huge benefits already in the triage of patients. Say you've got many patients inbound with CT scans of the lungs and other things. You need to say, *"This one is more of an emergency. This one can wait."* Or say you've got a lot of information you need to put together in one single data point to help you prepare a therapeutic plan for your patients. That's the first generation of problems being solved – that's about knowledge structure.

The next generation is about bringing information that you wouldn't see otherwise. Say you've got a brain scan. In that brain scan, there's a lot of information. With the naked eye and even a fancy calculator, you wouldn't be able to unlock all that information. You have to use ML in large-scale analytics to start to get anything out of that. That capability to help a doctor to see things more clearly and make better decisions is on the way.

So, that's one level that excites me when I look at the future of AI in the medical field: getting past those teething issues and the first-gen problems and moving into these second-gen applications.

When I think about the AI field overall, I think it's quite interesting. I came across something a couple of days ago. The latest ChatGPT was trained and optimized, and part of that optimization actually required a lot of human steps, one of which was reinforcement learning from human feedback. That's something that fascinates me. We are essentially taking the human brain as a black box and saying, *"Well, we're not going to try to do anything with that black box. We're just going to take an input, which is a prompt, and then we're going to generate content that humans find appealing and compelling."* You essentially optimize against human bias, which in a way puts the human at the center of those systems.

I found this enlightening. You're an ML practitioner. You know how AI systems work, and I'm sure you've read about how the brain organizes itself. There's a phenomenon that we believe in neuroscience called **emergence**. Emergence means that when you have systems that are complex enough, properties come out of that system that you haven't hardcoded. The big fear about AGI is that the system will create emerging properties that are suddenly going to take over humanity, or at least compete for the same resources and therefore create conflict. But when you realize that the system has been optimized for humans, it places that context of emergence into a different space.

> There are certainly emergent properties in the whole GPT system, but that bad property is not emerging. It's not suddenly creating something to oppress you – it's actually optimized against that.

This falls into a theme that I find very exciting – the use of AI and ML to understand ourselves better. There's a wonderful paper in the press right now. You can find it on bioRxiv (`https://www.biorxiv.org/content/10.1101/2022.11.18.517004v2`). It shows that people have used stable diffusion to convert the activation in visual cortices into high-resolution pictures. So, you basically take a recording of the brain, use a system, and what comes out on the other side is an image of what you're seeing. That's a pretty impressive use of a human-to-machine interface. It doesn't require any kind of chipset or anything. It just takes your recording. Obviously, it's limited to a task list, but still, what it outlines for me is that there's a future use of AI that is more about understanding ourselves.

I'm sure over time we'll also build protection against the misuse of AI. You can have a system that professors can use to see whether an essay has come from a student or ChatGPT. We're going to see more and more of those protective systems coming through. If you assume that they're proficient and they can stop the misuse of AI, then we're going to get into a space where we'll be able to interface with machines more and more so that we actually understand our own functioning even better. If you understand your own functioning better, then you can manage it in a way that is positive and better for the brain. So, that's the part that excites me most about the future of AI.

Summary

With expertise across the start-up, consulting, and academic sectors, it was enlightening to hear Stephane's advice and insights on several key topics.

One of the insights he shared is something that sounds simple but isn't commonly talked about: that technological literacy is needed as a foundation for data literacy, as data can't be disassociated from the systems that generate and capture data.

Organizations can suffer from the "curse of data," as Stephane put it, where the data is scattered across various systems and can't easily be combined and reconciled. This can make it challenging and inefficient to unlock the value in the data. It's important to be strategic in ensuring that the data is fit for purpose and accessible for analysis, with appropriate governance and technology to support its use.

When scaling solutions from the proof-of-concept stage to production, it's important to be aware of any constraints that may exist in the production environment relative to the development environment. There are also several principles, according to Stephane, that need to be considered:

- Break projects into bite-sized pieces
- Ensure communication and knowledge sharing between teams
- Create small, cross-disciplinary teams who own an end-to-end solution

Stephane also reiterated the common view throughout this book that experiments sometimes fail – it's part of the process. However, there is value in identifying approaches that don't work.

I encourage all readers to take a look at the FDA guiding principles for good ML practices that Stephane referenced. These guidelines contain principles that apply equally well to ML governance within organizations as they do to medical device development.

18

Intelligent Leadership with Meri Rosich

Meri Rosich is an experienced and recognized CDO, with leadership expertise spanning multiple industries and sectors – having led digital transformation journeys for some of the world's leading organizations, including Amex, Samsung, and Visa. She is currently the CDO of Standard Charter Bank.

I was eager to understand her recipe for leading large-scale data initiatives and increasing data and analytics maturity in an organization, and how to become a successful CDO.

I also wanted to learn how she develops and sustains a positive data culture under her leadership.

Becoming a chief data officer

Alex Antic: I'd like to get a sense of how you entered the field. How has your career progressed over time? How did you become one of the most respected senior leaders in our field? I'd love to hear your story.

Meri Rosich: When I reflect on my career journey, I realize that the past 25 years have been a fascinating evolution of data, the internet, and artificial intelligence. During the 90s, AI was still very theoretical and limited to a few people in academic circles, where we would submit thick, complex papers at academic conferences. However, the rise of the internet brought new possibilities for data usage, and that's how I found my way into the field.

My PhD in virtual reality opened up doors to data and coding, and I entered the private corporate world of internet companies, starting in Europe and then in New York. While working with data in New York, I quickly realized that it wasn't enough to merely work with it; you had to create value and execution from it. Working with strategy consultants gave me insight into strategic planning and futuristic views of the industry, inspiring me to understand how to build value-driven use cases.

This realization led me to pursue an MBA at London Business School, and from there, I embarked on my corporate journey in London, Hong Kong, Singapore, and Tokyo, with most of my career in technology organizations, including telecom and payments. Throughout my career, I have been inspired by solving problems that have never been tackled before – big challenges that require big data and uncharted territories.

Overall, my passion for data, innovative thinking, and solving complex problems has helped me become a respected senior leader in the field. It has been a privilege to see how the industry has evolved and to be part of the exciting changes that will shape the future of our field.

Improving diversity and inclusion

AA: I'm interested to hear your thoughts on diversity. There's a known lack of diversity and inclusion in the broader technology sector, not just the data field. What should we be doing to change things? How can we actually make changes in a pragmatic way?

MR: We are having today's conversation right after International Women's Day, so it's a topic that is very fresh in our minds.

Historically, there are a lot of different reasons why women have been less encouraged, less motivated, less attracted, and less incentivized to pursue technical education or technical degrees. It depends on both cultural and life stages. Different countries are at different stages. When you look at Japan, Korea, and India, the challenges are significant because of cultural expectations for gender roles. When you look at countries like China, it's always interesting to see there is a lot of equality, and it's because they come from a very different cultural background when it comes to education and gender roles.

When it comes to technology, all over the world you can see that women have been less inclined or less encouraged to take technical roles. Originally, you could expect that that was because there were certain risks in certain technical fields – whether it was travel or higher physical risk. But today, we know that there are almost more women in every single country graduating from technical universities than men. So, women enter at the same rates but they don't rise at the same rates.

Then, one of the challenges that we see is that even if companies implement hiring processes and rules, most organizations will not start a hiring process unless there's enough diversity in the candidate pool. So, unless you have at least one woman in your shortlist of candidates, you can never end up hiring a woman.

Even if there are now certain expectations of the hiring process, there are still many different challenges, and organizations need to be better at supporting not just women's but everybody's needs so we can have research papers and highly qualified, high-performing teams with women involved.

I'm an advocate for women in tech initiatives, and I believe that we have to go beyond that. We've been advocating for women in tech for the past decade.

> We're ready to move on to not just diversity but also inclusion, which means simply having women is not enough – they need to have a seat at the table.

It doesn't matter how strong your voice is; it doesn't matter how right your point of view is.

> I still find myself in meetings where I share a point of view and it's dismissed by men, who then express exactly the same point of view and suddenly it is accepted.

There's still bias, and there are still a lot of challenges that women face, and we need to tackle them one by one. Some are systemic, and we are all working together against them, from government organizations to non-profits, corporates, and the public sector. I think we all are more aware today, and education is a big part of it, but there are other elements, which are our own biases.

For anyone who has kids and has boys and girls, even if they've grown up with a very equal mindset and try to be equal, it's very hard sometimes to either pick the right toys or the right clothes. There are some elements that are still systemic, which we need to fight one by one.

> It's everybody's job these days to make sure that we have diversity and inclusion for everyone.

Discussing the high failure rates of AI projects

AA: We often hear of high failure rates for AI projects. I'm sure you've heard of these supposed numbers from Gartner – around 85 percent.

Assuming you think that rate is accurate, what are some of the common causes of failure you've seen across different industries? It depends on how we define failure, but what do you think are the common challenges and pitfalls that most people are facing?

MR: When it comes to evaluating the success rates of AI initiatives, we need to think beyond traditional IT frameworks. Therefore, the high rate of "failure" is not necessarily indicative of the ineffectiveness of AI, but rather a natural part of the process of experimentation and learning.

> Measuring success in AI is more akin to a scientific approach, where experimentation and continuous tweaking are necessary until the desired outcome is achieved.

It's important to acknowledge that AI is an experimental area and not all use cases require AI. As with start-ups, not all experimental ideas succeed, and that's okay. We need to factor in the exploration, learning, and time that scientists need to understand the problem deeply before they can effectively apply AI.

> It's important to recognize that not every problem requires AI, and sometimes, a simpler solution, such as a spreadsheet, is more appropriate.

Becoming a data-driven organization

AA: For organizations that are beginning their journey into becoming data-driven, going down the path of building analytics maturity, and using data to truly drive their decision-making process, what do you think are some of the key things they should do?

MR: I mentor many young people entering the data discipline, and I've noticed that the role of CDO is still a relatively new position. Often, individuals in this role come from technical, CIO, or governance backgrounds, particularly in industries such as banking and finance, where data protection, standards, and compliance are required by regulators.

When it comes to strategically valuing data within an organization, it's crucial to understand its role in the business.

> Data is like oxygen, essential for survival but not owned by anyone. It's critical and ubiquitous. The CDO's role is to ensure that the best quality data is accessible to all, much like oxygen for medical devices.

Data is a horizontal area and is used for the most high-value use cases in any organization.

In order to execute the role of the CDO effectively, the most important aspect is people. While companies are legal frameworks for revenue and taxes, the people within them are the most important.

> A company's culture and people should be prioritized, with a focus on having the right skills for the job.

If your organization is migrating to the cloud, for example, it's essential that your team is trained and familiar with the chosen platform.

Otherwise, the skills gap will delay progress by several months. Retention is also key, and companies should avoid neglecting their existing employees while focusing on recruitment and engagement.

> Leaders should have a clear understanding of what data scientists do and embrace neurodiversity to retain their top talent.

The first pillar to achieving success in data is building a culture and team that prioritizes people.

> Data scientists possess superpowers such as making connections in data, understanding complex problems, and having extreme focus. They require empathetic leadership to retain their talent – a quality that's rare in today's business world.

The second pillar is the technology infrastructure. It's crucial to identify your target state and determine what's missing. If your organization already has top-of-the-line cloud-native data products in partnership with research, then everything is in place. If not, you need to map out what's missing and work toward it, investing in the necessary technology to accelerate progress.

The last pillar is the use cases. Organizations need to ensure they have the right value-driven measures in place and that data is fueling the value of the company. This can be challenging, as it requires asking difficult questions about the accounting side of outcomes, but measuring the usage of certain data can enable a data-driven approach to business decisions.

> My recommendation for organizations is to always bring in experts to complement their teams.

There are people who are passionate about value creation within the data science discipline, and their expertise can ensure that everything the team does brings out the value that stakeholders expect.

AA: I see organizations that have some capabilities. They've invested, they've gone down this path, they've onboarded some data scientists and engineers, they have some alignment to strategic goals, they've identified the value element, and they have some of the right infrastructure in place. But what I often see is that they can get hung up on moving products from development to production. Do you have any advice for organizations that may be stuck in that gap and trying to understand how to do it well? Is there any advice that you can give them to develop scalable products?

MR: Scalability can be a challenge due to poor product design.

> Many POCs fail to scale because they were not intended to do so, often due to incomplete or insufficient data.

A small-scale idea may work well in a limited space but not necessarily everywhere else. The key is to start with an end-to-end product vision that considers the problem to be solved, not just the model. This approach avoids starting with data that may be specific to a particular location or region, which limits the scalability potential.

It's essential to identify the problem first and then determine what needs to be solved. For example, if the goal is to address anti-money laundering for 150 countries, certain elements must be in place to achieve scalability. However, focusing on the top three countries may be suitable for some organizations if that's where the primary risks are. To make POCs more scalable, it's crucial to design products that go end to end using the appropriate tools.

Scalability is a critical element of the problem to be solved, and it needs to be designed the right way. It's not enough to have a solution that works on a small scale; it must be scalable to achieve success.

> Starting with the problem, not just the data, is the key to building products that can scale across different regions and use cases.

Establishing an effective data culture

AA: One thing you touched on earlier is the importance of an organization having the right data culture. In your opinion, what does an effective data culture look like, and, more importantly, how do you actually establish one?

MR: Establishing a strong data culture can be a daunting task. However, regulated entities tend to have good data cultures due to accountability and standards they need to meet to comply with regulations. It would be great to see every industry adopt this level of accountability. Some governmental bodies, such as the EU, have already set examples that we can learn from. To achieve this, a regulatory framework needs to be in place to guide organizations on what good data culture looks like.

Leaders play a crucial role in shaping culture by modeling good behavior and setting high standards.

> In fact, culture can be defined as the worst behavior a leader allows in their organization.

This underscores the importance of leaders in creating a strong data culture.

Personally, I have been fortunate to work in organizations that prioritize good management practices, such as American Express, where I started after completing my MBA at London Business School. It is crucial that young people entering the workforce are exposed to high standards in their first job. This will set a foundation for high standards that will be followed throughout their careers. Conversely, if one's first job exposes them to low standards, it becomes easy to accept subpar practices without questioning them.

Therefore, it is vital that we prepare the next generation with a strong foundation in data culture to enable them to thrive in data-driven organizations.

What makes a good data leader?

AA: Speaking of leadership, what do you think makes a good data leader, such as a CDO or equivalent? What do you think are some of the key characteristics, attitudes, and skills that they need?

MR: A good CDO is someone who recognizes the limitations of their own knowledge and is skilled at delegating tasks to their team – trusting them to execute. This requires letting go of control and accepting that it's impossible to have all the answers or oversee everything. It can be particularly difficult for those with a strong technical background, who are used to always having a solution.

> As modern CDOs, we need to empower our teams and delegate responsibility.

While we should always be available to offer guidance or support, it's important to recognize when our team members need help and provide them with resources such as code coach clinic experts.

As CDOs, we must acknowledge that there is always more to learn and that things are constantly changing. We need to maintain an open mind and be willing to look outside of our immediate environment to find the best tools, systems, and solutions to problems. Unfortunately, many individuals in technical roles are trained to think in strictly black-and-white terms, so we must foster creativity and critical thinking in our teams.

Often, people are afraid to speak up about complex issues, which can lead to low-quality work and bad governance. We need to start with the idea that if something isn't easy to understand, we shouldn't push it forward. This applies to acronyms, technology, data, and code. If we can't explain it in simple terms, it will only create confusion and won't provide any value or be actionable. Let's aim for clarity and simplicity in all aspects of our work.

AA: How important do you think enterprise-wide data literacy is, especially in the senior ranks?

MR: Data literacy can be intimidating, but it's important for people to keep learning new things. However, it's important to understand that data literacy doesn't necessarily mean being able to write complex equations or code. Rather, it means having the critical knowledge to understand what something means and how it's used.

As a CDO, my role is to see the big picture and identify what pieces are missing from the puzzle. This involves understanding the data flows, capabilities, and infrastructure of a product or service. It's about identifying when something is about to fall through the cracks and addressing it before it becomes a bigger problem.

To be successful in this role, a good CDO needs to have a strong understanding of the organization's maturity level and the right teams in place with the right skills, infrastructure, and use cases. It's like putting together a puzzle, where every piece needs to be in the right place at the right time.

So, instead of thinking that CDOs are math geniuses, it's important to recognize that the role is about having a holistic understanding of data and its applications within the organization. With this understanding, a CDO can help ensure that the organization is making the most of its data and avoiding potential pitfalls.

The importance of data storytelling

AA: Apart from hearing a lot about data literacy, there's also a lot of talk about data storytelling. How important do you think that is at different levels, and how do you develop the necessary skills?

MR: Storytelling is an essential skill for human communication. Without effective communication skills, we cannot fulfill our basic functions in life. When it comes to data storytelling, it means providing a data point that enables a story to be told.

> The power of storytelling lies in making something complex understandable.

For instance, a book on sustainability that mentioned, "*If you close the tap when you're brushing your teeth, you save an Olympic-sized pool of water*," is an example of great storytelling.

With data, storytelling is about taking something complicated and making it simple to understand. When discussing data, I use storytelling that is relatable across cultures, such as food. Everyone eats and has food preferences, regardless of where they come from. Therefore, I use analogies such as apples, oranges, and tomatoes. They grow on different trees or plants, on different farms, and are sorted by size before being packaged as products. The external data can be compared to spices that are purchased from external sources. By using a relatable analogy, we can make data processes understandable to everyone.

Basic data storytelling is an essential skill that we have been using to communicate across various industries. It's not a different skill; we just need to ensure that people understand the message. It's about making sure that the point gets across, and storytelling can be a powerful tool to achieve that.

Making AI ethical and trustworthy

AA: With increasing awareness around ethical and trustworthy AI, how does an organization establish a responsible AI capability? A lot of organizations have the practices, processes, and frameworks, but they find it difficult to actually put any of them into practice. What advice can you offer?

MR: All organizations have a responsibility to establish internal standards that reflect the relevance and materiality of their data to decision-making processes. For instance, if a hospital uses a triage system, a wrong decision could cost a patient's life, while a travel agency sending a recommendation to visit Rome instead of Athens carries a lower materiality impact. Thus, each organization must identify its level of materiality and impact, and establish standards and processes accordingly.

> Many companies are implementing AI systems to gain a competitive advantage, such as personalizing marketing offers. However, it is essential to make the systems explainable and traceable.

For example, if I receive a credit offer or a loan denial, the company should be able to explain why I received that specific offer based on my browsing history or search queries. This type of decision carries a higher materiality impact, and it is crucial to be transparent and accountable to the customers.

It is crucial to balance the need for approval processes with innovation. Automation tools such as bias identification and remediation tools can help set up checks for AI models. The responsible AI frameworks and models developed by most governments by 2023 have also added maturity to the AI space compared to just a few years ago. Therefore, companies should not be worried about implementing AI tools but should consider the risk and reputational impact of their choices.

However, some decisions require human intervention. For instance, for candidate selection, it is vital to have a human in the loop unless the AI engine can provide full explainability on why a candidate has been prioritized or not. Companies must also review and reframe their risk management practices beyond bias to include reputational risk. These are critical questions that every organization must address.

Advice for aspiring data scientists

AA: For people starting out in the field of data science today, what advice can you offer in terms of the skills and attitudes that they need to develop to have a successful career in the field?

MR: I would strongly recommend that individuals adopt a growth mindset to keep up with the rapid pace of change in the industry. Rather than focusing on obtaining a specific certificate, it's more beneficial to continuously learn and stay up to date on new developments, tools, and research. This can be accomplished by regularly dedicating time to learning and exploring new ideas. I suggest setting aside one day a week to look at what's new in the industry and stay current with key vendors, players, and toolkits.

To prioritize learning, I encourage people to protect their Fridays and use that time to finish emails, plan for the following week, and most importantly, learn and collaborate with peers. This is an opportunity to gain insights into what others in the industry are working on and to share your own knowledge and expertise.

> By investing in your own learning, you can stay ahead of the curve and ensure that you're well equipped to handle whatever challenges come your way.

AA: How important do you think formal university studies are relative to online learning?

MR: I believe that, in today's world, where knowledge is available everywhere, it's important to adopt a self-driven approach to learning, regardless of whether it's university-based or not. When I started university in the 1980s, it was the only way to learn, but now, learning is ubiquitous. People need to be selective about how they spend their time because there's too much knowledge available, and they need to make the best choices to prove their skills. Certifications such as PhDs and degrees are not as important as having the skills and being able to implement them.

However, with the overload of information available, curation is more important than ever. People need to be careful about what they choose to learn and how they learn it. They should be able to make the best choices because time is scarce. There is an infinite number of podcasts and push learning materials, which can be confusing and take you in the wrong direction. It's important to choose wisely.

To prove their skills, people need to implement them in their careers.

> It's essential to make career choices that enable us to execute the skills we have learned.

For instance, it's important to showcase that you have implemented projects on specific technologies, products, or services and achieved results. By doing so, you can demonstrate your skills to potential employers or clients.

Looking forward

AA: What excites you the most about the future of AI?

MR: I find it both exciting and scary that AI is everywhere these days. On the one hand, it's great to see technology becoming more mainstream and accessible, but on the other hand, I worry about the potential risks of using constantly evolving learning systems.

While we're all comfortable with software and code and know how to fix it when it breaks, the same cannot be said for learning about AI systems. These systems can evolve and change rapidly, and it's hard to predict where they'll be in just a few years. As a result, we need to be very mindful of the capabilities of AI and the potential implications of using advanced, continuously learning technology systems.

It's not just about ChatGPT or other AI tools; it's about the broader implications of AI for society and our daily lives. We need to invest in more research to better understand the potential pitfalls and how to mitigate them. Only then can we fully harness the power of AI while minimizing the risks.

AA: Speaking of ChatGPT, how does it affect the data science scene, including the field of NLP?

MR: ChatGPT is a remarkable technology that has the potential to shape the future of AI and data. As someone who is deeply interested in the world of data, I find it fascinating to think about how ChatGPT could change the way we interact with machines and process information. In particular, I'm excited about how ChatGPT will affect the data science scene, especially in the field of NLP. With the ability to understand and interpret natural language, it's much easier to process large amounts of text data. This has the potential to benefit industries such as healthcare, finance, and marketing, where understanding and processing text data is critical. It has also raised interest in large organizations where more funding could now be allocated to its research and deployment.

AA: What are some of the ethical implications for organizations, and society more broadly?

MR: It's important to consider the ethical implications of this technology for organizations and society more broadly. One of my main concerns is the potential for bias in the data that ChatGPT is trained on, which can lead to discriminatory outcomes. Additionally, we must be mindful of the potential for misuse, such as using ChatGPT to spread misinformation or engage in malicious activities.

AA: How can data scientists/AI experts leverage ChatGPT to benefit their organizations and their own careers?

MR: Despite these concerns, I believe they can leverage it to benefit their organizations and their careers in many ways. For example, ChatGPT can be used to develop chatbots that can answer customer inquiries, saving time and resources. It can also be used to analyze and summarize large amounts of text data, making it easier to gain insights from it.

AA: What are your predictions for the future of LLMs and their broader use?

MR: Looking ahead, my prediction is that LLMs such as ChatGPT will become more prevalent and accessible in the coming years. As the technology continues to develop, we will see LLMs being used in more industries and applications, making it easier for people to interact with machines in a more natural way. However, we must also be mindful of the ethical implications and work to ensure that LLMs are used for the greater good.

Summary

My conversation with Meri was both insightful and thought-provoking. With a wealth of experience and expertise, she is well placed to provide advice on how to develop a successful, sustainable, and scalable data science capability.

When discussing the oft-quoted high failure rates in AI projects, Meri was aligned with the common consensus throughout this book that data science and AI projects are scientific in nature, and as a result of their experimental nature, failure is part of the process. It's important to understand that if we define failure as the project not delivering the expected business benefits, then there are a multitude of non-technical reasons for "failure" in this context. To help mitigate failures due to scalability issues, her recommendation is to start with the problem – and not the data – to build solutions that are truly scalable and aren't limited to the dataset used for a POC or small-scale solution.

I really like Meri's analogy that data is like oxygen – rather than the well-known "data is the new oil" analogy. Thinking of data like oxygen, and thus not being owned by any one business unit, but rather being critical to the survival of the organization, helps people understand why it's important to identify and remove data silos and other obstacles in the way of sharing data internally.

Meri's insights on the idiosyncrasies of data scientists, and the fact that some are neurodiverse – and thus require empathetic leadership – highlights the need for leaders to develop an understanding of workplace adjustments that may be required to support neurodiverse staff.

Meri offers valuable and pragmatic advice for increasing an organization's analytics maturity, and suggests focusing on three pillars:

- **People and culture**: Build a culture and team that prioritize people

- **Technology infrastructure**: Identify your target state and determine what's missing

- **Use cases**: Ensure that you have the right value-driven measures in place and that data is driving value for the organization

We discussed how to tackle the challenges and biases that limit diversity in the field. Meri's personal stories are unfortunately not isolated incidents and are reiterated by other women interviewed for this book. Her advice is simple yet powerful: it's everybody's responsibility – not just senior leaders' – to ensure diversity and inclusion.

Given the rapid rate of development in the field, Meri's advice for all data scientists is to invest in continual learning and to stay up to date with new developments, tools, and research. Her key advice for aspiring data scientists is to make career choices that enable you to execute the skills you've invested in developing.

For technical professionals looking to advance to leadership roles, her advice is to learn to delegate tasks and accept that you won't always have the answers or be in control. This can be challenging for some, especially when you're used to being the "smartest person in the room" – from a technical perspective.

19

Teaming Up with Dat Tran

Dat Tran is an experienced data and technology leader who understands the importance of an entrepreneurial mindset. He is the CTO and co-founder of Priceloop and former head of AI at Axel Springer.

I was keen to understand his principles for building high-performing cross-functional teams and how to become a successful leader in the field.

I also wanted to hear his views on leveraging Agile and Lean practices when developing and scaling data science projects.

Entering the industry

Alex Antic: I'd like to begin by understanding your career trajectory. How did you get into the field and become the leader that you are? Were there any pivotal moments or people that inspired you or anything that happened in your childhood or during your career?

Dat Tran: My path into this field was not straightforward. I didn't study computer science.

I actually studied business and also mathematics, and once I graduated, I was a little bit lost as to what I wanted to do. I was into investment banking during my undergraduate studies, but after doing a couple of internships, I knew it wasn't for me. I was already very technical, and although I didn't formally study computer science, I learned computer science very early on in my childhood. I got my first computer when I turned seven. I learned to code by myself, and my first programming language was Turbo Pascal.

During my master's, I majored in operations research. I don't know if you know about the field, but it's partly about optimization in a business context and other related problems: logistics, supply chains, and so on. I wanted to do something in this field, and luckily, a consultancy at the time was active in that area.

When I entered the consultancy, machine learning in industry was still new, especially in Germany. I met a few colleagues already in the field, and they recommended that I do a course. It was pretty new at the time. A lot of people did this machine learning course, by Andrew Ng, on Coursera. I thought, "*Wow! There are a lot of things here that I already know from university and that I enjoyed, especially the* **Operations Research (OR)** *side.*" I thought, "*This is something that I can be interested in and still make a career out of it.*"

AA: It is a famous course. I'm sure many people became data scientists by doing his course.

DT: He inspired many people to go into the field, many of whom didn't know what you can do with data science in industry. At the time, you learned about linear regression and support vector machines. Then, you would think, "*OK. What can you do with that after university?*"

Discussing the high failure rates of AI projects

AA: You've been in the field for a while now; you've managed teams and done a lot of fantastic work that I've followed through LinkedIn. There are often reports of high failure rates for AI projects. What do you think are some of the common challenges that organizations face that could contribute to these failures?

DT: I've managed quite a few teams at different organizations, and they were pretty successful from my perspective. I know people for whom it's been tougher, though.

The most common pattern I've seen is companies making the wrong investment or doing the wrong thing at the start of their AI or data science project. They start by hiring a lot of people because that is what is common. They want to build a new capability, so they hire a lot of people. They allocate a very big budget to that.

But what happens is that you have a lot of new people with no industry experience. A lot of researchers get hired, and it makes sense. Researchers are very important in AI, but to be successful, you need more than that. You need to understand that machine learning is all about data in a company context.

Usually, when companies hire machine learning people, they just hire them and say, "*Now you're here – do something. Create those models.*" But the companies usually don't really have data. They'll say they have – they'll tell me, "*We have a lot of data. We've just been waiting for someone to come here and unlock our treasure.*" But usually, what they have is not clean, it's not in the right format, and it's not really useful. At the end of the day, you need to use the data somehow in a production system. So, there needs to be a business model around it, which most companies don't have.

Another dimension that comes into this is that they don't have the engineering infrastructure required but have a big budget for the data team. They hire a bunch of machine learning researchers and engineers, but then they don't adequately invest in the tech side or the business side. It's a one-sided investment where priorities are not aligned. So then, those machine learning teams are treated as an isolated area where those fancy researchers and engineers are allocated a lot of money. But that isolated area can't do anything useful. I've seen that lead to a lot of failures in machine learning projects.

AA: I couldn't agree more. On the engineering infrastructure side, you're right – I've often seen the same things through my consulting business where companies, from the outset, don't adequately invest in data infrastructure. IT is seen as a cost center, not an area that drives business outcomes.

On the topic of engineering, how important do you think practices such as Agile are? Would you advocate for Agile to be applied both in engineering and data science? How do you think it's best leveraged in the broader data science ecosystem?

DT: I'm a big fan of Agile, and I recently gave talks about a concept I call **Lean AI**. It's something that I came up with because this is a mindset that I have always wanted to bring into organizations and cultures.

In Lean, you have the concepts of iterations, and fast cycles, and you fail fast. In machine learning, if you think about the cycle, it's very similar. When you want to solve a problem, you usually have a problem statement: "*Hey, I want to build a classifier to identify sentiment,*" for example. Then, you review the data. Once you've done that, you might do feature engineering, create a model, and then evaluate the model. Once that is done, you want to use it somewhere in a production setting.

Obviously, the wrong thing is to spend a year on the cycle. A year on the cycle means that you never go through the entire cycle once. I've seen this problem in many company settings, and also team settings where they have this problem. They spend a year plugging in new models and trying to improve core metrics without actually deploying to users.

That's why I'm a big fan of Agile, which means for me, the first model can be the most basic baseline model that you can have. Obviously, it depends on the problem. If you think about self-driving cars, you cannot deploy a baseline model unless you have a system around it. So, you would say, "*Hey, it's a baseline model. It will not cover everything. You have to be at the wheel.*"

You need a **minimal viable product** (**MVP**) to release so you can understand the problem you need to solve quickly. Most of the time, companies have projects just because they think they're fancy. So, there are many projects, from customer lifetime value to price prediction. But sometimes, there's not a lot of value in that, and companies don't realize it because they work on a project for such a long time, and then once it's released, they realize, "*Oh, it's actually not really useful for us. But now we have been working for a year on it, so we cannot abandon it.*" That is really risky because, in AI, you actually want to move fast, break things, and then see whether things are useful or not.

AA: **Do you think that works with model development on the data science side and integration into the broader end-to-end pipeline through engineering? Have you ever seen issues where people try applying those more rigid methodologies (Scrum or Scaled Agile, for example) to data science, which may work for a more process-driven approach such as engineering? Have you seen Agile methodologies work for both disciplines?**

DT: On the engineering side, things are clearer. It's deterministic. In a machine learning project, it's a very stochastic approach, in a way, because you don't know where the data fits or what patterns are in the data.

I grew up with both sides, so that's why I bring these approaches to data science/machine learning teams, and I think it works. Even though it's stochastic, you still have a deterministic component. The deterministic component is that you have to decide at some point whether a model is worth putting into production.

It's the same thing with software. Software is also not completely deterministic. You can write as much code as you can. You can have as many tests as possible, but bugs still happen when you release software. You cannot cover all aspects because, at the end of the day, software is logical, but humans are not. There are many things that a human can do to harm the technical side. Therefore, even with software, at some point, we must say, "*OK. This is good enough for me. I have the confidence to release it.*"

You need to have the same mentality for machine learning and data science as well because, most of the time, people are just afraid to release. In many conversations that I have had, people have said, "*What is a good metric? What is a good value for an accuracy metric for a model? Do I really need to reach 99 percent in a business setting?*" If you have a model, for example, that predicts correctly 80 percent of the time, and the remaining 20 percent of the time predictions are wrong, you may just need to accept it. You need to look at the risk. For a self-driving car, 80 percent accuracy is obviously something I would not release, but if it's a churn model, I would say it's fine. You have to accept that even at 99 percent accuracy, you have a 1 percent chance that your prediction is wrong. So, just accept it and calculate the money you would lose from inaccurate predictions, but it's better than waiting for another year or two before you really learn anything.

Setting up for success

AA: We've talked a bit about where organizations can go wrong when trying to build up a data-driven capability. If you were to go into a greenfield setting, how would you try and get things right from day one, in terms of the people, strategy, culture, and infrastructure? Then, assuming they have data and a legitimate business problem, what would be your key steps to ensuring the journey begins correctly?

DT: This is a very tough question. I get asked this a lot by companies, and unfortunately, there's no completely right solution for everyone.

> Usually, what I recommend is that you need to hire one key person to drive the journey.

Obviously, there are so many things around that, such as when to hire the right people. But you usually need someone who understands a good data-driven culture. You need to know what talent you need to hire. You will need to understand a lot of job descriptions. Many of those job descriptions have evolved, and the road's become clearer, but many times, you cannot distinguish between a **business intelligence** (**BI**) analyst and a data scientist in poorly written job descriptions.

Sometimes, there's no boundary between a data engineer and a machine engineer because people don't really understand the difference. You need someone who understands the differences: job description, job title, role, and location.

> You need someone who can also think about use cases, talk to different business areas, understand and identify problems they can solve with data, and then inspire people across different business functions.

Teaching is a very important thing. At the end of the day, everyone breathes data. You might go to a marketing department, and they tell you, "*We have a great idea for how to use machine learning for a problem, and we need you to give us some guidance.*" But that kind of buy-in takes a bit of time. In the organizations I've come across, things are very siloed. There's no key person who can drive that kind of change to build a data-driven capability. There's usually a decision by committee, or the organization will see machine learning models in competitions and decide that they need that kind of thing. They then just give HR the task of hiring people in the data area to see what comes out.

AA: I really like your focus on having the right keystone person who understands enough to really push the data capability forward; that's pivotal.

DT: It's also a chicken/egg problem. You might want a great data journey for your company, but it's very hard to find that person. The reality is not all companies are successful in their data journeys. Good people in that area are very rare.

AA: One thing I've observed throughout my career and in my reading, especially lately, is that organizations have trouble transitioning from proof of concept and development into production. What are some of the key ingredients to smoothing out that transition to making it efficient and reproducible?

DT: We talked about this a bit already with Agile.

I have worked for smaller and bigger companies, and what I have learned is that bigger companies need to improve at doing proof of concepts. That relates to culture, to not having an agile, risk-taking culture. Having an Agile, risk-taking culture is very important to me. One of the key ingredients is the right people with the right culture around them.

I previously worked for a media company, Axel Springer. It's one of the most traditional companies here in Germany. Despite that, we still managed to put a text-to-speech model into production, and it's used by one of the bigger news media outlets, which Axel Springer owns. We managed it within a year. So, the question is, "*How did we manage to do that within a big enterprise?*"

Obviously, when I joined Axel Springer, I couldn't say, "*Just give me a big enough budget.*" I needed that budget to hire the right people with the right mindset. That's people who are not only Agile but also have the right mindset for wanting to learn about DevOps and being interested in everything. I needed people who could talk to business people, understand a problem, and help with the engineering and software side.

The software side involves questions such as, "*Where do we need to deploy it? How do we work with the software so that it is useful? How do we work with product and design to get a compelling machine learning product?*" You need people who can get buy-in too. With the right people and bringing those elements together, moving from the proof-of-concept setting to the production setting was pretty easy.

Establishing a good data culture

AA: You just touched on data culture, which is something I'd like to dive into. It's something that I think many organizations – both large and small – can struggle with, especially if they don't have the right person at a senior level to advocate for a positive culture or who knows what it looks like. So, in your view, how do you establish a successful data culture in an organization? What are some of the key ingredients to having it work in practice, not just on paper?

DT: I can only speak for myself again.

I don't think there's a textbook that will tell you how to follow the recipe so that you will become successful in your culture. It depends on the company and the context.

What I always try to do is to inspire. Inspiring people means reaching out and creating presentations, ranging from basic introductory presentations about machine learning and AI that give business people an understanding of what it's all about to technical presentations for the software side. While I was at Axel Springer, my day-to-day work consisted of going to the subsidiaries and giving presentations, whether introductory presentations or more advanced ones.

There are other things that I see as really important in terms of inspiration. For instance, writing blog posts about real-world solutions to show that these things really work. My team was always working very hard and very fast to make things work, and people were interested when things were used in production by our customers. They got inspired: "*Wow! You did that? My department could also use that.*" Then, people start to talk about it. You create a viral kind of marketing within the company, where people suddenly say, "*Wow! That is really cool. They automated this and that,*" or, "*Our customers are using this to do this and that to increase our performance.*" People become interested.

> You need to be an influencer within the company.

Once you are, people will naturally become interested. Then, the culture will come, but it takes time. Many of my peers don't really invest in it. They stay in their area and manage their team without going out in the field. Inspiring people is hard work, and people don't realize it needs to be done.

AA: Inspiring and educating people at all levels is fundamental to building a good culture. I completely agree.

Being a data leader

Some of the specific attributes you have described are – in my mind – those of a great data leader. I'd love to hear your thoughts: what makes someone a standout chief data officer (CDO) or senior data leader in this field? What do you think are some of the key attributes they need to have?

DT: You need great communication skills, and communication skills are not just being able to explain model X but showing how you can use data within the business setting.

You also need to be a great listener. You really need to listen to the business and understand what they want to do. You need to be able to listen, identify the gaps, and work out how to fill them to provide business value.

Those are the key ones. Obviously, you need to be analytical and adaptable as a data person. If one of the approaches that you're taking is not working, you need to know when to select a different approach.

I also always like people to be entrepreneurial. By entrepreneurial, I mean being self-driven and comfortable making decisions by yourself. If you have that mentality, you will inspire people.

You'll also not be afraid to fail. Being entrepreneurial means failing a lot: a lot of ups and downs. Many companies that are going through a data transformation encounter a lot of ups and downs. It's not a straightforward journey where everything is fine. It's a lot of hard work. Being entrepreneurial or having someone who has an entrepreneurial mindset really helps on the journey.

AA: Is there any specific advice you can share with data scientists or engineers who think, "*I want to eventually move up into a management position and climb the corporate ladder?*" I've seen that sometimes it can be a difficult transition from doing hands-on technical work every day to suddenly having to let go of it. In addition, managing people, expectations, and politics can be a new challenge for technical people. What do you think people should do beyond focusing on communication and entrepreneurship? Are there any other things that they should study and practice? Should they learn more about the human side to become good data leaders and managers?

DT: I don't think this question is just related to data science. It applies to any field if you're moving from a technical specialization role to management.

You need to ask yourself if you want to do it. I've seen many people move from a very technical or specialized field into management because they thought they needed to do it because it was their track. Unfortunately, that's also what companies sometimes advocate: "*Once you have worked in a specialized field, you need to become a corporate leader to progress and have a higher salary.*"

But luckily, it's not always like that anymore, though there are still many companies where you cannot stay in your specialized field and still earn well.

If you decide to go into management, you must learn how to communicate. You also need to have a different mindset because you need to deal with politics. It's a tough thing.

Other than that, I also recommend that people who go into this field mentor people for a long time. They should take care of juniors and early to mid-career technical specialists. Why? Because once you become a leader, you need trust. You need the trust of your people. They need to follow you. To establish that, you need to have already built a mentor relationship over time so that once you are there and managing, people don't say, "*Oh, this guy is giving me another task – how annoying,*" but rather, "*I respect that person a lot so I will follow them and work very hard to make that person successful because if they are successful, I'm also successful as an employee.*"

That is something that I've always thought is good: mentoring people to build trusted relationships in the organization on the way to management.

Discussing data storytelling

AA: We have talked about communication and how important those skills are. We hear a lot about data storytelling these days. There are a lot of blog articles on it, and there are a lot of people who teach it in courses. How important do you think it is at different levels (depending on your career trajectory) for an individual, and how do you develop those skills, apart from just saying to someone, "*You need to work on your communication skills?***" What would be your advice there?**

DT: There are different things that you can do.

Everyone will tell you, "*Just do a course in communication.*" There are a lot of people giving training on how to communicate well and how to do presentations. But for me, I advocate for something different, though I don't know whether it is possible for everyone.

When I create my teams, from the most junior to the most senior engineer, I make it a rule that everyone gives a presentation after finishing a project – not just internally to the team but also other groups in the company. Within the company, we have different formats, such as a show and tell. I always encourage everyone to give a presentation about their work.

In the beginning, it was pretty funny for the juniors because they had little experience in giving talks. Especially when you come from a technical field, it's not like you go out and give talks regularly, but doing presentations helped them to build some security and confidence.

One other thing that helped me quite a lot is my writing skills. At school, I really enjoyed writing in general. I'm very good at writing and telling a story. I use that skill in my job to teach people about good storylines – how you start them, what the middle point should be, and what to do for the finishing touch.

So, if people want to go into this field, I think writing will help them. You can start with fictional stories, where you need to think about the story's structure and what a good start, climax, and end look like, and then use that skill to create compelling data stories.

AA: That's excellent advice. I used to write a lot of blog posts, and I really felt that helped me to finesse targeted stories that I wanted to tell to different audiences. And, of course, I'm writing this book!

Hiring team members

When employing data scientists and engineers for your teams, what specific skills and attitudes do you look for, especially if you're trying to decide between two people? How do you know who to pick? Of course, there's experience and many other factors, but in general, how would you encapsulate the key things you're looking for in technical employees?

DT: Making hiring decisions is always the hardest thing because what you get is a piece of paper. You have to review that paper, which is difficult, and you could make a mistake.

When I'm shortlisting resumes, I like it if the tasks the candidate displays are concise and interesting. If someone works in deep learning, I want to understand what exactly they do. If they've done something open source, I can see the work and its quality. That is very important for me. I don't like bad code.

When we do coding challenges, I really like to see that people use either double quotes or single quotes, not a mixture. I want the line distances to make sense in the code because that shows me a lot about how they work.

It even shows in the CV. For example, if someone writes a CV in LaTeX, I know that that person puts a lot of effort into formatting things, whereas if someone just puts a CV together in an hour, it does show. I can already tell the difference in passion between those people.

That said, you can have someone with a poorly written CV who turns out to be pretty good. It's very hard because you need to decide at some point.

AA: What about when you're interviewing? What do you look out for? What are some warning signs or things that stand out to you?

DT: It's similar to the coding challenge.

> I ask very detailed questions in interviews. For example, if a person did a machine learning project, I like to know whether that person really understood what they were doing.

You get a lot of people who have just started in this field, even if they've come from university. They have some interest in machine learning and just did one textbook-like Kaggle challenge: put data in, transformed it, and then produced some metrics. When I dig deeper and ask questions such as, "Why did you use this metric? What is the difference between this metric and that metric?" those people struggle because they haven't thought about it.

It's like mathematics. There's not just one solution to a problem but many solutions. I look for people who are creative and not stuck with one textbook solution but have a toolset. What we need in machine learning, in general, is a toolset. You need to understand what tool you need to solve a given problem. Many people know only one way, and that is not what I am looking for. I always look for people who know many different ways to solve a problem.

AA: How important do you think formal university studies are, as opposed to doing online training or on-the-job training in a self-paced fashion? Do you have a preference for one or the other?

DT: I don't have preferences. I've hired a couple of people who didn't study computer science. One of the best engineers and researchers who has published stuff in machine learning didn't come from this field but from physics. I've had people from a math background who are already very technical.

In general, I wouldn't say that universities prepare you for what we do anyhow. When I came into the field, there was no machine learning course at the university here, but there was a little bit of machine learning within computer science. It was part of computer science, not its own course, and I'm not sure if you really need that. For me, machine learning, in general, is basic algebra and analysis. If you finish your first year in math, you're prepared. You can understand machine learning quite well if you have basic calculus and basic algebra.

Advice for beginners

AA: What advice would you give to data scientists looking to develop a successful career in the field? What should they focus on beyond just technical skills, and what do you think are some of the challenges they may face throughout their journey?

DT: Giving advice is always challenging for people in this field. There is the same generic advice that I would probably give to anyone, but it also depends on luck at the end of the day.

What I saw as a good pattern was doing capstone projects: doing projects in machine learning to learn how things work. Then, if you really want to go into this field, either you're lucky (meaning you are in a corporation that already has machine learning teams, and you can ask for a transfer to that department), or you could do an internship where you can showcase that you have the skills and can build your practical experience.

Those are the two options for people to grow in this field. What I've observed so far, which is very challenging, is people doing paid courses and then trying to get into the field. It's very challenging because, in the data science/machine learning world, it's very competitive. Everyone is looking for senior people and not juniors. But everyone who wants to get into the field is a junior and that's why it's very challenging to do these courses and apply to companies. It is very difficult to get a job with that approach.

AA: I think a lot of graduates tend to focus on big-name companies, especially in the tech fields. They are focused more on the name rather than finding an organization that gives them a chance to learn from senior staff and that teaches them proper coding techniques and how to actually

apply data science in practice. When I mentor or teach students, I try and steer them toward what is more beneficial for their career development rather than what necessarily looks best on their CV.

DT: That's the thing. You shouldn't think short-term. You should think about the long-term aspect of a career. Obviously, the big brands will pay you more, but you will be stuck at some point with your career because you haven't learned the skills necessary to grow further. Also, at a bigger corporation, you don't get the full attention of someone who would mentor you the way you could in a smaller setting. Your start will be good, but overall, you will harm your mid to long-term career.

Looking to the future

AA: Last question: what excites you the most about the field of AI and the trajectory that we're on, beyond ChatGPT?

DT: What excites me the most (and I've been with this field for a long time now) is that it's constantly changing. Some people don't like it, but I like it.

> I like that every week – every day, actually – there's something new: someone trains a new model and solves a different problem in a different way.

Models are getting faster. That is pretty cool because you can use models in settings where you could not use them before.

Also, how machine learning is applied in different fields is very exciting. When I started, it was just a thing for nerds. Now, many people use it. You have business users using ChatGPT and being amazed by it, but if you know about machine learning, you know it's just a normal thing, and you know how it works. There's also the impact of AI on the arts. The popularity and use of ChatGTP is driving even greater AI adoption. In general, I don't think there have been any significant advances – LLMs have been a thing for quite a while. I must admit that I see the development from a more skeptical perspective – as "good" models are becoming larger and computationally more expensive. We need a shift in paradigm to create more efficient models. Also, data used to train these models is another topic that could fill an entire chapter of this book. For example, what data is used to train ChatGPT and similar models that is eventually also used for coding?

When I started my career, there was a kind of gamble: am I going into a field that will stick around, or will I go into a field that will just have another winter? For me, the gamble paid off because AI is here and here to stay. It will not disappear.

AA: Definitely. And who would have thought when we started our careers that our parents would use this technology daily, such as predictive text on Gmail, and facial recognition on our phones?

DT: Yes! It is amazing. On my iPhone, there are so many places where machine learning technology is used, from sorting photos, to Azure asset translation, to even just writing an email.

Summary

I enjoyed tapping into Dat's experience in leading machine learning teams to better understand his approach.

For organizations commencing their journey to becoming data-driven, Dat recommends onboarding the right key/foundation hire to help lead the change. This person needs to understand both the business and technical domains to understand what problems need to be solved and which solutions may be suitable.

It's also important to look for data professionals who are adaptable and interested in everything from business problems to AI, from DevOps to engineering.

Dat's focus is on an Agile approach – beginning with an MVP to help you quickly understand the problem you're solving and what the solution will look like. Then, in production, you'll learn whether the model is useful in practice – you will fail fast and learn in the process if it's not.

For data leaders, he strongly encourages a focus on being inspirational – and becoming an influencer within your organization. Successful projects in production will create a kind of viral marketing opportunity within the organization. Data leaders also need to be good listeners and have an entrepreneurial mindset, including not being afraid to fail, a common theme throughout this book.

For data professionals considering the transition to management, make sure you want to make the move; don't just follow the traditional career progression path.

It's become a given that communication skills are fundamental for all data scientists. Dat's approach to helping his staff develop their skills is to get everyone on his team to give a presentation upon the completion of a project. This not only helps them develop their communication skills but also builds awareness and knowledge sharing across the organization.

Given the importance of attention to detail in the field, Dat focuses on indications of a candidate's attention to detail when shortlisting and interviewing. He also likes to see candidates who are creative, who have a toolset of solutions, examples of good coding practice, and with well-written CVs.

20
Collective Intelligence

When I set out to write this book, my intention was to collate the learnings and wisdom from some of the industry's most successful and renowned data leaders, and to share some pragmatic advice on how to *do data science right*.

After many hours of conversations with these data leaders, I feel excited about the potential for good using AI – and concerned about the potential harms. One thing is clear: the discipline of data science will continue to increase in relevance.

It was inspiring to hear about how each of the interviewees overcame the challenges they faced – hopefully, some of their stories will save readers from learning some lessons the hard way! My plan has always been to offer clear and actionable advice and recommendations that you can apply to your own organization and career. In this chapter, we will distill the invaluable information from each of the interviewees into a set of key themes, as follows:

- Entering the field and becoming a successful data scientist
- Becoming a CDO and senior data leader
- Developing an effective data strategy
- Establishing a strong data culture
- Becoming data driven
- Ethical and responsible AI
- Data literacy
- Scaling your data capability
- Structuring and managing data science teams
- Avoiding AI failure
- Measuring success
- Storytelling with data
- Predicting the future of AI
- Striving for diversity and inclusion

The recommendations that follow not only reinforce each interviewee's key takeaways as captured in the summary section of each chapter but also include my own insights and advice.

Entering the field and becoming a successful data scientist

It's been fascinating to watch the field of data science develop over the past decade. This has been primarily driven by an increase in awareness of the value that data analytics can offer.

People who are looking to transition into the field of data science and build a career often ask me for advice on how to go about this. Based on the conversations in this book, and my own views, we can offer the following recommendations.

To become a data scientist, you first and foremost need to develop a strong foundation in the core technical skills required to do the job. These core skills are even more important if you plan to pursue a career in research. The technical skills required to perform data analysis, and develop machine learning models include the following:

- Mathematics and statistics:

  - Linear algebra

  - Calculus

  - Optimization theory

  - Frequentist and Bayesian statistics, including probability theory

- Computer science:

  - Programming, including learning from different families of languages, such as scripting languages (for example, R and Python), object-oriented and/or functional languages (for example, C#, Scala, Java, and Go), and data-specific languages (SQL).

  - Become familiar with the platforms and tools that are commonly in use, as well as the ones that are currently high on the hype curve. Becoming familiar with at least one of the major cloud environments (AWS and Azure tend to be the most common at the time of writing) is important. Other things to become familiar with include the concepts behind such tools as Spark, Hadoop, and Snowflake. Obtaining a level of DevOps skill is often useful, and it is highly regarded by potential employers.

  - In many roles, you will need support from IT to set up the specialized environments you need for data science work. You might find that the IT people in your organization are used to supporting software developers, but not (those weird and wacky) data scientists, who want to access real (in other words, production) data in a dev environment. It helps to be able to speak the same language as IT people – to have a basic understanding of hardware, platforms, software engineering principles, and cybersecurity principles.

In addition, you need to develop a solid foundation in the most common machine learning and deep learning algorithms and approaches, starting from linear and logistic regression and decision tree variants, and moving to reinforcement learning, neural networks, natural language processing techniques (including large language models), and graph analytics.

Critical and logical thinking skills are imperative, as are problem-solving skills. Looking for ways to continually strengthen them is highly encouraged.

Fortunately, the unrealistic expectation of being a *unicorn* and being able to do it all as a data scientist is less common than it used to be (in organizations with mature analytics capabilities, at least). Instead, focus on your strengths and interests. For instance, if you enjoy working with unstructured data and have a passion for understanding human language, then consider developing a career as an NLP specialist.

Whether you decide to go down the path of formal and traditional university studies to acquire these skills, or instead choose a more flexible self-learning approach via online courses, is a personal choice based on many factors, including finances, time, and family commitments. Either way requires significant commitment and continual learning, given how rapidly the field is evolving.

Interviewees had different views on the question of whether they prefer to see potential job applicants with university qualifications versus self-learning and online courses. There is no consensus.

Such technical skills are necessary, but not sufficient, to develop a successful career as a data scientist. Beyond the technical, effort must also be placed on developing other relevant skills and aptitudes, including the following:

- Communication skills, including storytelling skills
- Domain/business knowledge
- Stakeholder/relationship management skills
- Curiosity and enjoyment of exploration and experimentation
- Flexibility and adaptability
- A team-player attitude

What emerged clearly in the interviews was the need for aspiring data scientists to find ways to stand out and excel in such a competitive field. Showcasing your work regularly is a key way to achieve this. This could include writing blogs, posting on LinkedIn, maintaining a repository with your portfolio of projects, presenting at conferences and industry events, and doing demonstrations for your stakeholders and clients in your organization.

As I've previously written, there are two other elements that are beneficial to developing a career in data science – and any career really: networks and mentors.

Networks and mentors

You can read what I have previously written on these topics online: networks (`https://impartiallyderivative.com/2018/02/16/why-you-need-to-put-the-work-into-networking-as-a-data-scientist/`) and mentors (`https://impartiallyderivative.com/2018/02/15/the-secret-to-a-successful-data-science-career-1/`).

Networking provides an important opportunity to land your next (or first) role. As your career progresses, your network continues to grow, and you establish a name for yourself. If your reputation is good, a shift will occur in how you land roles. Rather than applying for jobs, you'll increasingly become headhunted, and people in your network will reach out to you to join their teams or suggest you for other roles. And don't forget the importance of internal networking. As Cortnie Abercrombie suggests, "*Go out and start networking and politicking with as many of the senior leaders within your company as possible. Start forming relationships and understanding what drives them and what's most important to them.*"

It's also a great way to mingle with like-minded people, hear what your peers are doing, and learn about the problems they're facing and how they're solving them. Mentors can be integral in helping you deepen your technical expertise and understanding of your business or research domain. This is why having two mentors – a technical mentor and a business mentor – can really help advance your career. They also play a leading role in helping you grow your network.

Petar Veličković (*Chapter 6*) had some specific tips for people who want to build a career in research:

- Don't just read research papers, but learn by implementing the techniques yourself. Authors will note the limitations of their work in the paper, and Petar encourages early-career and aspiring researchers to see whether they can overcome some of the limitations with their implementations. In addition to the learning aspect, this helps build a valuable portfolio of research.

- As an introduction to the world of research, Petar suggests presenting and publishing your work at workshops associated with the leading ML/AI conferences. The reviewers accept a high proportion of submitted papers and provide valuable feedback.

Of course, your career doesn't stop once you've made it into the field. Leadership offers a new range of challenges.

Becoming a CDO and senior data leader

Given the increasing understanding of the value that data and analytics bring to an organization, it was only a matter of time before data and analytics entered the C-suite. The **Chief Data Officer (CDO)** and **Chief Data and Analytics Officer (CDAO)** are relatively new entrants to the C-suite, and as such are less standardized in terms of roles and responsibilities. What is even less well understood is exactly what it takes to become a CDO or senior data leader.

Some of the fundamental attributes and aptitudes needed are listed here:

- Given that many leaders become a CDO or senior data leader without having a background or training in data and analytics, it can be challenging to understand not only what is needed but also how to develop the necessary skills. One of the most fundamental skills to acquire is data literacy, which can be defined as the ability to "speak" and reason with data. It certainly helps to have had hands-on experience with data and analytics, but at the very least a solid conceptual understanding of the technology – and the practice of analytics – is required. Jason Tamara Widjaja (*Chapter 10*) is of the strong belief that leadership is not a generic role in itself – and it's not agnostic of the team. He believes that to be a senior data leader, you want to have some grounding in data.

- For those coming from a hands-on technical background, it is first important to think about whether a move into management and giving up hands-on work is really what you want to do. More organizations are creating *individual contributor* pathways, so the move into management is no longer the only option for developing a career that allows you to have maximum impact.

- According to Cortnie Abercrombie (*Chapter 1*), the realities of any senior leadership role necessitate political savvy in a business context – and this is especially true for CDOs. This includes having *data charisma*, the ability to influence and negotiate with data. To be effective, Althea Davis (*Chapter 12*) notes that senior data leaders need to be seen as trusted advisors in order to win the support of colleagues and the broader organization.

- Dat Tran (*Chapter 19*) believes that an important part of leadership is being able to inspire people – to inspire and educate people at all levels.

- One of the best insights I've come across to help articulate the fundamental elements of a CDO are Maria Milosavljevic's 5 Hats: The Visionary, The Bridge, The Regulator, The Scientist, and The Engineer (*Chapter 16, Figure 16.1*). In line with Cortnie's advice, I'm tempted to add a sixth hat – The Politician! I strongly urge anyone who is new to the role – or looking to become a CDO – to read through Maria's details of each hat and determine how to apply them to their own role.

- Regardless of the seniority of the leader, an understanding is needed of the diverse skills required that enable a large-scale data science project to be successful, and the key roles each plays in the end-to-end pipeline of a data project.

- The relevant experience and skills are needed to find the right people to lead your data teams. Appropriately skilled and experienced advisors can assist with this if the expertise isn't available internally.

In the next section, we'll discuss how leader need to develop the overall strategy for an organization.

Developing an effective data strategy

A data strategy is integral to setting the data and analytics vision for an organization. It is typically developed, set, and advocated for by the CDO, and needs to be driven from the top down.

The data strategy needs to be one component of the broader organizational strategy – and there needs to be clear attachment points between each of the components, and into the broader organizational strategy. Organizational strategy is composed of four key sub-strategies:

- Business strategy
- Analytics strategy
- Data strategy
- Technology strategy

All these sub-strategies work together to achieve organizational goals.

It's important to have relevant support and clarity for what's within the scope of the data strategy and what is beyond it, so that everyone understands what the strategic outcomes are and the integral role they play in their accomplishment.

Establishing a strong data culture

The culture of an organization is one of the most important elements for success. We all immediately recognize a good culture when we experience one, but it's a concept that is difficult to articulate.

Maria captured the essential elements when she said, "*Culture is not just about a piece of paper, strategy, or policy; it's actually about implementation. It's about building data culture into the DNA of the organization, making sure that people can see its value.*"

One aspect of an effective data culture is that it supports change. Becoming data-driven often requires transformational change, and each successful data project also generally results in a change – often a change to the way people work. Being able to embed this change is therefore a crucial part of reaping the business benefits of data projects. The success of this ultimately lies in senior leadership and support. It needs to be driven from the top down.

Try to anchor your projects with change management, the business needs, and the creation of business value. An integral part of this is helping people trust the solutions that you're developing for them. One strategy for helping create this trust is to be able to quantify value via metrics, and Kshira Saagar's **Accessibility**, **Intelligence**. and **Reliability** (**AIR**) **metric** for benchmarking a data capability is a fantastic approach (*Chapter 4*).

Becoming data-driven

A data-driven organization has data at the core of the business – it is used to inform decision-making by providing a quantitative evidence base and to drive the creation of business value. Ultimately, being data-driven requires decision-makers to be willing to change their minds based on the objective evidence provided by the data and analysis – which also requires them to be data-literate.

Key recommendations include the following:

- See data as a team sport: business, technology, and analytics need to be integrated to deliver value. Maria advised that the data function exists to support the business – and in return, it requires business buy-in and support. Support included the resourcing of data science projects with SMEs, and being prepared and resourced to run the appropriate activities to embed the solution delivered by the project. The responsibility for making sure business readiness is in place before starting a project likely lies with IT/data science if we want projects to succeed. As Kathleen adds, *"Analytics can't provide value in a vacuum."*

- Funding for data science functions came up in a number of the interviews. IT is typically seen as a cost center, and sometimes data science is managed in the same way. Data science is not a support function, however, as some parts of IT are; rather, it exists to provide business value. Costs and expectations need to be closely managed from the start. As Christina Stathopoulos warns (*Chapter 14*), the right talent is going to cost you, and as Charles Martin mentions in *Chapter 5*, data science is not cheap. It needs to be thought of as a long-term strategic investment.

- As part of managing the costs, *bootstrapping* the data science function was recommended – where data work needs to immediately start providing an ROI to fund the next piece of work.

- When commencing your data journey, it's easy to get excited by the hype and to focus too much on the technology. Rather, your focus should be on clearly identifying your business problems and then trying to determine the simplest and best solutions. As Kirk Borne suggests (*Chapter 8*), *"Think big, start small."* This could mean investing in a small proof-of-concept to validate an approach or to secure the required funding.

- As Christina stresses in *Chapter 14*, stakeholder buy-in is imperative to developing a successful data and analytics capability – so spend effort in understanding their problems, and how you can help them solve them.

- It's also easy to fall into the trap of thinking that more data equals more value. Ensure that the data is fit for purpose – both from a business perspective (does it help solve the business problem?) and from a technical perspective (is it clean and suitable?).

Ethical and responsible AI

The use of AI will play an increasing role in shaping society and the way we live and work – and AI technology brings with it considerable risk. The scale at which AI can operate is fundamentally different to the scale at which most individual humans can influence. For a simple example of this principle, consider the scale of decisions that an individual hiring manager can make (decisions that are inevitably influenced by the biases of the hiring manager), compared to the scale of decisions that an AI candidate selection tool can make (also inevitably influenced by biases). The collection and use of data within AI systems also presents risks and ethical questions.

A Google search will quickly reveal that it is currently difficult to even arrive at a meaningful description of what ethical and responsible AI seeks to achieve. Somewhat ironically, ChatGPT provides a solid attempt, stating that it seeks to protect the rights and interests of individuals and societies. I would add that it should seek to promote trust through embedding principles such as transparency, explainability, and accountability.

Many organizations are moving toward developing AI ethics frameworks. Edward Santow commented in *Chapter 3* that emerging research shows that most ethical frameworks include high-level, vague platitudes that in practice yield no discernible results.

Organizational will is an important factor in whether such frameworks are effectively implemented. A number of interviewees commented that compliance levers are an effective driver for ensuring organizations prioritize such implementation. It's important, however, to also be clear-eyed about organizational drivers such as profit. The documented behavior of Cambridge Analytica and the data use practices of Uber were raised as cautionary examples.

Despite the challenges in defining and implementing ethical AI, there are some practical points we can take away from the insights provided. The first step is to perform a risk assessment on each AI system and particularly look at the materiality of the decisions made by the system (perhaps using a traffic light system to categorize the impacts of AI systems on human lives, as Cortnie suggests).

For any decisions that will have a significant impact on someone's life, it is important to have a human in the decision-making loop. For all systems, Ed advised that it's important to have in place a series of checks by humans so that errors in AI systems can be identified and addressed to minimize potential harm to individuals.

Cortnie raises the idea that culturally, everyone in an organization (not just the data scientists) needs to be both empowered to ensure, and responsible for ensuring, that AI solutions are ethical and responsible. Jon Whittle (*Chapter 11*) points out that design patterns, guidelines, and assessment frameworks need to be developed to support technologists who are building the systems. This is the layer that translates AI ethical frameworks into something that is usable and implementable.

The concept of explainability for AI systems can be a bit vague in practice. Ed gives great, clear advice on this point, stating that AI systems should yield explanations that are "*at least no worse than a human explanation.*" Specifically, two levels of explainability are required:

- A layperson's explanation
- A technical explanation

Ed also points out that some of the issues being cast as ethical issues are not new and are already addressed by existing legislative instruments such as the Privacy Act. This highlights the interplay between the technical, philosophical, and legal aspects of developing and managing ethical and responsible AI. I agree with Jon's views that universities need to take the lead on teaching the ethics of technology across the curriculum – rather than it being a standalone course – to ensure that the technologists developing AI have a solid foundation in it.

Data literacy

Data literacy is important because it ensures that everyone in the organization speaks the language of data. It can also help demystify AI, particularly for senior executives, and help them better understand what AI can and can't do – which also empowers them when being courted by solution vendors. It also helps senior executives better understand the risks, threats, and opportunities of AI, and enables them to better manage procurement, implementation, and oversight (such as governance) functions. In addition, being data-literate helps you ask the right questions – which, as Kshira says, is a fundamental skill.

It also provides exciting career opportunities to people from other disciplines who can supplement their skill set with data science micro-credentials. The growth in data-literate workers – especially those with deep data literacy and data analytics knowledge – helps facilitate strong working relationships between workers and data scientists and data engineers to create cross-disciplinary capabilities.

Apart from data literacy training for staff, and the creation of cross-disciplinary teams between business domain experts and data experts, it's important to democratize your data by making it available to all staff, as Christina mentions. No one business unit owns the organization's data – but rather they are its custodians, and the onus is on them to share this important strategic asset with all staff to help discover insights and support better decision-making. It's up to the CDO and senior leaders to facilitate data democratization by breaking down silos.

Scaling your data capability

The transition from development and proof-of-concept to a scalable production solution can often be a challenge for an organization. Some helpful suggestions include the following:

- Having a plan for productionizing and maintaining models early on – including looking beyond the development stage, and thinking about the entire data and analytics life cycle. Scalability needs to be considered from the beginning of a project.

- Having the right skill sets for moving from research and development to production and operationalization. Paradigms such as MLOps and ModelOps can play an integral role in helping monitor data and concept drift – and ensuring the models are ethical – at scale. It's important to have data engineering and DevOps skills in your data science team to help with the transition from development and proof of concept to production.

- As Jason says, as soon as you are working at a certain scale, you then need specialists – such as project managers – to help lead the transition into production.

- Althea suggests that the data life cycle needs to be anchored into the **Software Development Life Cycle (SDLC)**.

- Data science and engineering need to be thought of as a continuum throughout a project's life cycle, and not two distinct steps. At the beginning of a project – and especially during the development stage – data science plays a leading role, but as the project progresses and nears production, engineering will then come to the fore, upon which the data science function is scaled down.

The structure and management of teams is also a vital topic.

Structuring and managing data science teams

There are some unique aspects that need to be considered when building and leading data science teams. Data scientists tend to be motivated by curiosity and the challenge of solving complex problems.

It's also important to understand that data professionals may not wish to follow traditional career trajectories and move up into leadership roles. For many, becoming less "hands-on" means losing the technical skills that they've invested in developing, and enjoy focusing on. Thus, it's important to offer those staff an alternative path, where they can still progress and grow but don't have to give up what they love doing. It was refreshing to hear from Christina that many tech companies now offer career paths for technical experts, often termed **individual contributors**, for those not wanting to pursue the traditional management path. This is an important initiative to not only help address staff retention issues that many are facing but to also support staff to develop into roles with maximum impact.

In my experience, technical people who are interested in pursuing the management pathway are relatively rare. Those who are can play an important role in team leadership succession planning and should be supported through the transition as they develop their management skills. I know from my own experience, as well as that of colleagues, that it is not an easy transition!

The question of where an analytics capability should sit in an organization is one that involves balancing the various trade-offs of different structures within the organizational context. Angshuman Ghosh (*Chapter 15*) and Kshira lean toward a hybrid, federated model – where the function is centralized and centrally funded (at least for larger companies). The central team is responsible for bringing in the right talent. Then, the data specialists are distributed to work with local teams. They solve problems for them, gather domain knowledge, and deeply understand their needs.

As for the question of where this central data team should sit, Kshira suggests that they be placed close to the ultimate key decision-maker for the organization. For a small tech company, this might be the CTO, for example. This will not only help in gaining much-needed support for staffing and resourcing the capability but also in turning the findings into actionable outcomes that are aligned with the strategic goals of the organization. It's crucial to have a business sponsor who wants a solution to the problem you are solving.

In addition, note the following:

- Building a high-performing team is like a jigsaw puzzle – you need to focus on people and how they all fit together as a team, complementing one another's skills and personalities, rather than having preconceived ideas of fixed roles. A cross-discipline team, including business, data science, and engineering skills, is often the best approach.

- Cortnie advises hiring people you trust and proactively assisting them in developing new skills. Kshira uses a rule of thumb that there should be 7 data people for every 100 people in the organization.

- When interviewing data and analytics staff, Kshira suggests focusing on three key areas:

  - A willingness to learn, unlearn, and relearn things

  - A continual learning mindset

  - The ability to use logic and first principles to solve problems – especially in new domains

- Executives need to understand how analytics is integral to delivering business value, as opposed to being seen as something that's distinct and isolated from the broader business. Kathleen stresses the importance of visibility for the analytics team to ensure it remains relevant and funded. This is where data leaders need to get out there and tell stories about data.

- Senior technical SMEs also play an important leadership role in an organization. Jason advocates for leaders to build teams that have a diversity of personality types, strengths, weaknesses, likes and dislikes, and skill sets in order to build the most productive team possible.

Avoiding AI failure

A Gartner prediction in 2018 led to numerous articles stating that, "*85% of AI and machine learning projects fail to deliver.*" It's hard to be certain of the original source of this statistic because some poetic license appears to have been taken in interpreting it. This, however, hasn't stopped this statistic from being referenced often when discussing the relatively low success rates of AI projects.

> **85%?**
>
> The ironic misinterpretation may come from the line, "*Gartner predicts that through 2022, 85 percent of AI projects will deliver erroneous outcomes due to bias in data, algorithms or the teams responsible for managing them*" (`https://www.gartner.com/en/newsroom/press-releases/2018-02-13-gartner-says-nearly-half-of-cios-are-planning-to-deploy-artificial-intelligence`).

Why has this statistic of questionable accuracy captured so much attention? As many I interviewed pointed out, failure is a normal part of the scientific process – and more broadly, most new ideas fail (think of the failure rate of start-ups). In addition, we are generally aware of the high failure rates of large IT projects. For those who aren't in the field, the difference between AI projects and IT projects isn't readily apparent. I know of a number of CEOs who find that it's the large IT projects that keep them awake at night.

Defining failure and, conversely, success (see the next section, *Measuring success*), is important. As Jason puts it, "*It's very common for use cases to fail but, for me, there's no shame in failure.*" The focus on high failure rates in AI may in fact be perpetuating an unrealistic expectation that project failure rates should be close to 0%. Dat observes that fear of failure can inhibit data teams from releasing their models into production in a timely manner.

There are, however, some common challenges faced by organizations when implementing AI and building data science capabilities. Our experts offer the following advice to avoid these pitfalls and challenges:

- Kathleen outlines four common pitfalls that organizations can struggle with:

  - The business thinks they need more and more data, and more reports

  - The data is not clean and in an appropriate state for use

  - A linear waterfall approach restricts the speed at which analytics can add value

  - Thinking that analytics is separate from the business

- Data needs to be fit for purpose – and obviously high-quality. The AI models created are only as good as the data they're trained on.

- Cultural issues can hinder success. These need to be identified and addressed. An issue such as lack of data sharing between business units is ultimately the remit of the CDO and all senior leaders in the organization who have the power to effect change.

- Alignment is needed with the requirements of stakeholders/clients/customers to ensure that solutions meet their specific needs. Cortnie advises trying to work out what they actually want and need from the following:

  - What they tell you

- What they tell others

- What they won't admit, such as their career motivations and political machinations

    Cortnie also suggests working out who you need to get on board as key sponsors, and figuring out how to attach and align your work to their priorities.

- Lack of prioritization of the myriad of business opportunities can stymy progress and success. Once business requirements are captured, it's important to prioritize high-value projects that are aligned with the strategic goals of the organization and sufficiently resource them.

- Enterprise-wide data literacy is fundamental. All staff need to be empowered to innovate, reason, and question with data. If data is like air, as Meri Rosich (*Chapter 18*) refers to it, then all staff need access to it for the organization to function at full capacity. Restricting data to those functions that need it most can result in unintended failures.

- There is often a lack of understanding among senior executives and leaders as to how data science works. For one, you can't simply apply software engineering practices and management paradigms to data science. These tend to be rigid processes that are not directly suitable for *science* projects and can hinder the development and productionizing of data science models.

- Organizations need to be committed to investing in data science and AI. As Charles says, "*You have to have money, for one thing. Science is expensive and slow.*" As data science and AI are not commoditized like IT, data science projects can't be time-limited in the way IT projects are. Neither can you assume that all data scientists are equal. For certain roles, you need specialist skills and domain knowledge – which reinforces the importance of having people with sufficient technical knowledge in leadership roles.

- Nikolaj Van Omme (*Chapter 9*) offers some very helpful advice on how imperative it is to ensure that the solution fits into the way that the business operates, rather than just focusing on a brilliant technical solution in isolation. I've seen many data scientists get caught out by this, and not focus enough on how the business will actually use the solution, and the constraints it needs to operate within. This is why it is just as important to understand the domain and the people as the solution itself.

Measuring Success

Measuring success with data science provides an evidence base to create trust and generate support for the function. Various metrics can be used – which ones are most useful will depend on the business context.

It's very important to get metrics right. The wrong metrics will drive unwanted (and often unexpected) behaviors – this often occurs when it's difficult to measure the actual outcome you're after, so a proxy metric is used. For example, putting undue focus on a metric such as time to insight (one component of which involves measuring how quickly new data sources can be made available to analytics) could lead to corners being cut when it comes to data governance.

Here is a list of recommendations on how to measure value, and what considerations need to be made:

- Cortnie offers some valuable advice for designing business metrics that measure progress against strategic goals. In my experience, this is context-specific, as designing good business metrics in the public service sector, for instance, can be more challenging than in the private sector. One reason is that sometimes the outcome you're optimizing for is a second-order effect for which you may not have suitable data – such as providing actionable and timely intelligence to another government agency.

- Dollar value metrics are often appropriate in the private sectors (percentage increase in profits, for example), however, in the public and not for profit sector, profit is not applicable, and a reduction in costs is often not the key business driver.

- Jason categorizes the value that can be gained from analytics into three buckets: *insights*, *automation*, and *new capabilities* (see *Figure 10.2*). It may be useful to create metrics specifically for each bucket.

- According to Charles, measuring model accuracy is really about measuring business value, and how to do that needs to be thought through carefully from the outset of a project, as it varies from one business customer to another.

- Organizations sometimes use the Gartner Analytics Maturity Curve to benchmark their analytics capability. Kathleen replaces the curve with the concept of an ecosystem, rather than a linear path. Jason raised a similar point, in that the business value is the measure of its maturity, not whether it is descriptive or prescriptive. Organizations require an ecosystem of different model types and not a progression over a maturity curve.

- The human, subjective metrics are also important. Kathleen described one way she measures the value she brings: "*I want the business to talk about the work I've done for 5 minutes, and then talk for 25 minutes about what they can do now that they know something they didn't before.*" I think this is a simple but powerful concept that all data leaders can use. Similarly, Nikolaj's measure of success is simple: does the solution make people happy?

- Nikolaj stresses that the human element of AI is vital; in particular, the importance of understanding how the solution will actually be used. This is something that needs to be understood and factored into the solution from the beginning. More broadly, the problem to be solved includes non-technical and cultural factors – the wider context in which a solution will need to operate. My advice is simple: don't forget about the people using the solution, and who the solution may ultimately impact.

Getting people on board with your vision is also important, as we discussed in terms of data storytelling.

Storytelling with data

In discussing the factors that make a successful data analytics capability, the interviewees talked about the importance of engaging stakeholders and fostering collaboration throughout the organization. A key method for achieving this is through storytelling. Christina calls it *"the last mile of analytics"* because it is the technique you use to pitch your idea, project, or your findings and ultimately influence your audience to take action. Our interviewees provided great advice, such as:

- Christina stressed the importance of being able to tell a story with your data, and her three tips are as follows:

  - Have a deep understanding of what you're talking about

  - Break it down into clear and simple terms

  - Craft a story around each key message to draw in the audience

- Know your audience, avoid technical jargon, and use visuals. Christina also suggests using Grice's conversational maxims, which she summarizes as follows: be brief, be clear, be relevant, and be true.

- Meri uses metaphors and similes when storytelling that are relatable across cultures, such as food analogies.

Developing your storytelling skills is also a good career move. Christina points out that it can help you stand out and grow in your career because there is a skills shortage in effective data storytellers.

Predicting the future of AI

It's fair to say there was a range of views on the future direction of AI from our interviewees. Different interviewees, reflecting their different research backgrounds, are putting their money on different technologies for leading us to **Artificial General Intelligence** (**AGI**). Nikolaj believes hybrid approaches, such as the combination of ML and OR will be successful, while Jon sees neuro-symbolic AI being the focus of the action in the next couple of decades. Petar believes that if you truly want to mimic what a general intelligence system does, you need to do reasoning over some kind of graph structure.

Petar's prediction is that we may achieve AGI within our lifetimes. He believes that in the not-so-distant future – even in the next decade – we could have the first working example of something that might convince broad groups of people that AGI is coming.

Igor Halperin (*Chapter 13*) believes that AI will not compete with human intelligence but rather will provide helper tools. Stephane Doyen (*Chapter 17*) highlighted some of the exciting potential advances in AI for providing such helper tools in medical applications. He said, *"The prospect of doctors being able to access vast amounts of medical data, enabling them to deliver personalized treatments tailored to each patient's unique needs is thrilling."* Similarly, Stephane's research work will give a better understanding of how the brain functions and also how it *dysfunctions* – such as in the case of mental illnesses.

Jon mentioned exciting work being led by CSIRO, working with indigenous rangers and using AI to help them manage invasive weed species on country, as well as work in the Great Barrier Reef to track the crown-of-thorns starfish (a starfish that poisons the reef).

ChatGPT and generative AI are currently dominant within the hype cycle. Some of the interviewees shared their views on this technology – conversations that highlighted both ways that generative AI can help us in our work and some deeply unsettling insights into the potential negative consequences for society.

Kshira has picked up on how existing negative social consequences resulting from AI systems and broader technosocial trends are likely to be both amplified and accelerated by the rise of generative AI.

Generative AI and LLMs learn from relatively few examples. This means that the bias of existing AI systems will only be further amplified, depriving affected groups even further, and potentially perpetuating a self-reinforcing cycle of discrimination and disadvantage.

Social media is already a powerful tool that can be used by anti-social actors, large corporations, and state-sponsored actors to generate misinformation and manipulate behavior. Generative AI is likely to make it difficult for citizens to identify content that is generated by humans versus machines, reducing trust in all content. A post-truth and post-trust world is likely to amplify social divisions and unrest.

As a reader of this book, it's likely you are either in the data science space, are a data science function stakeholder, or would like to build a career in data science. Ed believes that generative AI technology is too dangerous to be let loose without appropriate controls being put in place. He's talking specifically about the use of these tools within organizations, but the same principle also applies more broadly. I believe it is up to us, those with the kind of skills represented by the experts interviewed in the book, as well as its readers, to start work on identifying appropriate controls. Like data, it will be a team sport.

Striving for diversity and inclusion

Actively striving for diversity within the data science field and data science teams is not only the right thing to do; it's critical given the work data scientists do.

Christina put it well when she said, "*We need to ensure that the teams that are working on these data applications and algorithms are diverse because a lack of diversity in the team can end up instilling unconscious biases directly into the algorithms themselves.*"

Hiring is one necessary component of this. Cortnie put it like this: "*We need to strive to hire melting-pot teams who are as diverse as the people the models could impact in society.*" Putting out a job advertisement and hoping to get diversity in applicants is not enough – you need a multi-layered strategic approach that includes the following:

- Actively developing the diversity in your network
- Analyzing the diversity in the applications received for previous job rounds, and working out how to reach and attract the groups that are typically under-represented

The diverse people you have hired also need to be empowered and supported to speak up and raise concerns about unconscious bias both in the workplace and in the data and algorithms.

One point that was raised by a number of the women I spoke to for this book was the common experience of being in meetings, and putting forward a point of view that, as Kathleen puts it, *"lands with a thud."* Then, the same thing is repeated by a man and is heard and applauded. These examples of bias tend to be invisible to those who don't experience it. A great approach was taken by one of Kathleen's colleagues, who would question this behavior by asking *"So-and-so, I just heard Kathleen say the same thing. Are you saying something different or are you reinforcing what you just heard her say?"*

As Meri aptly puts it, we're ready to move on to not just diversity but also inclusion. Simply having women present is not enough – they need to have a seat at the table.

Maria mentioned the axiom, *"You can't be what you can't see."* Although a number of the women interviewed commented that they don't naturally like being in the limelight, they mentioned pushing themselves to be visible, knowing that their experiences serve as inspirations to the younger generation. *"Use your voice, be seen, be heard, and be present"* was the exhortation from Althea Davis.

The changemakers

When she first reviewed this book, my partner (and contributor to this book), fellow senior data science leader, and technical editor Dr Tania Churchill proclaimed that it was *"a bit like having access to the wisdom of 18 mentors,"* and that *"the diversity of backgrounds, disciplines, and industries they each come from gives the reader a breadth of view"* that she and others don't normally have access to. I couldn't have articulated my vision for this book any better.

From aspiring data scientists to veteran data leaders, achieving the impact you seek is no easy task – and it can feel lonely at times. My hope is that this book will serve as an invaluable resource for when you need some guidance on how to do data science right, and reassurance that you're not alone in the challenges you face.

At those times, as you flip through the pages in this book – and take in the personal stories, advice, and insights from these prominent leaders – you'll be reminded that we're a tribe of change-makers and you're an integral player in this game of ours. After all, data is a team sport.

Index

A

AAA framework 62
Abercombie, Cortnie 5
  career trajectory 5, 6, 7
  data culture views 16-25
  data strategies, designing 25-34
  diversity and leadership discussion 7-11
  ethical approach to data,
    implementing 11-16
actionable insights 288
AIOps 139
AIR 60
AI Truth
  URL 5
AlphaFold 71
analytics carer
  pursuing 116
Artificial General Intelligence (AGI) 341
  graphs, using for 107, 108
ASD Essential Eight 277
Association for Computing
    Machinery (ACM) 202
attention mechanism 103
augmented reality (AR) 149
Australian Institute of Machine
    Learning (AIML) 206

Australian Public Service (APS)
  future of AI 281, 282
Avogadro's number 231

B

Bengio, Yoshua 155
Borne, Kirk
  AI products, failing, reasons 143, 144
  data-driven transformation, recommending,
    to organization 136-139
  data scientists, managing 141, 142
  data science, teaching 147, 149
  effective data culture, building 145-147
  future of AI, predicting 149, 150
  teams, structuring 139, 140
  views, on getting into data science 133-135
business as usual (BAU) 126
business intelligence (BI) 65, 216, 318

C

ChatGPT 173, 174, 193, 312
Chief Data and Analytics
    Officer (CDAO) 330
Chief Data Officer (CDO) 5, 11, 321
  becoming 330, 331

chief executive officer (CEO) 55, 153, 220
Chief Experience Officer (CXO) 181
chief financial officer (CFO) 55
Chief Information Officers (CIOs) 11
Chief Legal Officer (CLO) 28
chief technology officer (CTO) 55
Cognitive Science 284
Collaborative Intelligence Future
    Science Platform 208
Commonwealth Scientific and Industrial
    Research Organisation (CSIRO) 198
COMPAS program 46
compassion 142
computer vision 3
concept drift 138, 139
constant building paradox 57
consultative 142
Consumer Policy Research Centre 42
Craiyon 208
curve fitting 170
Customer Life Time Value (CLTV) 263

D

Dall-E system 208
Data61 42, 197
data access maps (DAMs) 65
data action loop 54
Database Administrators (DBAs) 6
data capability
    scaling 335, 336
data drift 138
data leadership 281
    Althea's four dimensions 216
    culture 281
    curiosity 281
    management 281
data literacy 309, 335

Data Management Association (DAMA) 217
data monetarization 225
DataOps 138
data science teams
    managing 336, 337
    structuring 336, 337
data scientist 294
    becoming 328-330
    data culture, establishing 332
    data-driven approach 333
    data strategy, developing 332
Davis, Althea
    artificial intelligence (AI) failing,
        reasons 225, 226
    data, obtaining 211, 212
    data service, establishing 218-2
    diversity and inclusion, increasing 212, 213
    projects, managing 223-225
    working, in consulting 214-218
deep learning (DL) 155
Doyen, Stephane 283
    AI solutions, developing for
        medical field 289-293
    data science 283, 284
    opinion, on becoming a leader 288, 289
    data-driven approach 288, 289
    establishing, at organization 295
    future of AI 298-300
    right team, building 296, 297
dual affordance 48
dueling datasets
    dealing with 123

E

econophysics 229
effective data culture
    building 145-147

emergence 300

environmental, social, and
    governance (ESG) 207

ethical implications, of AI 312

Excel sheet 64

expert systems 134

F

Facebook, Apple, Amazon, Netflix,
    and Google (FAANG) 76, 160

fast.ai 209

fear, uncertainty, and doubt (FUD) 136

fraud detection 3

freemium model 64

future of LLMs
    predicting 313

G

Gartner Analytics Maturity curve 121

GenderMag 204

generalized model 135

Generative AI
    challenges 49, 50
    future 194

Ghosh, Angshuman 257
    AI 257, 258
    data culture, building in
        organization 264, 265
    data-driven approach 260, 261
    data storytelling, importance 309
    data team, organizing 262, 263
    field evolution, watching 258, 259
    team members, hiring 266, 267

Google search 334

Google sheet 64

Gradient Institute 42

graph attentional models 103

Graphical Processing Units (GPUs) 103

graph neural network (GNN) 95, 98

graphs
    using, for AGI 107, 108

H

Halperin Igor 229
    AI, making explainable 234
    AI, planning for 235
    AI hype, navigating 235
    future 237, 238
    machine learning (ML), applying,
        in finance 232, 233
    role of education 236, 237

heterophilous 103

heteroskedasticity 67

homophilous 102

Hopfield associative memory network 71

hype cycles 135

I

individual contributors (ICs) 250, 336

International Joint Conference on Artificial
    Intelligence (IJCAI) conference 201

inverse reinforcement learning 237

Ising model 231

J

Jira board 64

L

Large Language Models (LLMs) 68
Lean AI 317

M

machine learning (ML) 3, 153
  applying, in finance 232, 233
**machine learning operations
    (MLOps) 58, 138, 139**
Maley, Kathleen 113
  data-driven approach 118-122
  data culture, establishing 12
  diversity 116, 117, 118
  dueling datasets, dealing with 123
  future of analytics 130, 131
  roadblocks, overcoming 124-126
  views, on pursuing career in
      analytics 113-115
Martin, Charles 71
  AI expert, becoming 71-73
  AI explainability 91-93
  companies, advising on AI roadmap, 75
  data integration 86, 87
  data projects, failure reasons 78-84
  impact, measuring 85, 86
  limits, finding of NLP 89, 90
  research and consulting, balancing 73-77
maximum entropy methods 230
micro-education platform 134
Milosavljevic, Maria
  analytics, obtaining 269, 270
  data-driven approach 178-181, 273-276
  data leadership 281
  diversity and inclusion, considering 270, 271
  data science projects, failing reasons 280

data culture, establishing at organization 295
ethical AI 277-279
future of AI, in APS 281, 282
Minimal Viable Product (MVP) 18, 317
ModelOps 139
**mutually exclusive, collectively
    exhaustive (MECE) 275**

N

**National Criminal Intelligence
    Fusion Capability 278**
**Natural Language Processing
    (NLP) 71, 91, 158**
  limits, finding 89, 90
neural network (NN) 156
Next-Generation AI Graduates 200
nondeterministic polynomial time (NP) 156
No One Right Answer (NORA) 193
normalization 67

O

observability 138
Omme, Nikolaj Van 153, 154
  AI ethics, in organization 169-172
  data-driven, becoming 160, 161
  future of LLMs, predictions 173-175
  integration, of ML and OR 155-160
  leadership views 165, 166
  progress of AI, assessing 154, 155
  project setup to succeed 161-165
  skills, for starting out in data 172, 173
  success, measuring of projects 167, 168
operational research (OR) 153
overfitting 135
oversmoothing problem 100
oversquashing phenomenon 100

P

point of function optimization 170
pretendotype 81
prompt engineering 194
proof of concept (POC) 58, 124
Public Interest Advocacy Centre (PIAC) 37
PyTorch 72

Q

quality assurance (QA) 291
quantitative analyst (quant) 229-231

R

Return on Investment (ROI) 17
roadblocks
    overcoming 124-126
Rosich, Meri
    AI ethical and trustworthy, creating 310, 311
    data-driven organization, starting 306-308
    data leader, qualities 308, 309
    data scientists, guidance for 311, 312
    data storytelling 310, 311
    diversity and inclusion, improving 304
    effective data culture, establishing 308
    failure rates of AI projects, discussing 305

S

Saagar, Kshira 53
    AI future 68, 69
    path 53, 54
    data-driven approach, implementing 55-62
    data storytelling, significance 65, 265
    entering, into data science industry 67
    leadership, discussing in data culture 62-64

Santow, Edward 37
    AI pathways, developing 37-41
    ethics, applying 42-44
    impact on society, considering 45-48
    generative AI, challenges 49, 50
scientific machine learning (SciML) 233
semi-empirical quantum
        chemistry methods 73
senior data leader
    becoming 330, 331
singular value decomposition (SVD) 80
Software Development Life Cycle
        (SDLC) 224, 336
Stathopoulos, Christina 241
    changes, observing in data science 243, 244
    data science leader, becoming 241-243
    data storytelling, using 251, 252
    diversity and inclusion, increasing 245, 246
    fundamental skills, of data science 252, 253
    interviews, for data scientist role 253, 254
    leadership, progressing into 254, 255
    organization, advising 246-248
    projects failure, reasons 248, 250
stock-keeping unit (SKU) 58
subject matter experts (SMEs) 223
Suspect Target Management
        Plan (STMP) 38

T

Tamara Widjaja, Jason 177
    AI projects, failure reasons 184, 185
    data culture, establishing 186- 280
    data-driven approach, starting 178-181
    data governance, importance 187-189
    data science 177
    data science projects, managing 182-184
    entrants, advising to 191, 192

future of 298-300

leadership discussion 189, 190, 284-287

realistic expectation, communicating
    to clients and partners 185, 186

views, on ChatGPT 193, 194

views, on future of Generative AI 194

Technical and Further Education (TAFE) 38

TensorFlow 72

Tensor Processing Units (TPUs) 103

Theano 72

total cost of ownership (TCO) 61

Tran, Dat 315

beginner's advice 324, 325

data culture, establishing 320

data leader, selecting 321, 322

data storytelling, discussing 322

failure rates, of AI projects 316-318

future, of AI field 325

high-performing cross-functional
    teams , building 315, 316

success steps 318, 319

team members, hiring 323, 324

traveling salesman problem (TSP) 156

U

unique selling proposition (USP) 207

V

value

automation 180

insights 180

new capabilities 180

Veličković, Petar 95

advice, for getting into research 110, 111

AI research 95, 96

gap between academia and
    industry, bridging 108-110

graph neural networks, applying 102-104

graphs, for AGI 107, 108

machine learning, with graph
    networks 97-101

research boundaries, pushing with
    machine learning 104-107

virtual reality (VR) 149

W

WeightWatcher model 71

research and consulting, balancing 73- 77

Whittle, Jon 197

AI, in Australia 206, 207

career 197-199

career, in data science or AI 209

future of AI, predicting 208, 209

knowledge, for developing
    responsible AI 201-206

leadership discussion 207

research, translating into
    real-world impact 199, 200

X

XGBoost 285

Packtpub.com

Subscribe to our online digital library for full access to over 7,000 books and videos, as well as industry leading tools to help you plan your personal development and advance your career. For more information, please visit our website.

Why subscribe?

- Spend less time learning and more time coding with practical eBooks and Videos from over 4,000 industry professionals

- Improve your learning with Skill Plans built especially for you

- Get a free eBook or video every month

- Fully searchable for easy access to vital information

- Copy and paste, print, and bookmark content

Did you know that Packt offers eBook versions of every book published, with PDF and ePub files available? You can upgrade to the eBook version at packt.com and as a print book customer, you are entitled to a discount on the eBook copy. Get in touch with us at customercare@packtpub.com for more details.

At www.packt.com, you can also read a collection of free technical articles, sign up for a range of free newsletters, and receive exclusive discounts and offers on Packt books and eBooks.

Other Books You May Enjoy

If you enjoyed this book, you may be interested in these other books by Packt:

Network Science with Python

David Knickerbocker

ISBN: 978-1-80107-369-1

- Explore NLP, network science, and social network analysis
- Apply the tech stack used for NLP, network science, and analysis
- Extract insights from NLP and network data
- Generate personalized NLP and network projects
- Authenticate and scrape tweets, connections, the web, and data streams
- Discover the use of network data in machine learning projects

The AI Product Manager's Handbook

Irene Bratsis

ISBN: 978-1-80461-293-4

- Build AI products for the future using minimal resources
- Identify opportunities where AI can be leveraged to meet business needs
- Collaborate with cross-functional teams to develop and deploy AI products
- Analyze the benefits and costs of developing products using ML and DL
- Explore the role of ethics and responsibility in dealing with sensitive data
- Understand performance and efficacy across verticals

Packt is searching for authors like you

If you're interested in becoming an author for Packt, please visit `authors.packtpub.com` and apply today. We have worked with thousands of developers and tech professionals, just like you, to help them share their insight with the global tech community. You can make a general application, apply for a specific hot topic that we are recruiting an author for, or submit your own idea.

Share Your Thoughts

Now you've finished *Creators of Intelligence*, we'd love to hear your thoughts! Scan the QR code below to go straight to the Amazon review page for this book and share your feedback or leave a review on the site that you purchased it from.

`https://packt.link/r/1-804-61648-6`

Your review is important to us and the tech community and will help us make sure we're delivering excellent quality content.

Download a free PDF copy of this book

Thanks for purchasing this book!

Do you like to read on the go but are unable to carry your print books everywhere? Is your eBook purchase not compatible with the device of your choice?

Don't worry, now with every Packt book you get a DRM-free PDF version of that book at no cost.

Read anywhere, any place, on any device. Search, copy, and paste code from your favorite technical books directly into your application.

The perks don't stop there, you can get exclusive access to discounts, newsletters, and great free content in your inbox daily

Follow these simple steps to get the benefits:

1. Scan the QR code or visit the link below

https://packt.link/free-ebook/9781804616482

2. Submit your proof of purchase
3. That's it! We'll send your free PDF and other benefits to your email directly